THE NEW
SEED-
STARTERS
Handbook

Other Books by the Same Author

Vegetables Money Can't Buy
Working Wood (Written with Michael Bubel)
Root Cellaring
The Adventurous Gardener
The Country Journal Book of Vegetable Gardening
The Seed-Starter's Handbook

THE NEW
SEED-
STARTERS
Handbook

BY NANCY BUBEL

Illustrations by Frank Fretz
Photographs by Alison Miksch
& Rodale Press Photography Staff

Rodale Press, Emmaus, Pennsylvania

Printed in the United States of America on acid-free ∞, recycled ♻ paper

Distributed in the book trade by St. Martin's Press

Book Design by Linda Jacopetti

Library of Congress Cataloging-in-Publication Data

Bubel, Nancy.
The new seed-starters handbook/
by Nancy Bubel; illustrations by Frank Fretz; photographs
  by Alison Miksch.
        p.   cm.
  Bibliography: p.
  Includes index.
  ISBN 0–87857–747–5   ISBN 0–87857–752–1 (pbk.)
  1. Seeds. 2. Sowing. 3. Gardening. I. Title.
SB121.B83 1988
635′.0421—dc19
                             87–27698
                              CIP

4 6 8 10 9 7 5   hardcover
        10   paperback

Grateful acknowledgment is made to the following publishers for permission to reprint copyrighted
material:

John Wiley & Sons: *Handbook for Vegetable Growers*, by J. E. Knott. Copyright © 1957 by J. E. Knott.
Reprinted by permission.

Lea and Febiger: *Vegetable Growing*, by J. E. Knott. Copyright © 1955 by J. E. Knott. Reprinted by
permission.

Macmillan: *The Years in My Herb Garden*, by Helen Fox. Copyright © 1953 by Macmillan. Reprinted
by permission.

Sierra Club Books: *The Unsettling of America*, by Wendell Berry. Copyright © 1977 by Wendell Berry.
Reprinted by permission.

Van Nostrand Reinhold: *The Beautiful Food Garden*, by Kate Gessert. Copyright © 1983 by Van
Nostrand Reinhold. Reprinted by permission.

Grateful acknowledgment is also made to J. F. Harrington, Department of Vegetable Crops, University
of California at Davis, California, for permission to reprint previously published tables; and to Dr. H.
Garrison Wilkes for permission to reprint excerpts from his article "The World's Crop Plant
Germplasm—An Endangered Resource" from the February 1977 issue of *The Bulletin of the Atomic
Scientists*.

For My Parents,
Milton and Grace Hangen Wilkes, and for our children and their families:
Mary Grace, Anthony, and Nicholas; Greg, Tish, and Ansel.

We lose our health—and create profitable diseases and dependences—by failing to see the direct connections between living and eating, eating and working, working and loving. In gardening, for instance, one works with the body to feed the body. The work, if it is knowledgeable, makes for excellent food. And it makes one hungry. The work thus makes eating both nourishing and joyful, not consumptive, and keeps the eater from getting fat and weak. This is health, wholeness, a source of delight. And such a solution, unlike the typical industrial solution, does not cause new problems.

The "drudgery" of growing one's own food, then, is not drudgery at all. (If we make the growing of food a drudgery, which is what agribusiness does make of it, then we also make a drudgery of eating, and of living.) It is, in addition to being the appropriate fulfillment of a practical need, a sacrament, as eating is also, by which we enact and understand our oneness with the Creation, the conviviality of one body and with all bodies.

Wendell Berry
The Unsettling of America

Contents

Acknowledgments

In referring to our debt to those who have gone before us, a wise man once said, "We stand on the shoulders of giants." This is especially true in gardening, where we have been both enlightened by the experiments of learned men and supported by the faithful efforts of obscure dirt gardeners who have saved and selected seed and cared for the soil over the centuries.

In preparing this book, I received generous help—in the form of correspondence and printed matter—from Dr. R. J. Downs, Phytotron Director at the North Carolina State University at Raleigh; Dr. Robert F. Fletcher, Extension Specialist in Vegetable Crops at Penn State University; Dr. O. A. Lorenz, Chairman, Department of Vegetable Crops, University of California at Davis; Dr. Raymond Sheldrake, Professor of Vegetable Crops at Cornell University; Dr. H. Garrison Wilkes, Botany Department, Boston Harbor Campus of the University of Massachusetts; Dr. Jay S. Koths, Department of Plant Science, University of Connecticut; Dr. Betty Ransom Atwater, Director, Ransom Seed Laboratory, Santa Barbara, California; and Richard Grazzini, Manager of the H. G. German Seed Co., Smethport, Pennsylvania. Dr. H. Garrison Wilkes, Dr. J. F. Harrington, and Dr. James Edward Knott have kindly permitted me to quote from their work.

In addition, Dr. R. Gregory Plimpton of Atlantic and Pacific Research, Inc., has supplied me with much helpful information. Kent Whealy, originator of the Seed Savers Exchange, Forest Shomer, Director of the Abundant Life Seed Foundation, and Lawrence Hills of the Henry Doubleday Research Foundation have answered my inquiries in a spirit of kind cooperation and allowed me to quote from their writings. Articles by Dr. Jeffrey McCormack and Dr. Mark Widrlechner, published in Seed Savers Exchange Handbooks, have been very helpful.

Also, visits to the following growers were especially helpful: Bertha Reppert, Rosemary House, Mechanicsburg, Pennsylvania; Fairman and Kate Jayne, Sandy Mush Herb Nursery, Leicester, North Carolina; Cyrus Hyde, Well-Sweep Herb Farm, Port Murray, New Jersey; Janet Urban, Botanist, Bowman's Hill Wildflower Preserve, Washington Crossing, Pennsylvania; and Bob Hyland, Education Director, Longwood Gardens, Kennett Square, Pennsylvania.

In addition, I learned much from phone conversations with Roger Kline, Extension Associate, Department of Vegetable Crops, Cornell University; Dr. John Gerber,

University of Illinois; Dr. Jim Austin, Park Seed Co.; Pam Dwiggins, Research Botanist, National Wildflower Research Center, Austin, Texas; Dr. Arthur O. Tucker, Delaware State College; Greg Edinger, Naturalist at Bowman's State Hill Wildflower Preserve, Washington Crossing, Pennsylvania; Rob Johnston of Johnny's Selected Seeds, Albion, Maine; Dick Meiners of Pinetree Garden Seeds, Gloucester, Maine; Colleen Armstrong of The New Alchemy Institute, East Falmouth, Massachusetts; and Shepherd Ogden of Cook's Garden Seed Company, Londonderry, Vermont.

My editor, Claire Kowalchik, always had the reader in mind. I'm ever so grateful for her careful questions and perceptive suggestions, which have helped immeasurably to make *The New Seed-Starters Handbook* a better book than the original.

I'm grateful, too, to my husband and children, who good-naturedly endure my absentmindedness when I am "with book." Special thanks to my son, Greg (who now has his own family garden) for his perseverance in typing the whole much-spliced and much-corrected manuscript of the original edition.

Introduction

Since 1957, when we tried (and failed!) to grow radishes in a window box outside our third-floor apartment in Philadelphia, my husband, Mike, and I have been learning about growing vegetables. We got off to a slow start. I had only vague memories of the Victory Garden my parents grew for a few years in our New England backyard, and an insistent yearning to begin a garden of my own. Mike had grown up on home-raised vegetables—a surprisingly limited variety of them, though, mostly grown from seed his mother had carefully saved from one harvest to the next: cabbage, potatoes, beans, carrots, beets, dill, cucumbers, and sunflowers. Later, during a stay in Germany, he had taken a course in gardening at a free university of sorts. Mike wanted a garden, too.

When we moved into our first house, we bought digging forks and shovels almost before the ink was dry on the deed. We started that first year with tomatoes and beans and many flowers. It took us several years to progress to planting a garden that we could eat from all summer, but by 1970 our vegetable garden was carrying us year round. Today we hardly ever buy a vegetable. Our year-round supply of vegetables is due in large part to starting seeds early indoors and making continuous outdoor plantings of varieties of food chosen for quality and ease of storage.

Along the way, we've learned a lot about how and when to start different vegetables. Trial and error has helped. Older relatives and neighbors have been generous with advice and lore, and we never start a new gardening year without thinking fondly of our different mentors and how much they have given us over the years. Reading about gardens and gardening, seeds, plants, soil, and insects has given us a framework that often supports relationships between what we've observed and what we've been told. Mostly, though, we've muddled along, taking longer than it now seems we should have to see and use the full potential of a piece of ground and a packet of seeds.

That's why I decided to write the first edition of *The Seed-Starter's Handbook*— to help other gardeners make that jump from dabbling to self-sufficiency sooner and more easily than we did. At the same time, I hoped that experienced gardeners would find in it some insight into possibilities never considered, into alternatives and experiments in areas of gardening where the final word has yet to be written.

You know how it is when you put the phone down. You then remember all the *other* things you had intended to say. So it has been with this book. From the time the first edition was published, I've wished that I had included more about growing other garden flowers and wild plants from seed. In the ten years since I sent off that first manuscript, I've visited many gardens, asked a lot of questions, talked to numerous researchers, tried growing many new kinds of seeds, and read thousands of pages of gardening research, lore, and advice. Some of my learning grew out of my failures—with stubbornly dormant linden tree seeds, light-seeking scarlet sage seeds, and wildflower seeds that wouldn't germinate without prechilling.

The result is *The New Seed-Starters Handbook,* which contains new sections that detail the principles and techniques for starting garden flowers, wildflowers, herbs, trees, and shrubs from seed. This material has been combined with the chapter from the first edition on growing vegetables and garden fruits from seed to create a handy encyclopedia section. No book of this size can include every possible plant in each of these categories, but I've tried to give you a selection of the best. Although this is a handbook, not a textbook, I've included Latin names for the plants under discussion, just so we're all sure we're talking about the same plant, and for your convenience in ordering seeds. Common names are charming but often vary locally. Latin names are accurate worldwide.

Apart from the encyclopedia section and several additional chapters, the other new material in the book has been inserted into the existing text via new words, paragraphs, pages, and sections. These additions and revisions represent an eight-month distilling process during which I evaluated and interpreted a whole file of new material accumulated since the day *The Seed-Starter's Handbook* was published. Not a chapter remains untouched. To be honest, gardening friends, rechecking all the facts, summarizing old material, adding newly discovered findings, and writing new chapters has been more difficult than writing the original manuscript. I'm hugely relieved that it is done, but also delighted to have had the chance to improve my original work.

Consider this book a manual of procedures, giving you the step-by-step how and when of various planting techniques. Look, too, in every chapter, for the principles on which these techniques are based, and count on finding at least a few open-ended questions that might challenge old suppositions or suggest new growing frontiers.

There's no one right way to do most of these things, you know. A good many workable options are open to you in planting seeds. I've tried to suggest the range of possibilities. The choice is up to you.

Let this book be smudged. Let it be marked. It's meant to be used. I hope it will make a difference in your garden, and ultimately, on your table.

Once again, I wish you joy in planting seeds. The seed-starter works, always, at the edge of a mystery. Though we may take it for granted, we are part of that mystery, along with the fragility, the resilience, the dependability of the green world. Happy planting!

A borage seedling.

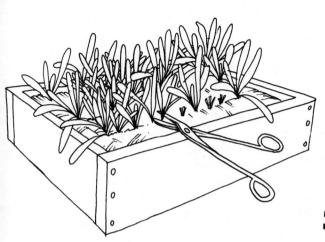

Section One

Starting Seeds Indoors

To own a bit of ground, to scratch it with a hoe, to plant seeds and watch the renewal of life—this is the commonest delight of the race, the most satisfactory thing a man can do.

Charles Dudley Warner

1 Why Start Your Own?

Security and adventure might be considered opposites in some situations, but the gardener who raises plants from seed can experience both. Security—that confidence in the future that springs from one's own ability, forethought, and preparation—and adventure—the soaring sense of "anything is possible" and "there are so many interesting things to try"—are well known to those who grow new varieties, experiment with new methods, and dabble in plant breeding and seed saving.

Skill in raising vegetable plants from seed is the very cornerstone of gardening independence. Choice of seeds and careful handling can bring you not only earlier harvests, but better vegetables. You can select varieties of food known to keep or process well so that the winter season, for which we gardeners are always planning, will be a time of abundance. Likewise, good eating will be yours all summer long from the selection of fresh vegetables you've planted for their superior quality.

The Reasons

I suspect that I'd continue to raise my own seedlings even without a good excuse, because I enjoy the process, but when I stop to think about it I realize that there are all kinds of good reasons for nurturing one's own plants from seed.

Earlier Harvests. You can get a much earlier start in the garden, and therefore put fresh food on the table sooner, when you've grown flats of cabbage, tomatoes, eggplant, and peppers indoors for setting out when weather mellows. The sooner you can begin picking from your garden, the greater your yield for the year.

Greater Variety. Varieties of plants offered by commercial seedling vendors represent but a tiny fraction of the possibilities open to you as a gardener. Buying started plants severely limits your options for raising vegetables of special flavor, insect or disease resistance, or extra nutritional value. If, for example, you want to grow Juwarot or A+ Carrots (extra high in vitamin A) you'll have to start with seeds. Looking for special gourmet foods like globe artichokes, watercress, or Japanese melons? It's back to the seed catalogs. Peppers that are hot, but not too hot? Start your own Hungarian wax plants. Delicious, mild, sweet Golden Acre or Jersey Queen cabbage? You need to start those with seeds.

Stronger Seedlings. Seedlings you've grown yourself *can* be super seedlings. If you do all the right things at the right times, you'll have the best that can be grown, and you'll know that your plants have well-developed roots growing in good soil that hasn't been hyped with chemicals. You can even plant organically raised seeds to give your plants an extra start on excellence.

Healthier Seedlings. By raising your own plants, you minimize the chance of introducing soil-borne diseases to your garden. Club-root and yellows that affect cabbage, along with anthracnose, tobacco mosaic, and wilt, are some examples of

2

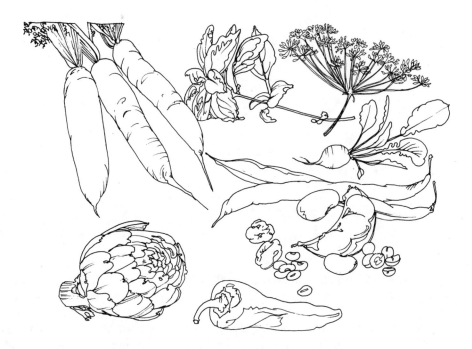

Some varieties of vegetables and flowers can only be grown from seed.

plant diseases you may avoid importing if you grow your own. Of course, you must use uncontaminated soil, and, especially in the case of mosaic in pepper, eggplant, and tomato seedlings, avoid handling tobacco around the plants.

Cost Saving. You'll save money. Well, maybe. Certainly for the price of a dozen greenhouse tomato plants you can buy a small handful of seed packets, each of which will give you plants to share or to sell, or extra seeds to save for the following year. So many interesting plants can be grown from seed, though, that once you start raising seedlings you might find that you tend to put some of that saved money back into seeds of other kinds. But since you're likely to eat even better as a result, you may well consider that you're still far ahead.

Satisfaction. Creative satisfaction ought to count for something, too. From settling a well-chosen seedling in a pot of its own carefully prepared soil and watching it grow greener, sturdier, and leafier, to picking and eating the peppers, eggplant, or other nutritious food the mature plant finally bears, you've been involved all the way, and you can see that your skillful care has made a difference.

Enjoyment. At the very least, planting seeds indoors is a good cure for the winter doldrums—those bleak, cold days when February seems like a permanent condition and you feel you simply must do something to nudge the season into turning. Choose your earliest plantings judiciously, though. You don't want them to be past their prime when you set them out in the garden. Onions, chives, peppers, certain wildflowers and perennial plants, and houseplants like coleus and geraniums are good candidates for beginning the season.

2 First the Seeds

You have in your hands an array of seed packets and perhaps a few jars of seeds you've saved yourself. You're anxious to plant, to get a start on the growing season that still seems far away. Take a minute now, if you will, to be aware of what the seed really is before committing it to the soil. Dry, flaky, hard, smooth, warted, ridged, powdery, or wispy, these distinctively shaped particles may look as lifeless as the February garden patch. Don't be deceived, though. Seeds don't spring to life when you plant them. Seeds are alive.

Often symbols of beginning, seeds are living guarantees of continuity between generations of plants. Inside even the most minute, dustlike grain of seed is a living plant. True, it's in embryonic form, possessing only the most rudimentary parts, but it lives, and it is not completely passive. At levels that we can't see, but laboratory scientists can measure, seeds carry on respiration. They absorb oxygen and give off carbon dioxide. They also absorb water from the air. Seeds need a certain minimum amount of moisture within their cells in order to make possible the metabolic processes by which they convert some of their stored carbohydrates into available food. Thus they maintain their spark of life—dim though it may be—until conditions are right for them to complete their destiny as germinated plants.

The Botanical Facts

By strict botanical definition, a seed is a ripened fertilized ovule containing an embryonic plant and a supply of stored food, all surrounded by a seed coat. In practice, though, gardeners use many seeds that are actually fruits (the mature ovary of a flower containing one or more seeds). A kernel of corn is really a seedlike fruit. Carrot, dill, and fennel seeds are, technically, dry, one-seeded fruits.

Seeds, then, are completely self-contained. Within the boundaries of the hard, dry coat that protects them, they possess enough food energy to carry them through their dormancy and into their first few days as seedlings. They have all the enzymes

4

Begin by gathering your stored and bought seeds.

they'll need to convert this stored food into a form their tissues can use, and they carry within their cells the genetic information that directs what they will be and when and how.

Let's look at the bush bean as an example—not a typical seed, perhaps, but one in which it is easy to see the parts and their arrangement that are common to all seeds. The good old garden bush bean is the favorite of botanists for this purpose because its size and structure make it possible for us to see clearly how it is formed.

If you soak a bean seed in water for a few hours, the hard outer coat will slip off easily. The bulk of the bean that you now see is composed of the cotyledons, the two identical fleshy halves that comprise the "meat" of the seed. Cotyledons are rudimentary leaves. Unusually large and thick in the bean, they contain stored fat, carbohydrates, and protein. Both cotyledons are attached to a rudimentary stem, and they curve protectively over a tiny leafy bud. The root tip, at the other end of the seed, will elongate into the first root of the plant when the seed germinates. Any seed, no matter how tiny, wispy, or irregular, will possess these features: cotyledons (sometimes one, more often two), a brief stem, a leafy bud, and a root tip.

Most seed-bearing garden vegetable plants are dicots; that is, they possess two cotyledons. Beans, tomatoes, celery, cabbage, and other vegetable seedlings are

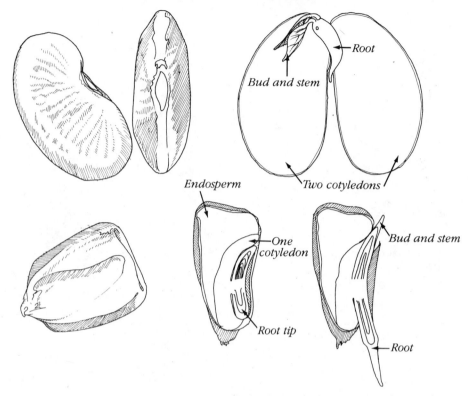

Cross sections of a bean seed, or dicot (top), and a corn seed, or monocot (bottom).

dicots. When they germinate they send up two "wings." Monocots, with just a single cotyledon, are represented by the grass family (Gramineae), which includes corn, wheat, rye, and other cereal crops. These seeds send up the familiar, single grasslike spear of green.

In many seeds, some of which are important food crops in their own right, the stored food is not contained in the cotyledons, as in the bean, but in a layer called the endosperm, which surrounds the embryo. This part of the seed varies in different species. It may consist of starch, oil, protein, or waxy or horny matter, but whatever its form, its function remains the same—to nourish the seed from the time of its maturity on the parent plant until the beginning of the next growing season, when conditions will be favorable for its success as a plant in its own right. Often, of course, the tasty endosperm is what we're after when we raise the crop, as with buckwheat, corn, wheat, and rye.

Just to keep the record straight: Cotyledons of some plants may synthesize nourishment needed by the seed; others may both make and store food for the embryo. If you find that fact surprising, reflect that there is more—much more—that we don't yet know about seeds.

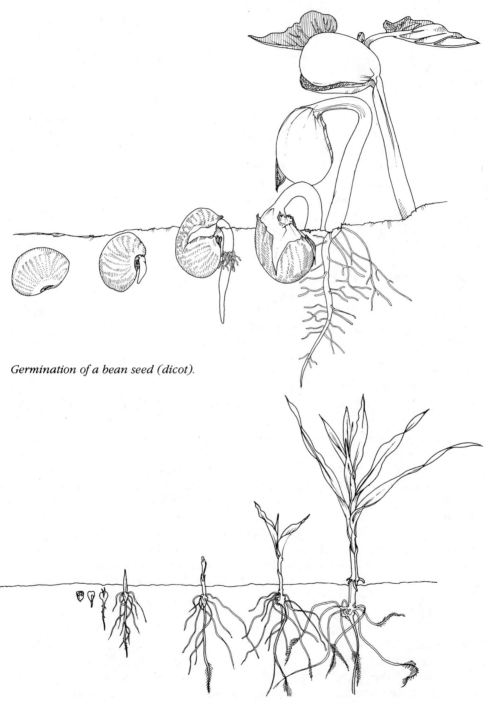

Germination of a bean seed (dicot).

Germination of a corn seed (monocot).

Dormancy

We're just beginning to appreciate, too, how much is still unknown about dormancy in seeds. If you've ever tried unsuccessfully to start a row of lettuce in midsummer heat, you have an idea of how a dormant seed behaves. It refuses to germinate, even if otherwise viable, when it lacks the right temperature, moisture, and oxygen supply that would ordinarily favor germination. Even though conditions might seem favorable for germination, such as those that occur in midsummer, they might not be right to induce germination in certain seeds.

It can be annoying to miss a seeding date for a certain crop and be unable to plant it later in the summer; however, the ability of seeds to remain dormant, in varying degrees, has contributed to the survival of seed-bearing plants as we know them. Certainly a plant that needs 90 days of warm weather to mature will be doomed to failure if it sprouts as soon as it matures at the end of the summer, shortly after the first frost. Lettuce, likewise, has less chance of success under random conditions when sown in hot, dry soil than it does in the moist, cool surroundings that promote its quick growth. Dormancy, then, is a protective device, designed to assure the continuity of the species.

A seed may be dormant because its embryo is still immature, its seed coat is impermeable to water or to gases, its coat is too unyielding to permit embryo growth (although this is rare), or because of a metabolic block within the embryo. Often, more than one of these factors operates at the same time.

Breaking Dormancy

As a gardener, it is often in your interest to try to break dormancy in certain kinds of seed. Since you intend to give the plant special care and optimum conditions, you can often do this and get away with it. For example, if you are anxious to raise a fine bed of lettuce to eat with your midsummer tomatoes—a real mark of gardening expertise—how do you give your lettuce seeds the message that it's all right for them to sprout?

Studies of seeds in research laboratories have furnished valuable clues to the interruption of dormancy. The period is relative, not absolute. In fact, there seems to be general agreement among a number of scientists who have studied this phenomenon that the whole question of seed dormancy must be considered a matter of balance between the growth-promoting and growth-inhibiting substances that are found in all plants.

So how do you tip that balance in your favor? Here is what the experts have discovered.

Temperature. Chilling the seeds often breaks dormancy.

Light. Subjecting the seeds to light—even a dim continuous light or a sudden bright photoflash—will sometimes help, especially with lettuce. Germination depends on the total amount received. The dimmer the light, the longer the necessary exposure.

Red Light. Exposure of some seeds to red light (660 nanometers) promotes seed germination. Experiments with lettuce bear this out. However, far-red light (730 nanometers) has been found to inhibit seed germination. Practically speaking, this means that seeds that are difficult to germinate will often do better under fluorescent plant lights (see chapter 10). Some seeds won't germinate when shaded by leaf cover, probably because the leaves filter out helpful rays while allowing the inhibiting far-red light to reach the seeds.

Dormancy is seldom a problem with vegetable seeds, except for some heat-sensitive seeds like lettuce and celery. Beans, mustard, and many other vegetable seeds never go dormant. Carrots and parsnips do need a month of afterripening following harvest before they'll germinate. Seeds you've purchased have had time, of course, to undergo any necessary afterripening in the months between collection and planting. Dormancy is not an uncommon problem when trying to germinate seeds of wildflowers, trees, or shrubs.

History of Seed-bearing Plants

We know very little about the origin of seed-bearing plants. Charles Darwin called it an "abominable mystery." Spore-bearing plants, like ferns, existed first, followed by gymnosperms, like the evergreens, which shed their seeds unencased. Discoveries in West Virginia of early seeds encased in cupules—incompletely fused seed coats—have pushed the date of known true seeds back to 360 million years ago. Angiosperms, flowering plants in which the seeds (encased in an ovary) are more fully protected, appeared during the Cretaceous period (roughly 100 million years ago). Eventually they dominated the more primitive spore-bearing forms of green life, probably partly because the well-developed seed embryo, clinging to the parent plant until thoroughly mature, had the edge over the naked, randomly shed seed.

Although we still know so little, we know enough of the internal workings of the seed to stand in awe at its variety, its toughness, and its practical simplicity.

3 Choose Your Medium

The first step in encouraging your seeds on their way to being the plants they really want to be is to prepare a starting medium that will nurture the seeds through the critical germination and seedling stages. The stuff to which you entrust your seeds should match the following description:

- Free from competing weed seeds, soil-borne diseases, and fungus spores.
- Able to absorb and hold quantities of moisture.
- Not so densely packed that vital air is excluded.
- Naturally derived and free from any substance you would not want to put in your garden.
- Noncrusting.

Since the physical conditions of their surroundings—temperature, moisture, air, and light—are more important to germinating seedlings than the nutrient content of the soil, these first mixtures in which you plant your seeds needn't be rich. In fact, it's better if they're not. But they should be light, spongy, and moist. It's a nice touch to start seeds in one of the special seed-starting media, like vermiculite, and then to transplant the seedlings into a potting soil mix that contains more nutrients. Some gardeners prefer to plant seeds directly in a potting soil mix. Many good commercial mixtures are available, or you can mix your own potting soil from the recipes listed later in this chapter.

I wouldn't recommend starting seeds in plain garden soil, because it tends to pack and crust when kept in shallow containers indoors. In addition, unsterilized garden soil may harbor fungi that cause damping-off, a disease that makes young seedlings shrivel and wilt at soil level and sometimes even interferes with complete germination. Soil in which such diseased seedlings have been grown must be pasteurized (see Heat Treatment later in this chapter) before another batch of seeds is planted in it. I find it easier to use fresh potting soil for each new seedling crop and to use the old soil for potting up young trees or rooted herbs.

Three Favorites

I've had good results planting seeds in each of the following three media.

Vermiculite. This is a form of mica which has been "popped" like popcorn by exposing it to intense heat. The resulting flakes are light and capable of holding large amounts of water. Vermiculite also absorbs nutrients from fertilizers and other soil components and releases them gradually to plants. Be sure to buy horticultural vermiculite, not the kind sold for the building trade, which is coarser, highly alkaline, and often contains substances toxic to plant roots.

A Mixture. Equal parts of vermiculite, milled sphagnum moss, and perlite combine to make another good starting medium. This provides a spongy, friable seedbed that promotes good root development. Perlite, despite its plastic appearance, is a natural product, a form of "popped" volcanic ash, and while I do not like to use it alone, as I do vermiculite in some cases, it promotes good drainage in seed-starting mixes. The moss must be milled sphagnum, which is very fine, not peat moss, which is too coarse for small seeds and tends to dry and crust.

Mix these three components together thoroughly before dampening. A large old tub or bucket set on newspaper makes the job easier and spillage less of a problem. When moistening this and all other soil mixtures, I like to use warm water because it is more readily absorbed. This seems to be especially important for mixtures containing sphagnum moss, which tends to float on cold water in a dustlike layer that takes ages to soften and swell. It is also a good idea to prepare your soil mixture several hours before you intend to plant seeds. Then, if you have added too

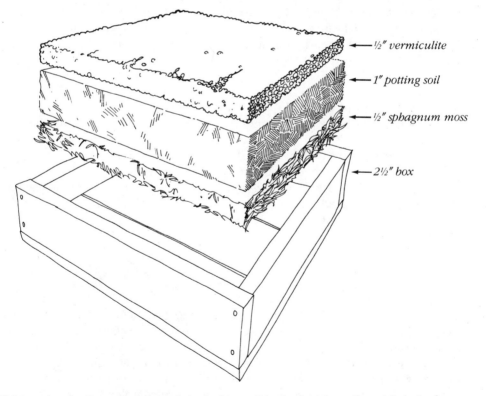

This growing medium, made with ½ inch of moss, 1 inch of potting soil, and ½ inch of vermiculite on top, produces healthy seedlings.

much water, you can simply pour off any that puddles on the upper surface, before planting the seeds.

A Three-Layer Arrangement. This medium is composed of a bottom layer—about one-half inch—of torn pieces of sphagnum moss that has been well dampened, then a one-inch layer of good potting soil, topped by a one-half-inch layer of vermiculite. The bottom layer of moss should not be finely milled. I use clumps of moss gathered in our woods. The air spaces trapped by the ferny fronds of the moss promote excellent root development, I've found, and the moss retains water, too. I tear up the wads of moss into small pieces so that individual plants can be removed more easily when it's time to transplant them.

You can also mix chopped sphagnum moss with other potting soil ingredients. Some evidence has shown that sphagnum moss exerts a mild antibiotic effect that helps to control bacterial diseases of seedlings. (A disease-causing fungus, *Sporotrichum schenkii,* has been found in sphagnum moss. Wash thoroughly after handling the moss, and wear gloves if you have any skin breaks on your hands.)

Alternative Media

There's nothing absolute about the mixtures that I use. Other gardeners have obtained good results with the following seed-starting media. What you choose may depend partly on what is easily available to you and on your personal reactions to the feel of the stuff.

Other options include:

- Equal parts of milled sphagnum moss and vermiculite.
- One part milled sphagnum moss and two parts each of vermiculite and perlite.
- Shredded moss alone (well dampened beforehand).
- Potting soil, either commercial or home mixed.
- A cube of sod, turned grass side down, which works well for the planting of larger seeds, like squash, melons, and cucumbers, and supports the seedling until it is planted out in the garden.

Feeding the Seedlings

Neither vermiculite, perlite, nor sphagnum moss contains the nutrients necessary to support plant growth. Seedlings growing in these substances, or mixtures of them, must be fed regularly until they get their roots in real soil.

Such liquid feeding, carried on over a period of weeks, amounts to hydroponic culture, with the growing medium serving only to hold the plants upright and to condition the roots. Seedlings benefit from the more complex interactions taking place in real soil, so they should eventually be transplanted to a growing medium.

Potting Mixtures

A seed-starting medium provides good physical conditions for seed germination. A potting medium contributes nourishment for the growing plant. There are all kinds of potting mixture recipes from which to choose. Here are some options, along with a few hints about their suitability or limitations.

The amounts in each recipe are given in parts. A part, of course, can be any measure of volume ranging from a teacup to a bushel basket, as long as the measurement used is consistent. I'd suggest making up more soil mix than you think you'll need while you have all the ingredients assembled. Mix the components thoroughly as you go, just as if you were making granola.

Cruso's Potting Soil

one part commercial potting soil or leaf mold
one part sphagnum moss or peat moss
one part perlite or sharp sand

This is the formula for potting soil that Thalassa Cruso recommends in her book *Making Things Grow.* For fast-growing seedlings, I like to change the proportions to two parts soil or one part soil and one part compost, mixed with one part moss and one part sand or perlite. As originally formulated, for houseplants, the mixture is a bit lean to support strong seedling growth.

Good and Simple Potting Soil

one part finished compost
one part vermiculite

Easy to mix. Moisten it well before setting plants in it.

Home-Style Potting Soil

one part finished compost
one part either loose garden soil or potting soil
one part sharp sand, perlite, or vermiculite, or a mixture of all three

This one can be made without buying any ingredients if you have access to sharp sand.

A Rich Potting Soil

one part leaf mold
two parts either loose garden soil or potting soil
one part compost or rotted, sifted manure

This potting soil is especially good for lettuce and cabbage transplants.

A Lean Potting Soil

one part either loose garden soil or potting soil
one part perlite
one part vermiculite
one part milled sphagnum moss
one part sharp sand

This is mostly physical support for the plants. Use this mixture for plants that should grow slowly, such as cactus.

Amended Potting Soil

four parts either loose garden soil or potting soil
two parts sphagnum moss or peat moss
two parts leaf mold or compost
two parts vermiculite
six teaspoons dolomitic limestone

The limestone helps to neutralize acids in the leaf mold and peat moss.

If you want to formulate your own potting mixture from what you have or can readily find, you'll probably be on safe ground if you include each of the following:

- Soil, preferably loam, for nutrients.
- Sand or perlite for drainage. Gravel may also be used in the bottom of a solid container but not as part of a soil mix.
- Compost, leaf mold, vermiculite, or moss for water retention.

Heat Treatment

Commercial potting soil should be free of harmful bacteria and fungi, but garden soil is, of course, teeming with organisms. Most of them are beneficial. I never heat-treat or otherwise "sterilize" my soil. Besides being a large nuisance, the process kills helpful soil organisms and, at temperatures above 180°F (82°C), releases dissolved salts that can be toxic to plants. Even at 160°F (71°C), some salt-releasing breakdown occurs.

If you decide to sterilize a batch of soil, the following methods will work. Plan a picnic supper or weekend trip when you do treat your soil, because the kitchen will smell awful!

Using the Oven. Before heating the soil, moisten it thoroughly so that small puddles form when you press your finger into it. Bake the soil in metal pans in a preheated 275°F (135°C) oven. Small amounts, a gallon or so by volume, should be ready within 30 to 40 minutes. Larger quantities, a half bushel or so, may need to

Table 1

Temperature Necessary to Kill Soil-Inhabiting Pests and Diseases

---

Pests	*30 Minutes at Temperature*
Nematodes	120°F
Damping-off and soft-rot organisms	130°F
Most pathogenic bacteria and fungi	150°F
Soil insects and most plant viruses	160°F
Most weed seeds	175°F
A few resistant weeds, resistant viruses	212°F

Source: George "Doc" Abraham and Katy Abraham. *Organic Gardening Under Glass* (Emmaus, Pa.: Rodale Press, 1975).

remain in the oven for 1½ hours. It is important that the soil be wet. The steam generated by the water penetrates the spaces in the soil. Dry soil takes much longer to treat and smells much worse.

A meat thermometer inserted in the soil will indicate when it has arrived at the proper temperature to kill offending organisms. Damping-off fungi, for example, die at 130°F (54°C), and a temperature of 160°F (71°C) kills most other plant viruses and pathogenic bacteria. Avoid overheating the soil, for you want to retain as many of the more numerous helpful soil microorganisms as possible. Soil that has been baked should be rubbed through a hardware cloth screen to break up the clumps before using.

Using the Microwave. Soil poured into plastic bags can be sterilized in a microwave oven. Do this in batches of up to ten pounds. Placing the bags in the oven for 7½ minutes will kill most damping-off and root-rot fungi. Some researchers report, however, that microwave treatment changes soil pH, cation exchange capacity, and mineral content.

Using Solar Energy. An easier method recently developed at the University of California uses solar heat. Fill gallon-sized black plastic planting pots with moist soil and then place each pot in a clear plastic bag. Tie the bag closed and set the bagged pot in a sheltered location where it will receive full sun. Two weeks of direct-sun treatment in warm weather should kill most disease organisms.

Using Boiling Water. Pouring boiling water over a flat of soil does not sterilize it, but kills many microorganisms.

Helpful Hints

Perhaps the following hints will save you a bit of trouble, too.

Collect Soil in the Fall. The time to collect good garden soil for spring seed starting is in the fall, before the soil freezes solid. Even if you're lucky enough to have a winter thaw at the time you're doing your planting and transplanting, the soil is likely to be mucky, and you may not be able to dig very deeply, either.

Make Leaf Mold. There's leaf mold and there's leaf mold. Make your own seasoned pile for use in potting mixtures. One year I scraped up a bushel of lovely, crumbly, woodsy-smelling leaf mold from the edge of our woods and put it into my soil mixture. All the seedlings growing in this "special" mixture died. I'm not sure whether the stuff was too acid or if it was toxic in some other way, but from now on I test such soil amendments on a few plants rather than gambling my whole crop of early seedlings.

Avoid Peat Moss. Watch out for peat moss, too. Personally, I don't like it. It's chunky, unmercifully hard to moisten, and it crusts when it dries. I would never use peat moss in a soil mixture when I could get milled sphagnum moss. If you must use it, wet it very thoroughly several hours before you need it. I'd caution you, too, against using wetting agents (surfactants) in potting mixes or irrigation systems. They tend to reduce soil water retention, and their effect on people who handle and consume them hasn't been tested. Even mild soaps like Basic H, when mixed with soil in highly dilute solutions, reduced the growth of vital root hairs.

Use Builder's Sand. Sharp sand, specified in the soil recipes, is coarse builder's sand. Sand brought from the seashore is too fine and too salty. Don't use it for plants. Even lake sand or sand scraped up by the roadside is too fine for our purposes. It will pack into a cementlike mass that kills plant roots. I know—I used it by mistake . . . once.

Make Your Own Mix. What would I use if commercial potting mixture ingredients like vermiculite and perlite were unavailable? I would mix equal parts of compost; good garden soil, rubbed through a screen; and torn moss. I would not use garden soil alone, for no matter how good the soil, it tends to pack and harden when used indoors in small pots.

4 Containers

Your seeds are ready. Your soil is mixed. Now it's time to comb closets, attic, and cellar for likely discards to hold the soil and provide a temporary home for your growing seedlings.

The ideal container for starting seeds should be no more than two to three inches deep. One that is deeper will only use up more potting soil. A shallow pan, though, will dry out quickly and limit root development. The soil holders you use for germinating seeds do not necessarily need to have drainage holes. You can prevent pooling of water around plant roots by choosing the right growing mediums to fill the containers (see chapter 3). Long, narrow flats fit neatly under fluorescent lights; however, don't use flats that are so long that they will be heavy and difficult to handle when filled with wet soil.

Types of Containers

I've grown seedlings in all of the following kinds of containers, for different reasons and with varying results. Each has its place. Some are more useful than others. Certainly it is preferable, all else being equal, to make use of some of the many throwaways that clutter up our daily lives, but there *are* times when a specially prepared device will be more effective than anything else. It's silly to make do, unless you must, when with a little effort and some scrap wood you could rig up a really fine seedling-care system. At any rate, here's my list of container candidates, briefly evaluated for effectiveness, cost, and useful life.

Eggshells and/or Egg Cartons. They are inexpensive but not really much good in the trials I've run on them. Neither the shells nor the egg carton sections can hold much soil, and they dry out fast. I think it's a mistake, too, to start a child gardener with one of these cute but impractical gimmicks. The learning child deserves good equipment that will help him or her to be successful. Something about planting seeds in eggshells seems to have captured the imagination of armchair gardeners, and every spring you'll see articles in magazines suggesting the method. Go ahead and try it if it appeals to you. Just don't expect great results.

Milk Cartons. Yes, they're free, good for one growing season, and they fit together well under lights. You can either halve a half-gallon milk carton the long way or cut one side out of a quart carton. If you don't buy milk, no doubt your friends, relatives, or neighbors would save their cartons for you. Even the little pint and half-pint cartons may be used for larger individual transplants.

Aluminum Disposable Pans. The loaf pans make fine seedling trays. They last for several years, work quite well, and can be had free if you know people who will save them for you. The shiny sides of the loaf pan reflect light back onto the seedlings. One of my best generations of eggplant seedlings was raised in an aluminum loaf pan on the kitchen windowsill. Pie pans are more of a last resort for plants, for they are too shallow to hold enough soil for proper moisture retention and root development. They're fine for sprouting seeds, though, if you transplant promptly.

Peat Pots. A tray of seedlings all tucked into peat pots looks neat and satisfying, but despite the assumption that plant roots will readily penetrate the walls of the pot and find the surrounding soil, I have not found this to be true in all cases. More than once I've uprooted a plant and found its roots pretty well confined to the peat pot. Also, unless grouped closely together, peat pots dry out fast.

Peat Pellets. These compressed-soil pots in netting seem to offer less resistance to plant roots than the peat pots, if you remove the net when you plant the seedlings out. However, even though the net has air spaces, it often appears to inhibit plant roots. The soil mixture of peat pellets sometimes contains a chemical fertilizer. Peat pellets are easy to use and neat, but they are expensive and can't be reused.

Plastic Gallon Jugs. Cut-down plastic gallon jugs have worked fairly well for me, but I prefer the long, narrow cut-down milk cartons, which seem to accommodate more seedlings and waste less space under lights. Plastic jugs are free, lightweight, last two or three years, and may be further recycled by saving the cutoff top to put over seedling plants in the garden for frost protection.

Shoes. Yes, shoes. I've tried planting melon seeds in cast-off shoes on the advice of an elderly gardener whose flourishing crops indicate he must be doing something right. The shoes were fun to use and they certainly make a stir when visitors noticed them, but I can't honestly say that they made any difference in the plants that grew in them.

Bricks. For another unconventional and more widely adaptable seed-starter, use an unglazed brick set in an open frame of scrap wood that extends 1½ inches above the top of the brick. On top of the brick, put an inch-thick layer of moist sphagnum moss or vermiculite, and plant seeds in it. Then set the framed brick in a tray of water. The water should always touch the brick, but should not cover it. The continuous supply of moisture, coupled with good drainage and aeration, encourages good seedling germination.

Market Packs. These are rectangular trays made of pressed fiber. I think they are preferable to peat pots. The packs are not cheap, but they last for several years, fit together well in storage, and allow plenty of depth for root development. Also, unlike the larger flats, they may be kept on a sunny windowsill.

Clay Flower Pots. Clay pots last a long time and are fine for special transplants. They are the best containers for starting watercress seeds, because they can be kept in a tray of water so that the soil is constantly moist. As far as I'm concerned, though, they are too costly and unwieldy for most transplanting. I use small plastic pots to start cucumber and melon seedlings.

Household Discards. Chipped enamel broiler pans, rusty cake tins, or lopsided dishpans make good containers for groups of transplants. Old refrigerator crisper drawers work well, too, especially for lettuce and other leafy crops grown under lights in winter, because they hold plenty of soil.

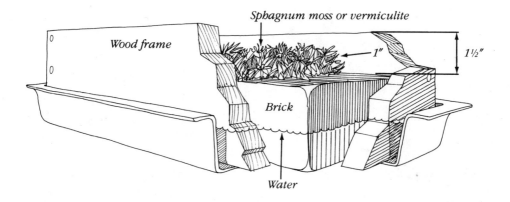

Sphagnum moss or vermiculite

Wood frame

1″

1½″

Brick

Water

The brick in this unique container absorbs water and keeps the growing medium moist.

Cottage Cheese Cartons. These cost nothing and outlast at least one use. I use them primarily for give-away seedlings, because in my setup the round shape of the containers wastes space under the lights. They do fit on most windowsills, though, which can be a boon to the gardener without lights. Incidentally, if you do set them at your window, you'll probably need to provide some protection—a tray or foil liner—to prevent spills and splashes from spotting your windowsills.

Flats. My favorite. When you raise your own seedlings, there is no such thing as too many flats. A flat is nothing more than a four-sided frame made of wood, about three inches wide, with slats nailed across one open side. The slats should be spaced one-eighth to one-fourth inch apart to allow for drainage. Most flats are rectangular, because that is the most efficient shape.

Six 12- by 16-inch flats will fit on each shelf of a 26- by 48-inch, double-bank fluorescent light cart. I wouldn't recommend making flats any larger than about 14 by 18 inches, or they'll need extra bracing to hold the weight of all that soil, and then they'll be too ungainly to handle. Flats are basic. Make as many as you can find time and materials for. You won't regret it.

Plastic Trays. Commercial plastic trays 2 inches deep and 18 to 20 inches long work well and last a long time, but they are costly. Poke a few drainage holes in them with an ice-pick.

Soil Blocks. These are better than peat pots because they promote good root growth without forming a barrier. They take time and special equipment to make, though. You need a soil-block mold, a metal device with a plunger that ejects blocks of soil from a mold; a tub for mixing; soil composed of two parts peat moss that has been rubbed through one-fourth-inch hardware cloth, one part garden soil, and one part compost; water; and a tray.

To make a soil block, wet the soil mix until it forms a slurry. Then pack a four-

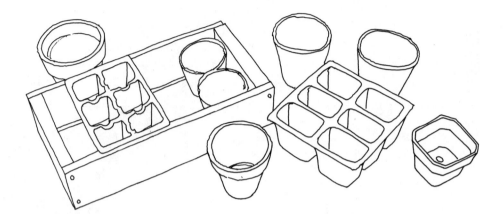

Any of a variety of containers, bought or found, can be used to start seedlings.

cube block maker with the wet mix, level off the top of the mold, and eject the cubes onto a tray. Plant a single seed in each block, or plant two and cut off the extra seedling if both germinate.

In my experiments with soil blocks, I found that they do indeed promote good root growth. For me, though, they had two disadvantages. The first was the time and trouble necessary to make the blocks. Secondly, the blocks must be very carefully and frequently watered from above so they don't disintegrate.

Speedling Flats. A more serious competitor for my favorite plain old rectangular flat is the Speedling flat—a molded plastic-foam tray 3 inches high, 13 inches wide, and 26 inches long. The plant-holding cells taper to a narrow open point at the bottom of each cell—an ingenious way to encourage a dense root ball that makes good use of the soil mix. In order to keep the roots confined to their spaces, the foam trays must be propped an inch or so above the surface to admit air. In my trials, foam flats produce excellent seedlings, but they're a bit more cumbersome to work with than conventional flats. They're durable, though; mine is still in use after four years.

Like most gardeners who can't resist starting enough plants for the whole neighborhood, my spring seedling setup is a motley arrangement of all the flats I can muster, plus representatives of most of the other kinds of containers mentioned above, which just about sums up my advice for other seed-starters. Use what you have, buy what intrigues you, but aim for a growing supply of long-lasting, inexpensive wooden flats.

To make soil blocks, pack a soil block mold with wet growing medium, eject the blocks into a flat, and plant a seed in each.

5 Sowing Seeds Early Indoors

I've been starting spring plant seeds indoors in late winter for some years now, and I'm still learning—still making mistakes, too. Once in a while the cats will knock over a flat of transplants. Then I must start all over again. Quite often I experiment with a new vegetable or flower, which shakes up the routine a bit, but in general I've worked out a pattern for this yearly ritual. I no longer worry about whether the seeds will come up. I know now that they will, if I keep them warm enough. I no longer start my tomatoes in early February, because I've learned it's better to set out plants that are in active growth rather than overblown specimens I've been holding for three weeks until the weather was right. Neither do I sow seeds as thickly as I once did, although this bad habit of the amateur gardener dies hard. It has to do, I suppose, with a lurking reluctance to trust those seeds. Only time and repeated sowing and harvesting can soften that inner uneasiness and lead to the respect for each seed that gives it sufficient room to grow.

My Timetable

It is all too easy when you have a severe case of the February blues and long for an early start on spring to get out the flats and the soil mixtures and start a whole batch of spring vegetables. Now, that's perfectly all right, if you can chalk the whole project up to experience or winter indoor recreation and start another batch later at the right time. It's also all right if you live in one of those benign climates where frost won't hit after early April. However, most of us would rather try to do it right the first time. So here's a little timetable to give you an idea of safe indoor starting times for your spring-planted vegetables.

Onions—12 to 14 weeks before the safe planting-out date (which is 4 to 6 weeks before the last frost)

Peppers—8 to 12 weeks before the last frost

Eggplant—6 to 8 weeks before the last frost

Tomatoes—6 to 8 weeks before the last frost

Lettuce—5 to 6 weeks before the safe planting-out date (which is 4 to 5 weeks before the last frost)

Cole crops (cabbage, broccoli, collards, etc.)—5 to 6 weeks before the safe planting-out date (after danger of severe weather has passed, but while nights are still cold)

Cucumbers and melons—2 to 4 weeks before the last frost (but don't plant out until the weather is warm and settled)

When I raised my seedlings on sunny windowsills, I started them well within the range of traditionally recommended times given above. If anything, I started them a bit on the late side, since late-winter sun is still quite weak, and it's not until well into March that you can expect old Sol to do much for your windowsill plants.

The date you choose for starting your plants, then, will depend on the variety you're growing; whether you'll raise the plants in a greenhouse, on a windowsill, or under lights; and the average frost-free dates in your growing area. If you are unsure of when those dates are, check with your local county extension agent. The office is often in the county courthouse.

A Sowing Checklist

Now that we've settled the when of seed starting, let's go on to the how. Here's a checklist of seed-sowing steps:

1. **Gather your equipment.** Seek out and collect the flats or other containers, planting mix, seeds, watering can, newspaper, labels, and markers you'll need.

2. **Prepare a work space.** Choose a spot where you'll have room to knock things around. Spread a layer of newspapers over the area. It's easier to gather up and discard the papers than to mop up spilled potting soil.

3. **Review your seed packets.** Look over your packets of seeds and write out a label for each one before you plant the seeds. You won't believe, till you've done it, how easy it is to mix up different seed lots. And if you plant five kinds of pepper and eight kinds of tomato, as I do, labeling the seed flats is doubly important. A hot pepper seedling looks no different from a sweet bell pepper seedling, and you may want to plant out ten of one kind but only five of the other.

Labels may be wooden Popsicle sticks, cut pieces of venetian blind slats, strips cut from plastic jugs, or ready-made labels from your garden supply store. I've also simply glued scrap paper labels to the sides of wood flats (glue new labels right over the old ones next year) or clipped tags to plastic containers to identify my plants. If you do any kind of experimenting with different fertilizers, kelp sprays, chilling treatments, or soil mixtures, accurate labeling and record keeping are even more important.

4. **Prepare your flats.** Spread a layer of newspaper on the bottom of each flat or container that has drainage holes or slits in the bottom to keep the soil from sifting through. If the spaces between slats are wide enough, wedge in a "caulking" of torn shreds of moss. Trim the newspaper so that it doesn't stick up above soil level, or it will act as a wick and draw off soil moisture. Fill the flats with the seed-starting medium of your choice (see chapter 3) or with potting soil if you prefer, up to about

one-half inch from the top. A deep container, half-filled, will both shade the plants and interfere with ventilation of the seedlings.

Sometimes I plant seeds in plain vermiculite, omitting the moss and soil layers described in chapter 3. The seedlings that are started this way must either be transplanted sooner to a soil mixture containing nutrients or fertilized every four days or so.

Finally, firm the top surface of the planting medium, preferably with a flat object like a board or a brick. This prevents the small seeds from tumbling too far into the crevices.

5. **Prepare the seeds.** Certain seeds will germinate more quickly if given a little head start before sowing. Scarifying helps to hasten germination of hard-coated seeds like morning-glory or New Zealand spinach. Just nick the seed coats with a knife or file. Don't cut deep enough to damage the embryo.

Presoaking cuts several days of germination time for slow-to-sprout seeds like carrot, celery, and parsley. I've had good luck planting these seeds after having poured just-boiled water over them, draining when cool, and mixing with dry sand to avoid clumping. Some gardeners put such seeds in small muslin bags to soak overnight before planting. Other seeds like peas and beans will sprout sooner if presoaked for an hour or so in warm water. Presoaking isn't necessary for most seeds you'll be planting indoors, though. Check the listing of plants in the encyclopedia section at the back of the book to find out whether a particular seed should be presoaked.

6. **Plant the seeds.** You can either scatter the seeds over the entire surface— especially when using a small container—or you can plant in rows—usually a good idea when raising seedlings in flats. At any rate, give each fine seed at least one-eighth inch of space from its neighbor; medium seeds need one-half inch of room, and large seeds one inch.

Getting the seeds to go where you want them can be tricky, but the knack comes with practice. To disperse fine seeds, try sowing them from a saltshaker. Medium-sized seeds like tomatoes, which may come only 20 or so to a packet in the case of some hybrid varieties, may be placed exactly where you want them with fine tweezers. According to most of the pictures you see, the real experts sow seeds by tapping them lightly directly from the seed packet. I always feel that I have better control over distribution of the seeds by taking a pinch of them in my fingers and then gently rotating thumb and forefinger to gradually dislodge the seeds. If all this is new to you, you might want to practice by sowing a few batches of seeds over a piece of white paper so that you can check to see what kind of distribution you're getting.

7. **Cover the seeds.** Except for very fine seeds, which may be simply pressed into the soil, you will need to cover the seeds with soil. The few seeds that need light for germination are noted in the encyclopedia section. These should be simply pressed into a damp seed-starting medium and covered with clear plastic or glass. The rule of thumb is that seeds should be covered to a depth of three times their size. A one-eighth-inch seed, then, would have three-eighth inch of soil over it. I hardly ever put more than one-half inch of soil over any seed, and then only for larger seeds. For indoor sowing I simply spread fine, light soil or vermiculite over the seeds. This

last layer needn't be wet. If the planting medium on the bottom is well soaked, the top will soon become damp, too.

The good thing about presoaking the growing medium is that it's then not necessary to water the planted seeds from the top. If you've ever tried this procedure, I'm sure you've found, as I have, that it's a surefire way to relocate all your carefully spaced seeds. If for some reason you do prefer to water your seedling flats from the top, either use a gentle spray from a rubber bulb sprinkler with a flat, perforated nozzle head, or spread wet burlap over the flat and water through that. In that way you'll avoid the flooding that would carry seeds willy-nilly to the corners of the flat. Be sure the top layer of vermiculite (or whatever) is pressed firmly over the seeds so that the seeds are closely surrounded on all sides.

8. **Cover the container.** You can use any one of the following materials to cover your container:

- damp newspaper or burlap
- a scrap piece of used aluminum foil
- a plastic sheet or bag slipped around the flat
- another seed flat

You can build a stack of seed flats if they are roughly the same size. In fact, the soil surface will receive better ventilation if the flats don't fit exactly edge to edge. Watch for mold in this case, or if you cover your containers with plastic, and provide better ventilation if mold develops.

9. **Set the containers in a warm place.** Place all planted containers in a warm place where seeds can germinate. I've found the arrangements listed on the next page conducive to good germination.

Using Moss in Seedling Flats

Because it is easily available to me, I always spread a layer of torn moss (gathered from our woods) over the bottom of any container in which I start seeds, except when using all vermiculite. The moss serves as an insurance cushion, soaking up extra water so that it doesn't puddle around the roots. At the same time, it holds this moisture in reserve so that the flat doesn't dry out too quickly. If you use moss, it is important to tear it into small pieces, rather than leaving it in big wads, so that each seedling may be easily separated from the others at transplanting time. I thoroughly dampen the moss before adding the second layer. Moss should be dried for several months before it is used in seed flats; fresh green moss sometimes seems to inhibit plant growth.

Although most experts recommend punching drainage holes in all seed-planting containers, I have found that in many cases my seedlings have done quite well in containers without drainage holes, as long as I've put this cushion of moss on the bottom, and as long as I'm in control of the watering. Flats that are kept outside and that are rained on become badly waterlogged if they have no holes.

You can use commercially available sphagnum moss also, as long as it is not finely milled.

- Put flats in the corner behind a wood-burning stove, preferably not on the floor where cool air settles, but elevated two feet or so on a small stable or box.
- Set flats on a high shelf above a floor register. Watch for excessive drying, though.
- Set small batches of seeds on the pilot light on a gas stove.
- Set flats on top of a turned-on fluorescent light fixture, like a light cart setup with several tiers of lights, and they'll receive steady warmth at the ends of the tubes. Once you have the early plants (onions and peppers) sprouted and under lights, you can set the next batch of seeds that you want to germinate on the dark but warm tier above the lighted plants.

A Second Method

An alternative method of indoor seed planting, used by an accomplished gardener with whom I enjoy trading lore, goes like this. Fill a cut-down milk carton or plastic jug with starting medium, water it well, and plant the seeds. Enclose the container in a plastic bag and keep the whole shebang good and warm. After three or four days, punch drainage holes in the bottom of the container (use a pick or knife). Keep warm and covered, but allow to drain and check for mold until you see the seeds have germinated.

Presprouting Seeds

Ever since I picked up the idea from Dick Raymond's book *Down-to-Earth Vegetable Gardening,* I've been presprouting seeds of tender crops like cucumbers, squash, pumpkins, and melons. The first three are planted directly in the ground after they've germinated, and you'll find more information on dealing with them in the encyclopedia section. Melons bear earlier if pre-germinated and then planted in individual pots for a two-week head start indoors. (Some gardeners feel that cucumbers also do, but I've not been convinced of that yet.)

Presprouting melon and other cucurbit seeds has given me a much higher rate of germination than I was able to get when planting two seeds to a peat pot, probably because the seeds receive more constant warmth and moisture. I highly recommend presprouting seeds of those crops that need especially warm conditions during germination.

How to Presprout Seeds. To start the seeds on their way, space them evenly on a damp double layer of paper toweling (or several thicknesses of paper napkin). Be sure that no two seeds touch. Carefully roll up the towel, keeping the seeds as well separated as possible, and tuck the rolled cylinder into a plastic bag. If you label the rolls, you can put more than one variety in the bag.

Put the bag full of damp, rolled, seed-filled towels in a warm place. I use the

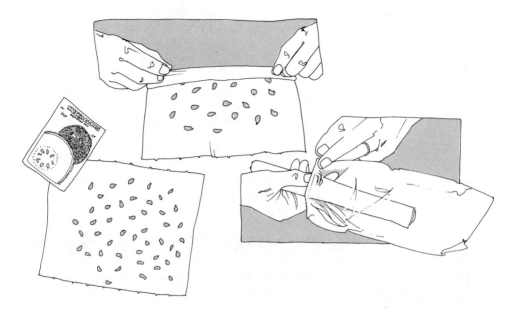

To presprout seeds, first spread them on a damp paper towel. Roll up the paper carefully and place it in a plastic bag. Set the bag in a warm place until the seeds have germinated.

warm top of a fluorescent plant light fixture or the top of my insulated hot water heater.

Check the seeds each day. Nothing will happen for several days, but germinating seeds do need a certain amount of oxygen, and the small amount of air that wafts in when you peek at the seeds will do them good.

Signs of Life. The first sign of germination in cucurbit seeds will be the development of the root. Be sure to remove your sprouted seeds from their incubator before the root hairs grow together and tangle. If one should grow through the towel—and this often happens—just tear the towel and plant the damp shred of paper right along with the seed.

Planting Presprouted Seeds. Plant the presprouted seeds in a good rich potting mix that you've scooped into individual containers and premoistened. Cucurbits don't take kindly to transplanting, not with those fleshy, sappy, easily bruised roots, so your plants will stay right in these containers until they go into the garden. Individual half-pint milk cartons work well, since the bottom may be easily removed and the whole plant clump set in the hill. I plant my sprouted cucurbits in small plastic pots, which are easily overturned to free the seedlings when it is time to set them out. Cover the presprouted seed root lightly but firmly with soil, and set the pots under lights immediately, or put them on your sunniest windowsill.

6 Germination

If you've ever lost track of a flat of germinated seeds, as I have, and discovered them too late—when the thready white stems had grown an inch before putting on pale little leaves—then no doubt you've also muttered ruefully, "Out of sight, out of mind." It *is* easy, in the busy spring rush, to overlook flats of planted seedlings tucked in out-of-the-way corners. For that reason, I like to keep my flats together in a place where I'll see them every day and remember to check on them.

The Process of Germination

What's going on in those flats while we wait? We commonly think of germination as being equivalent to sprouting, and it's true that the final test of complete germination is the emergence of a growing root or leaf sprout from the seed, yet the process by which a dry, dormant embryo quickens into tender, new green growth begins well before we have visible evidence of the new root or leaf.

The first step in the process of germination is the absorption of water by the seed. This is a necessary preliminary to the internal changes in the seed that trigger growth. The uptake of water by the seed (called imbibition by botanists) depends in turn on the content of the seed, the permeability of its outer layer, and the availability of the necessary amount of liquid. Seeds that contain a high percentage of protein imbibe more water than those that are high in starch. (Only under very acid or hot conditions, which don't exist in nature, will seed starch swell with water intake.) Seeds with hard coats, like morning-glories, will absorb water more readily if their hard outer shell is nicked with a file. Furthermore, the seed depends not only on the presence of moisture in the soil, but also on close contact with soil particles, to permit sufficient water uptake. The fact that a seed has absorbed water is not, by the way, proof of its viability. Even dead seeds can imbibe water.

As the seed swells with water, it develops considerable pressure, pressure that eventually ruptures the seed coat (which has already been softened by the surrounding moisture) and eases the eruption of the root. These are the physical effects of the seed's absorption of water. At the same time, internal metabolic changes are revving up life in the seed, changing its chemistry from neutral to first gear, you might say. As the seed tissues absorb water, food stored in the endosperm is gradually changed into soluble form, ready to be used as a component of new tissue.

In order, though, for the starches and proteins in the endosperm to dissolve, they must often be changed into simpler forms—the starches into simple sugars like glucose and maltose and the proteins into free amino acids and amides. The enzymes, necessary to split complex forms of stored food into simpler forms of usable food, are activated in response to the stepped-up metabolism of the seed. You will

28

remember that even dormant seeds carry on respiration. They take in oxygen and release carbon dioxide. The rate of respiration is markedly increased in the germinating seed and both generates and supports the many interacting internal changes in the embryo.

Enzymes, then, direct the breakdown of certain useful stored foods. Hormones, also present in the seed, control both the transportation of newly soluble foods to different parts of the seed and the building up of new compounds from the components of those that have been broken down into simpler forms. Pea seeds, for example, synthesize new compounds during the first 24 hours of the germination process.

The product of all this stepped-up activity within the seed is new tissue, originating at growing points in the root tip, the stem, the bud, and the cotyledons. This new tissue is formed in two ways: Cells already present in the seed grow longer, and cells divide to produce new cells, which then elongate.

Studies done on lettuce seeds, for example, show that cell division begins about 12 hours after germination has begun; the root cells show some elongation at about the same time. In corn, the first change to be observed is the enlargement of the cells, followed by cell division in the root as it emerges from the seed coat. Both kinds of tissue changes are necessary to the normal development of the seedling.

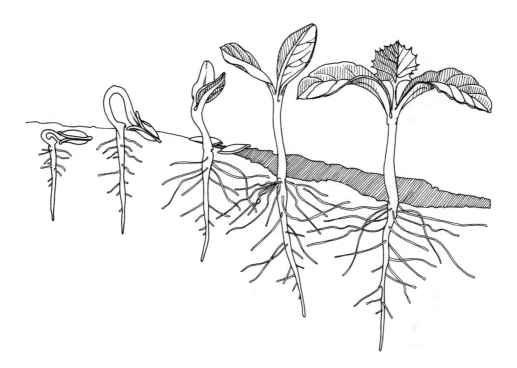

Seed germination.

Factors that Influence Germination

Many internal and environmental conditions influence the course of germination in seeds.

The Condition of the Seed. A shriveled seed that has been stored too long or under poor conditions will have a scant supply of food stored in its endosperm. The seedling that grows from such a seed, if it germinates at all, is likely to be weak and stunted.

Mechanical injury to the seed during harvesting or drying can injure the cotyledons, stem, or root tip, or produce breaks in the seed coat that admit microorganisms, which in turn deteriorate seed quality. The hormones that promote cell elongation are produced by the endosperm and cotyledons. Any injury, therefore, that interferes with the soundness of these hormone-producing tissues is likely to result in stunted seedlings.

The Presence of Water. Water must be available to the seed in amounts sufficient to start the quickening of respiration that leads to germination. However, few seeds will sprout if submerged in water. Some air must also reach the seed for it to absorb the oxygen it needs. Water serves several purposes in the germinating seed. Initially, it softens the seed coat so the root can emerge more easily. Then it combines with stored foods to form soluble forms of nourishment for the seed. As growth proceeds, it helps to enlarge new cells, as directed by the hormones, and serves as a medium of transportation to take soluble foods and hormones to parts of the seedling where they're needed.

Sufficient Air. Even quiescent seeds in storage need a certain minimum supply of air. The requirements of a germinating seed are more critical. Our atmosphere contains a mixture of gases, with the oxygen portion fairly constant at 20 percent. (Some seeds—certain cereals and carrots—have been shown to germinate more completely in an even richer oxygen concentration.) The oxygen taken in by the seed in respiration combines chemically with the seed's fats and sugars—a process called oxidation.

Seeds also need a certain amount of carbon dioxide in order to germinate, but they don't do well if surrounded by a considerable concentration of carbon dioxide. Cucurbits (squash, melons, pumpkins, and cucumbers) have seed membranes that admit carbon dioxide more readily than oxygen, so they are acutely sensitive to low-oxygen conditions. As gardeners, we can't manipulate the composition of the air that surrounds us, but we can make sure our germinating seeds are supplied with enough air by planting them shallowly, in a loose, friable medium, and by keeping the soil moist but not waterlogged, so some air spaces remain.

Temperature. In general, seeds need warmer soil temperatures during germination than they will need later when they've grown into plants. There *are* differences in heat and cold sensitivity, though, among the different species. Some seeds, like lettuce, celery, and peas, germinate best at low temperatures, while peppers, eggplant, melons, and others prefer more warmth. Extremes of heat and cold inhibit germination of most kinds of seed. There are, for most seeds, optimum temperatures at which they do best. This doesn't mean your seeds won't germinate at higher or lower temperatures; the drop in germination with less-than-ideal temperatures is gradual, not abrupt. Seeds that germinate best at a soil temperature of 75°F (24°C) will put forth some growth at 65°F (18°C), although it might occur later and might be less.

Some seeds, like dock, tobacco, and evening primrose, need alternating warm and cool temperatures in order to germinate. According to studies done with these plants, it's not the rate or duration of the temperature change but simply the fact of the change itself that affects them.

The most favorable germinating temperature for most garden seeds started indoors is between 75°F and 90°F (24°C and 32°C). That's *soil temperature*. Remember that whereas the air temperature in a room may be 70°F (21°C), a moist flat of soil set on the floor may be cooler unless kept near a source of heat. (See table 2 for temperature ranges at which different seeds will germinate.)

A soil-heating cable may be used to speed germination of some seeds that are more difficult to start. Applying bottom heat directly to the flat uses less power than heating the whole room or greenhouse. In most households, though, there is usually at least one spot—over the furnace or water heater, on top of the television set, near a wood stove or heat register, or on a pilot light—where seed flats can be kept warm during germination.

Light. A fair number of flower seeds and some tree seeds either require light for germination or germinate more completely in the presence of light. Most vegetable seeds are indifferent to the amount of light they receive during germination. We used to think that darkness was essential to germination, but recent studies don't seem to support that conclusion as a generality. Some flower seeds, like those of the pansy, germinate best in darkness, and germination of onions and chives appears to be retarded by exposure to light. A few vegetable seeds germinate more completely under some conditions when they receive some light. In the case of lettuce and celery, for example, light promotes more complete germination only when the temperature is higher than that at which these seeds usually germinate best. At the lower temperatures they prefer, exposure to light doesn't seem to make much difference. The lesson from this is clear: When putting in a late planting of lettuce or celery when the weather is warm, press the seeds into moist soil, covering them lightly, if at all, with fine soil—although you could also spread a few dry grass clippings over the row to prevent crusting.

Vegetable seedling flats can be covered with wet newspapers, damp burlap, or used aluminum foil. Flats of seeds that need light to germinate should usually be covered with clear plastic sheets, bags, or food wrap, except for seeds that are known to germinate within a week, which may be left uncovered.

Soil Conditions. Apart from the physical conditions of friability, aeration, moisture, and freedom from waterlogging, all of which promote germination, there are other conditions in the soil that may affect the outcome of seed planting.

Organic Matter. Soils containing a high percentage of organic matter along with many microorganisms may have a higher concentration of carbon dioxide than the surrounding air. This can retard germination, depending on the permeability of the seed coat to carbon dioxide. Seeds don't need a rich mixture to *start* germinating.

Table 2

Soil Temperature Conditions for Vegetable Seed Germination

Crops	Minimum (°F)	Optimum Range (°F)	Optimum (°F)	Maximum (°F)
Asparagus	50	60–85	75	95
Beans, Lima	60	65–85	85	85
Beans, Snap	60	65–85	80	95
Beets	40	50–85	85	95
Cabbage	40	45–95	85	100
Carrots	40	45–85	80	95
Cauliflower	40	45–85	80	100
Celery	40	60–70	70*	85*
Corn	50	60–95	95	105
Cucumbers	60	65–95	95	105
Eggplant	60	75–90	85	95
Lettuce	35	40–80	75	85
Muskmelons	60	75–95	90	100
Okra	60	70–95	95	105
Onions	35	50–95	75	95
Parsley	40	50–85	75	90
Parsnips	35	50–70	65	85
Peas	40	40–75	75	85
Peppers	60	65–95	85	95
Pumpkins	60	70–95	95	100
Radishes	40	45–90	85	95
Spinach	35	45–75	70	85
Squash	60	70–95	95	100
Swiss Chard	40	50–85	85	95
Tomatoes	50	60–85	85	95
Turnips	40	60–105	85	105
Watermelons	60	70–95	95	105

Source: J. F. Harrington, Department of Vegetable Crops, University of California at Davis.
*Daily fluctuation to 60°F or lower at night is essential.

Salt. A high salt content, found in some seaside soils, can block germination by drawing water from the seeds.

Calcium. Some seeds respond favorably to a high calcium content in the soil.

Leaf Mold. Leaf mold from the woods may contain germination-inhibiting substances. Beech tree leaves, for example, develop a compound that inhibits germination after they've been exposed for a winter. The fresh leaves do not contain this compound. Eucalyptus leaves also contain germination-inhibiting substances. As I mentioned in chapter 3, I once killed a whole batch of seedlings by planting them in a mix containing some perfectly lovely leaf mold I had scraped up at the edge of our woods. Leaf mold, on the whole, is great stuff, and I use it regularly in our garden and compost pile with no ill effect. But since many of these interactions are still little understood, I no longer collect my leaf mold for seedling mixtures in the woods, but rather save maple leaves from our yard trees for this purpose.

Table 3

Number of Days for Vegetable Seeds to Emerge at Different Temperatures

Crops	32°F	41°F	50°F	59°F	68°F	77°F	86°F	95°F	104°F
Asparagus	0.0	0.0	52.8	24.0	14.6	10.3	11.5	19.3	28.4
Beans, Lima	—	—	0.0	30.5	17.6	6.5	6.7	0.0	—
Beans, Snap	0.0	0.0	0.0	16.1	11.4	8.1	6.4	6.2	0.0
Beets	—	42.0	16.7	9.7	6.2	5.0	4.5	4.6	—
Cabbage	—	—	14.6	8.7	5.8	4.5	3.5	—	—
Carrots	0.0	50.6	17.3	10.1	6.9	6.2	6.0	8.6	0.0
Cauliflower	—	—	19.5	9.9	6.2	5.2	4.7	—	—
Celery	0.0	41.0	16.0	12.0	7.0	0.0	0.0	0.0	—
Cucumbers	0.0	0.0	0.0	13.0	6.2	4.0	3.1	3.0	—
Eggplant	0.0	—	—	—	13.1	8.1	5.3	—	—
Lettuce	49.0	14.9	7.0	3.9	2.6	2.2	2.6	0.0	0.0
Muskmelons	—	—	—	—	8.4	4.0	3.1	—	—
Okra	0.0	0.0	0.0	27.2	17.4	12.5	6.8	6.4	6.7
Onions	135.8	30.6	13.4	7.1	4.6	3.6	3.9	12.5	0.0
Parsley	—	—	29.0	17.0	14.0	13.0	12.3	—	—
Parsnips	171.7	56.7	26.6	19.3	13.6	14.9	31.6	0.0	0.0
Peas	—	36.0	13.5	9.4	7.5	6.2	5.9	—	—
Peppers	0.0	0.0	0.0	25.0	12.5	8.4	7.6	8.8	0.0
Radishes	0.0	29.0	11.2	6.3	4.2	3.5	3.0	—	—
Spinach	62.6	22.5	11.7	6.9	5.7	5.1	6.4	0.0	0.0
Sweet Corn	0.0	0.0	21.6	12.4	6.9	4.0	3.7	3.4	0.0
Tomatoes	0.0	0.0	42.9	13.6	8.2	5.9	5.9	9.2	0.0
Turnips	0.0	0.0	5.2	3.0	1.9	1.4	1.1	1.2	2.5
Watermelons	—	0.0	—	—	11.8	4.7	3.5	3.0	—

Source: J. F. Harrington, Agricultural Extension Leaflet, 1954.
Notes: 0.0 = Little or no germination.
— = Not tested.

Monitoring the Process

Keeping all these factors—water, air, temperature, light, soil conditions—in balance, while we wait for those first spears of green to show, calls for checking the seed flats at least once a day. The soil should be kept moist but not soggy. Air should be allowed to reach the soil surface at intervals, at least enough to prevent the formation of mold on the surface. Although I formerly surrounded each container of seeds with a plastic bag until the seeds germinated, I found that mold often formed on the soil because of poor air circulation. Now I simply cover the top of the flat, without surrounding it with a moisture-proof barrier. If you *do* find mold on the soil, chances are that exposing the flat to the air for an hour or so will take care of the problem.

Then, one day—often, in the case of peppers, just about when you'd given up hope—you'll notice little elbows of stems pushing through the soil surface.

Table 4

Percentage of Normal Vegetable Seedlings Produced at Different Temperatures

Crops	32°F	41°F	50°F	59°F	68°F	77°F	86°F	95°F	104°F
Asparagus	0	0	61	80	88	95	79	37	0
Beans, Lima	—	—	1	52	82	80	88	2	—
Beans, Snap	0	0	1	97	90	97	47	39	0
Beets	—	114	156	189	193	209	192	75	—
Cabbage	0	27	78	93	—	99	—	—	—
Carrots	0	48	93	95	96	96	95	74	0
Cauliflower	—	—	58	60	—	63	45	—	—
Celery	—	72	70	40	97	65	0	0	—
Corn	0	0	47	97	97	98	91	88	10
Cucumbers	0	0	0	95	99	99	99	99	49
Eggplant	—	—	—	—	21	53	60	—	—
Lettuce	98	98	98	99	99	99	12	0	0
Muskmelons	—	—	—	—	38	94	90	—	—
Okra	0	0	0	74	89	92	88	85	35
Onions	90	98	98	98	99	97	91	73	2
Parsley	—	—	63	—	69	64	50	—	—
Parsnips	82	87	79	85	89	77	51	1	0
Peas	—	89	94	93	93	94	89	0	—
Peppers	0	0	1	70	96	98	95	70	0
Radishes	0	42	76	97	95	97	95	—	—
Spinach	83	96	91	82	52	28	32	0	0
Tomatoes	0	0	82	98	98	97	83	46	0
Turnips	1	14	79	98	99	100	99	99	88
Watermelons	—	0	0	17	94	90	92	96	—

Source: J. F. Harrington, Agricultural Extension Leaflet, 1954.
Note: — = Not tested.

Flowers that Need Light to Germinate

We're accustomed to tucking seeds in under a thin blanket of soil when we plant them, but there are some flower seeds that should not be covered. For these seeds, exposure to light increases the permeability of their membranes, allowing oxygen to penetrate the seed coats more readily (as you remember, a germinating seed needs oxygen to support its quickened respiration). In most cases, it's a matter of degree. Seeds that need light to germinate may produce a few seedlings when covered (perhaps these are the ones that weren't quite covered after being sown), but many more will germinate if light reaches the seeds. In tests on cactus and tuberous begonia seeds, for example, 30 percent of the seeds germinated when covered, whereas 80 percent germinated when exposed to light.

The flower seeds listed on the facing page will germinate better in the presence of light. Just press them into the surface of moistened soil and cover the seed flat with a sheet of glass or with clear plastic wrap so that light can get through.

Flower Seeds that Need Light to Germinate
(Do not cover these seeds with soil.)

Annuals	*Perennials*	*Biennials*
Ageratum	Alyssum saxatile	Bellflowers
Begonias	Balloonflower	English Daisies
Browallia	Chinese Lanterns	Foxglove
Coleus (tender perennial)	Chrysanthemums	
Godetia	Columbines	
Impatiens	Edelweiss	
Kochia	False Rock Cress	
Lobelia	Feverfew	
Mignonette	Gaillardia	
Petunias	Maltese Cross	
Portulaca	Oriental Poppies	
Scarlet Sage	Primroses (except	
Snapdragons	Chinese)	
Strawflowers	Rock Cress	
Sweet Alyssum	Shasta Daisies	
	Sweet Rocket	

Flower Seeds that Often Germinate Best in Some Light
(Cover these seeds very lightly, if at all, with soil.)

Annuals

African Daisies	Nicotiana
Balsam	Stocks
Celosia	Tithonia (Mexican
Cleome	Sunflower)
Cosmos	Transvaal Daisies
Monkey Flower	(tender perennial)
(Mimulus)	Wishbone flowers

Flower Seeds that Germinate Best in Darkness
(Cover these seeds with ¼ inch of fine soil, well firmed down.)

Annuals	*Perennials*
Bachelor's Buttons	Delphiniums
Butterfly Flowers	Poppies (except
Calendula	Oriental)
Globe Amaranth	Shamrocks
Nasturtiums	Soapwort
Nemesia	
Painted Daisies	
Pansies (biennial)	
Phlox drummondi	
Salpiglossus (painted	
tongue)	
Sweet Peas	
Verbena	

7 What Seedlings Need

The most crucial time in the life of a seedling is the period just before it has broken dormancy. Each tentative-looking little plant sprout is rather like a baby. It must have its needs met immediately. If it finds only darkness when its cotyledons break through the soil surface, it will send up a pale, weak stem in search of the light it must have. If the soil in which it is growing dries out, it can no longer enter a holding stage like the dormancy it went through as a seed. There is no going back. It has started into growth, and if it is to continue, there must be moisture within reach of the roots and light on the leaves.

Light and Temperature

As soon as the plants have germinated, then, they must be given light, either from fluorescent tubes or from the sun by way of a house or greenhouse window. I try to catch my seedlings even before the whole cotyledon has emerged, when the seed is still just a sprout. This is easier to determine when the seeds haven't been deeply buried. Then when I put them under lights I know they'll have the stimulation they need from the very beginning. (For more about light sources, intensity, and so forth, refer to chapter 10.)

Temperature requirements of seedlings are not as critical as light, perhaps, but nevertheless are important. Once the plants are growing above ground, they need less warmth than is required for germination. The majority of vegetable plants that germinate most rapidly at 70°F to 80°F (21°C to 27°C) do well when grown at 60°F to 70°F (16°C to 21°C), with night temperatures about 10°F (5.5°C) lower. Exceptions are mentioned for individual plants in the encyclopedia section at the back of this book. Cool growers like lettuce and onions will still flourish when temperatures drop to 50°F (10°C).

In a severe winter, seedlings kept in an unheated room must occasionally be moved or covered to protect them from freezing. Undesirably high temperatures may be a more common problem. Although I keep flats of germinating seeds in the warm corner (70°F to 80°F, 21°C to 27°C) where heat rises from our wood stove in the kitchen, I move the flats to our cool solar greenhouse as soon as the seedlings are up. Plants grown indoors in warm rooms put on weak, spindly, sappy growth that is difficult to manage under lights and to prepare for the transition to colder outdoor temperatures. Start seeds warm and grow seedlings cool.

There is, in fact, some evidence that judicious chilling of tomato seedlings, at just the right time, promotes earlier and heavier fruiting. Dr. S. W. Wittmer of Michigan State University found that tomato seedlings kept at temperatures of 52°F to 56°F (11°C to 13°C), starting immediately before the seed leaves opened and continuing for 10 to 21 days, developed up to twice the usual number of flowers in the first cluster and often in the second. Since the position of the lowest group of flower buds on the plant is determined approximately a month to six weeks before blooming, the plant must be chilled during this early stage, before the opening of its first true leaves, in order to induce earlier flowering lower on the stem.

Other vegetable horticulturists report that peppers may also be induced to form early buds by chilling the plants in the seed-leaf stage. When I read these reports, it dawned on me that I'd been prechilling my pepper seedlings for years, though not through any planned program aimed at early bloom. The peppers *are* in bloom when I set them out in the garden in May, partly because I always start them extra early—in January—thus inadvertently chilling them at the right time.

Space

Soon after the seed leaves unfold, conditions often become crowded in the flat, unless you have spaced your seeds with mathematical precision. I know I usually need to thin mine, especially lettuce and cabbage. The best way to do this is not to yank the extra plants out of the soil, but to cut them off, using small embroidery or

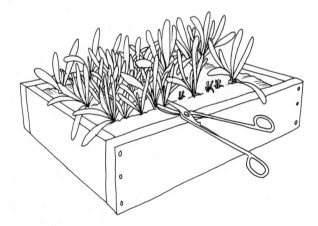

When thinning seedlings in a container, snip the tops. Pulling them can disturb the roots of nearby plants.

Give seedlings a little tender loving care.

nail scissors. A considerable amount of root growth often takes place even before the green leaves go into operation, and the roots of plants you want to save may be damaged by pulling out neighboring roots.

There is more going on in the roots of even the youngest seedling than most of us ever fully recognize. Roots not only anchor the plant and absorb nourishment, but are also responsible for maintaining the pressure that enables the plant to raise water, against the force of gravity, to its topmost leaves and stems. In many plants, roots give off exudates that help to define and sometimes defend the plant's territory. Roots of all plants synthesize many of the amino acids that control the plant's growth.

Roots are examples of "being and becoming," to borrow Aristotle's phrase, used perceptively by Charles Morrow Wilson in his book *Roots: Miracles Below.* Root growth is continuous. It does not stop, ever, as long as the plant lives. At the same time, root filaments are continuously dying. Root hairs, the tiny fibers that form the point of contact and exchange between the soil and the plant, constantly extend into new territory. They need moisture, but they also need air. A soil mix that is half solid matter and half pore space, with about half of the pore space filled with water, provides ideal conditions for root growth.

Superfluous seedlings are just like weeds. Thinning the seedlings, so that their leaves don't overlap, cuts down on competition for light, moisture, and nutrients and also helps to promote better circulation of air around the plant.

Seedlings also need water and nutrients, of course. Watering and fertilizing are discussed in chapter 9.

8 Transplanting into New Containers

There comes a time shortly after your seeds have germinated when you will want to transplant them into a deeper container with a richer growing medium. Some seedlings, though, should not be moved from their original containers until you plant them in the garden (see the box in this chapter).

Why Transplant?

Those seedlings that can be transplanted benefit in several ways.

Stimulation of Feeder Roots. Some fine roots are broken in the transplanting process. As a result, a new, bushier network of feeder roots is formed. If roots are very long and thready, as in the case of onions, I often prune them to one inch or so, at the same time pruning the top growth to correspond.

Room to Grow. Crowded seedlings become weak and spindly, and the lack of air circulation around them can encourage disease. Giving them more root and leaf space promotes the health and strength of the plant and ensures that the root ball of the plant will have sufficient protective soil around it when you cut the plants apart later to set them out in the garden.

Richer Soil. Although seedlings started in soilless mixes seem to be able to subsist on liquid feedings for quite a while, I like to get my plants into a good soil mixture so that they can take advantage of the micronutrients and the beneficial interactions in the microscopic soil life, limited as these may be in the small space of a wooden flat.

Easier Selection and Evaluation. Discarding a new young green sprout is always a painful process for me. In fact, I usually keep a small container of these extra plants by the kitchen window for a week or so after transplanting, in the hope that some visitor will adopt them. Selection, nevertheless, is a necessary step in raising the best possible nursery plants. No matter how sparingly I try to sow the seeds or how carefully I thin the seedlings, I usually have more seedlings in a starting flat than I

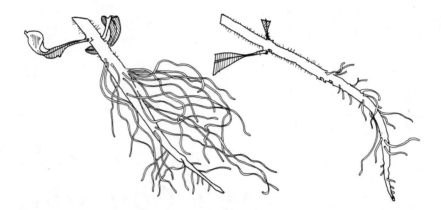

Good root growth (left) versus poor root growth (right) in a tomato seedling.

will have room for under lights or in the garden row, and so I choose, as I transplant, the seedlings that appear most vigorous.

I look for good green leaves that are symmetrical and well developed. Deformed cotyledons sometimes indicate early damage to the seed. Vegetable seeds that sprout early usually grow into vigorous plants unless they are retarded by waiting without light for the other, slower seeds to come up. There is, in fact, some evidence that early emergence is an indication of seed vigor and ultimate high yield.

Root growth is as important as top growth in evaluating seedlings. Look for a compact, well-developed root ball. A fringe of well-branched feeder roots will do far more for the developing plant than a single, thready trailing root. It is also difficult to transplant a long, single-filament root without tangling it, bruising it, or doubling it back on itself.

When to Transplant

By the time they have developed their first true leaves, your seedlings are ready to be transplanted. It's better to get the job done before the plant has a second set of fully developed leaves, because from then on the likelihood is that the stem will be longer and more easily injured, and the roots may be lengthy and trailing and difficult to trace through the soil mix.

Some accomplished gardeners prefer to do their transplanting when the seedling is even younger, just after the cotyledons have emerged completely, and the plant is standing upright in the soil. This method does have advantages. It permits you to save almost every plant grown from expensive seed, the kind that comes ten seeds to a $2 packet, since none are lost in thinning. It puts the plants into a richer soil mix at an early age and frees your seed-planting trays and mixes for the next wave of plantings.

I have used this method occasionally for selected flower and vegetable seeds, but generally I prefer to thin my seedlings in the cotyledon stage and let them develop their first set of true leaves before transplanting. This allows me to choose the very best ones from the flat.

How to Transplant

Transplanting is more than a technique. When done well, it involves respect for the young life of the plant, even a certain empathy that can sense the thrust and direction of the tender growing roots, the reach and promise of the unfolding green leaves. The plant wants to grow. It is, you might say, programmed to grow. Having set the process in motion by planting the seeds, we now have the opportunity to give each seedling the most careful treatment so that it will continue growing smoothly on its way to producing our food.

Let's suppose that you have a flat of seedlings that you want to transplant into a larger, deeper flat of richer potting soil. The following steps will guide you through the transplanting process:

1. **Prepare the container.** Your first step will be to prepare the flat. Flats or other containers used for growing seedlings should have drainage holes in the bottom. Spread a double layer of newspaper on the bottom to keep soil from sifting through the drainage cracks. Next, arrange a one-inch layer of torn pieces of moss on top of the newspaper. This is not absolutely necessary, but in my work with young plants I've found that those planted over a torn-moss foundation develop excellent root systems. If you have no source of fairly coarse natural moss or commercial unmilled sphagnum moss, I'd use plain potting soil, but I'd be sure to include some perlite or sand in the mix to promote drainage. Don't use a bottom layer of plain finely milled horticultural sphagnum moss. It's too fine.

Moisten the moss, pat it down flat, and fill the rest of the flat to within about one-half inch of the top with potting soil. Usually I moisten the soil thoroughly before setting the seedlings in their places, although at times I have put the seedlings into dry holes, watered them in, pulled fresh soil over the roots, and then watered again lightly. This unorthodox method has given me equally good results.

Plant Bands. Transplants may also be moved into plant bands, which should be wedged snugly together in a bowl or pan to prevent drying. To make plant bands, first take a piece of doubled newspaper about seven or eight inches long and four inches wide, and tape it around a jar two to three inches in diameter to shape it. While the paper is on the jar, fold in one end as you would a package, but don't tape it shut. Remove the plant band from the jar. Place your plant bands side by side in a pan—an old roasting pan works well—and put one-half inch of soil in the bottom of each band to anchor it in place before inserting the seedling. The folded-over

Transplanting a seedling into a new flat provides it with richer soil and more room to grow.

bottom, if it hasn't rotted away at planting time, may simply be peeled off before setting the plant in the garden.

    2. Prick out the seedlings. Most seed-starting gardeners soon acquire a favorite tool for this purpose. The miniature houseplant shovels that usually come as part of a set are too clumsy for most small, closely spaced seedlings. I use the slim, slightly pointed handle of an old salad fork. Wooden Popsicle sticks, pencils, and old screwdrivers can serve the same purpose. What you want is an instrument that will

Plant bands, used as containers for transplants, can be made by shaping newspaper around a jar and taping the sides together.

lift out the seedling while causing the least damage to its roots and those of its neighboring plants.

When seedling roots are compact enough to come up in a relatively dense cluster, it is often a good idea to water the old flat shortly before removing the seedlings from it so that enough soil will cling to the roots to help prevent transplanting shock. Seedlings raised in vermiculite, which generally falls away readily anyway, or those with long, extensive root systems, won't benefit from prewatering. By the way, it's better not to dip the roots of seedlings in water before planting them in their new location. This makes the roots cling together when, instead, they should be individually surrounded by soil.

Remove the seedlings one by one from their old quarters, planting each one before digging up another. Even a few minutes of air drying will adversely affect those delicate roots. Hold the seedlings by their first leaves rather than by the easily bruised stem, or support the root ball in your hand.

3. Replant the seedlings. You have several options here. You can, as suggested above, set the plants in already-moistened soil. A second option is to plant them in dry soil (over wet moss); then firm the dry soil around the seedling, and finish with a light watering to further settle the seedling.

Or, if you are putting the young plants into pots, you can tuck a wad of damp moss in the bottom, set the plant on it, half fill the pot with soil, water the plant in, and add enough additional dry soil to fill the pot.

If you're potting up a plant that has long roots, put some soil or moss in the pot and turn it on its side. Position the plant in the pot with its roots spread out on the growing medium, and gradually fill in with soil as you tilt the pot bit by bit back to its

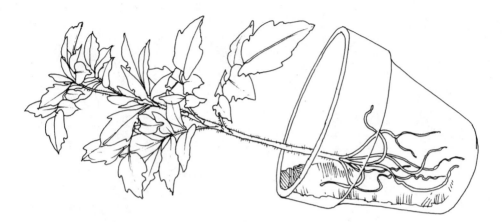

To pot a plant with a long root, lay the pot containing some growing medium on its side and spread the roots over the medium. Gradually lift the pot to a vertical position as you fill it with growing medium.

normal position. Instead of being coiled around each other, the roots will have more of their surface exposed to soil.

When transplanting, set the seedlings just slightly deeper than they were in their original container. Tomatoes can be planted as deep as the root will allow. Plants that form a leafy crown, though, like lettuce and cabbage, should not have soil pulled over or into the crown. When settled into place, the first leaves of young transplants should be at about the same level as the sides of the container.

Press the soil gently but firmly over and around the roots. Roots need a certain amount of air, which should be provided by a good soil mix that is not overwatered, but they also need close and immediate contact with soil particles in order to absorb necessary nutrients.

Except for onions, leeks, and chives, which grow spears rather than spreading leaves, transplants should be spaced at least two inches apart in the flat. Three inches is better for tomatoes, cabbage, and eggplant that won't be transplanted again.

4. **Watch your seedlings**. Check for signs of wilting. If they droop, even though you've watered them well when planting them, don't pour on more water. The roots are already doing all they can. Instead, help to balance the plant's moisture supply by arranging a tent of damp newspaper over the flat or by enclosing it in a large plastic bag or covering it with a large roaster lid or other protective layer that will not rest on the plants. Keep new transplants out of direct sun for a day. I put mine back under fluorescent lights half a day after moving them, as long as they show no signs of wilting. Wilted plants should be kept shaded and cool until they perk up, which shouldn't take more than a day or two at the most.

To Transplant or Not

Plants that take well to the transplanting process are generally improved by the experience. Other plants, because of their fleshy or deep roots or their special sensitivity to root insult, should not be transplanted. This group includes:

Vegetables
beans
cabbage, Chinese
corn
cucumbers
melons
pumpkins
root crops, except beets, turnips, and celeriac
squash

Flowers
balloonflower (or transplant early, before taproot forms)
California poppy
celosia
poppies (or transplant with care)
portulaca

Herbs
borage
burnet (transplant only when young)
caraway (transplant only when young)
chervil
coriander
dill (transplant only when young)

Trees
Nut trees should be transplanted as early as possible because they form deep taproots.

9 Growing On

Your young plants should be showing more new growth within a week after transplanting. The trick now is to keep them growing steadily until it's time to prepare them for planting outdoors.

Watering

It's important to remember, first of all, that soil in containers dries much more rapidly than the deeper, heavier soil you have in your garden. Wilting that lasts more than a day may retard the seedling, so it is best to check the flats every day to see whether they need watering. I usually water my transplants in flats about every third to fifth day, but I base the decision on the condition of the soil, not the time schedule. Short of letting the plants wilt, it is a good idea to let the soil dry out from time to time. Then the roots are stimulated to grow into the air spaces between the soil particles in search of water.

When to Water. Water is powerful stuff. It can nourish your plants, or it can kill them. Water is necessary to maintain the turgidity of the plant so that there is a continuous column of moisture in the cells. It is also indispensable to the intricate intracellular chemical processes that keep the plant growing. However, too much water ruins the soil tilth and spoils the plant. Soil that is continuously waterlogged has no air spaces to promote root growth and support helpful soil bacteria. In such a situation, other less beneficial organisms take over, rot sets in, and the roots suffer.

How do you tell when a flat needs watering? By touch, by feel, and by sight. Touch the soil. Poke your finger into a corner where it won't disturb any roots. If the soil feels powdery or dry, it needs water. When you've been handling flats for a while, you can judge by feeling the weight of the flat whether it has lost most of its water. A well-watered flat is heavier than a dry one. Other than dry soil surface, visible signs like drooping plants indicate a serious lack of water.

Water Temperature and Composition. Does water temperature and composition matter? I think it does.

46

Temperature. Our well water is cold, but I always use tepid water for seedlings of heat-loving plants like melons, cukes, peppers, eggplant, and globe artichokes. Water temperature is less critical for the brassicas (which include kale, cabbage, cauliflower, broccoli, and mustard), lettuce, onions, and such, which prefer a cooler environment; still, I try to temper extremely cold water to a more moderate 70°F (21°C) to avoid shocking the plant.

Chlorine. If your water is chlorinated, I'd suggest letting it stand overnight with as much surface as possible exposed to the air—in a bucket or dishpan—before using it to water seedlings. High levels of chlorine can harm plants and kill beneficial soil bacteria. Symptoms of chlorine damage include yellowing along leaf veins and, in extreme cases, curling of leaves.

Soft Water. Water that has gone through a water softener should not be used to water plants. It contains potentially toxic amounts of sodium.

Some gardeners who are good at planning ahead fill plastic jugs with water and keep them ready so that they always have room-temperature water available for their plants.

How to Water. Yes, there is more than one way to water your seedlings. Some expert seedling-raisers water their flats from the bottom. This practice supplies water quickly to the root zone and avoids excess surface dampness that can encourage damping-off disease. It is a messy business if you don't have a good setup for it, but if you have a large waterproof trough in which you can partially submerge a flat, or preferably two or three at a time, the method might be practical for you. Each flat must, of course, have drainage slits or holes in the bottom so that water can be absorbed. Leave each container of seedlings in the water until the surface of the soil starts to look dark.

Bottom watering is really more of a greenhouse refinement than it is a necessity. I've had no problem at all with top watering either seedlings or more mature plants. Watering from above carries nutrients down through the soil to the roots and makes it feasible to top-dress the plants with compost or rich soil if necessary. I often use the muddy water left from rinsing root-cellar vegetables to water my indoor seedlings.

Maintaining Humidity

Humidity—the amount of moisture in the air—is important to plants. They lose a great deal of moisture through their leaves when the air is excessively dry, as it can be in both centrally heated homes and in rooms warmed by a wood stove during cold weather. Plants suffer when humidity falls below 30 percent. They do best in the 50 to 70 percent range; higher humidity invites fungus and other disease problems. Misting plants helps, but only temporarily. The best way to increase humidity around potted seedlings is to set them in containers of damp moss or vermiculite, or in a container that holds an inch of pebbles in half an inch of water.

Potted seedlings that spend more than a month or two on this damp bed may send out roots into the moss or vermiculite. You can prevent this by setting a rack or some kind of improvised support between the damp bed and the plants.

A purple coneflower seedling shows healthy growth in a flat.

Draping a plastic tent over the indoor garden helps to hold humidity in, although it also cuts down on air circulation. During the winter, air circulates well in a heated house, because the decrease in air pressure caused by burning fuel prompts an intake of fresh air and keeps interior air in motion.

If you notice green algae or fuzzy mold growing on the soil surface, your plants probably have more water and less air than they need. Run a small fan in the area, remove any plastic covers, and hold off on water for a few days. You might also try sprinkling some powdered charcoal over the soil surface to correct the problem.

A good way to increase humidity around potted seedlings is to place them in trays containing an inch of pebbles and ½ inch of water.

Fertilizing

Fertilizer, like water, is a necessity that should not be overdone, or salts will accumulate in the soil to a toxic concentration. It is better to give frequent small feedings than occasional large feedings to very young seedlings. For the first three weeks, in fact, any fertilizer the seedlings receive should be half-strength rather than the full-strength mixture (1½ teaspoons to a gallon of water, for example, instead of 1 tablespoon to the gallon).

When to Fertilize. Newly emerged seedlings, still in the cotyledon stage, have absorbed enough of the seed's stored nourishment to get them well off the ground. I don't fertilize my seedlings until they've begun to develop their true leaves. Seedlings growing in nutrient-free vermiculite should receive about two feedings of diluted fertilizer a week. Later, when they have been transplanted to a richer mixture containing soil and/or compost, they can go on a ten-day to two-week fertilizing schedule.

Types of Fertilizer. You can choose from a few effective fertilizers to feed your seedlings.

Fish Emulsion. I use diluted fish emulsion, a source of trace minerals as well as the three major elements, nitrogen, phosphorus, and potassium, to feed my seedling plants. The usual dilution, one tablespoon to the gallon, seems to hold true for most of the commonly sold fish-fertilizer preparations. The labels of these bottles usually become hopelessly frayed and streaked before the contents are used up, so I'd suggest jotting down the appropriate dilutions for different plants while you can still read the label.

Sprout Water. Water drained from sprouting seeds, such as that I collect from the alfalfa and mung bean sprouts I raise in our kitchen, is valuable food for plants. I am convinced that sprout water helps to foster plant growth, although I can't cite any scientific studies as proof. In my one "controlled" experiment, two flats of Elite onion seedlings that I watered-in at transplanting time with sprout water grew thicker, sturdier tops than two untreated flats of the same variety in the same soil. They maintained their advantage until planting-out time. Since gibberellins (growth hormones) are known to be abundantly present in seeds, it makes sense to me to suppose that the water used to soak and rinse sprouts might well contain dissolved growth hormones.

Eggshell Water. Some gardeners save their eggshells and soak them in water until the brew develops an odor you wouldn't want in the house. (Leave them on the back porch or in the barn.) The resulting solution, diluted by an equal volume of water, is used to fertilize half-grown seedlings and other plants. Unless you're especially partial to essence of rotten egg, I'd use this on greenhouse, cold frame, and outdoor plants only, but it seems a good use of an otherwise neglected resource.

Providing Light

Plants need light, as discussed in chapter 10, in order to combine airborne elements, water, and soilborne food into the raw materials of growth. There's no sense in fertilizing a plant that receives insufficient light (less than eight to ten hours a day in the case of vegetable seedlings). The plant can't make use of the food.

If, due to conditions beyond your control, your plants don't receive as much light as they should, keep the temperature low so that new growth will be less spindly. At lower temperatures, plants can tolerate less light.

Plants do a large part of their growing at night. The side away from the light adds more new tissue than the side closest to the light. This is what causes them to stretch and lean in the direction of light. Plants grown at a window should be rotated one-quarter turn each day so that no one side elongates and leans too far in any one direction.

10 Light

Seedling plants may have all their other needs met—nourishment, air circulation, correct temperature, water, good growing medium, humidity—but without sufficient light, they will amount to nothing.

People are limited to using the sun's energy in indirect ways, but green plants have the ability to absorb energy from sunlight directly and to use that energy to make food they can store. Light striking a green leaf sets in motion the process of photosynthesis—the conversion, through the action of chlorophyll, of water and carbon dioxide to simple sugars and starches. Only the green parts of leaves carry on this process. White stripes, blotches, and leaf margins do not contain chlorophyll and so are not capable of photosynthesis. Red-leaved cabbage, coleus, ruby lettuce, and similar plants contain chlorophyll, but red pigment in their cells masks the green.

The Facts about Light

Light, although it may appear white to us, is actually a mixture of a rainbow of colors. A full spectrum of light includes the following color gradations, each of which has a different effect on plant life.

Green-Yellow Light. The chlorophyll in the plant reflects green-yellow light. Its effect on growth is thought to be negligible.

Orange-Red Light. This light stimulates stem and leaf growth.

Violet-Blue Light. Enzyme and respiratory processes are regulated by the violet-blue portion of the spectrum. In addition, this light encourages low, stocky growth.

Infrared (Far-Red) Light. This stimulates germination of some seeds, but can inhibit others. Its full effect still isn't completely understood.

Full-spectrum light, like sunlight, includes invisible ultraviolet rays, too. It is only the visible part of the spectrum described above, however, that provides the energy necessary for photosynthesis.

Making the Most of Natural Light

If you have a greenhouse, of course, your plants will receive the well-balanced light they need. The quality of sunlight that shines on windowsill-grown plants is the same as that in a glass-enclosed greenhouse, but the intensity is much lower, especially on cloudy and early winter days when the sun is low in the sky. You can boost the amount of light your windowsill plants receive by positioning shiny metal reflectors or boards painted with flat white paint behind the plants to bounce the light back onto their leaves. Use foil-covered cardboard, shiny cookie tins, or other household findings. The resulting arrangement may not win any interior decorating prizes, but it *does* get more light to the leaves. Try it with eggplant, a real sun and heat lover.

Another problem with raising seedlings on windowsills is that few houses have enough south-facing windows to provide a place for more than a few pots or flats. If you *are* limited to raising your seedlings at the window, choose the kinds of vegetables that most need an early start—main-crop tomatoes, peppers, and eggplant, for example—and sow lettuce, cabbage, and broccoli seeds in the open ground or in a cold frame. You can also raise two generations of indoor seedlings if you put your tomatoes and peppers in a cold frame about the first of May and then use your windowsills for the melons and cucumbers. Where space is limited, you can plant seeds of a hardier tomato like Sub-Arctic directly in the garden rather than raise your early tomato seedlings indoors.

Other places to put your seedlings where they will receive natural light include the following:

- A sunny corner in an outbuilding or barn; best in mid- to late spring in an unheated building.
- A roof garden, protected by a plastic tent, cloches, or some other covering, which can be effective but may be cumbersome to care for. A flat, black garage roof or house roof absorbs a lot of heat.
- Cold frames—see chapter 16.
- A solar greenhouse—see chapter 23.

Artificial Light

A few sources of artificial light are available to choose from, but fluorescent light seems to work the best.

Sunlamps. Sunlamps might sound like a good light source for seedlings, but any gardener who tries them will find that they are death on plants. The high concentration of ultraviolet rays in the sunlamp interferes with normal plant growth.

Incandescent Light. Incandescent bulbs produce red light, which alone makes the plant grow leggy. They also produce a great deal of heat in relation to the amount of light they give off. It is not practical to try to raise seedlings under incandescent lights. In lighting supply stores, you will see special incandescent bulbs with built-in reflectors designed for raising plants. These are used as auxiliary light sources in greenhouses or for houseplants. They must always be installed in a porcelain socket, and care must be taken to prevent cold water from splashing the bulbs when watering the plants, or the blown glass may break.

Fluorescent Light. The discovery that plants do well under fluorescent light has made it possible for many more gardeners to get a good early start on the outdoor growing season and to produce plants as good as any raised in a greenhouse. Plants grown under fluorescent lights develop excellent color and stocky growth. Fluorescent light comes closer than any other artificial illumination to duplicating the color spectrum of sunlight. In varying proportions, according to the type of the bulb used, these lamps emit light from the red and blue bands of the spectrum. The tubes give off more than twice as much light per watt of power consumed as incandescent lights.

Fluorescent tubes of various kinds give off different shades of light, including warm white, cool white, natural white, and daylight. The different kinds of fluorescent powder used to coat the inside of the tube account for the range of light quality available.

Reports of results from growing plants under fluorescent light at the North Carolina State University School of Agriculture and Life Sciences and at Cornell University, as well as from experienced nonprofessional gardeners, indicate the following:

- Plants do well under a variety of tube combinations.
- Special plant-raising tubes are not necessary for starting vegetable plants.
- Best results are often obtained by mixing tube colors: For example, using one warm-white and one cool-white or daylight tube in each fixture.

Cool-white tubes emit a bright bluish white light. Warm-white tubes have a faint tan or pinkish cast. The cool-white tubes are the easiest to find, but most hardware stores will order warm-white ones for you if they don't have them in stock.

Special plant growth tubes give more blue-red than green-yellow light. Their effect on plants is mostly cosmetic: The plants look good but do not necessarily grow any better than they do under cool-white bulbs, which are the most efficient. In addition, the special-purpose fluorescent tubes are more expensive and have a shorter useful life.

In the early days of fluorescent tube experiments, incandescent lights were thought to be necessary for flowering, but it is now known that they are unnecessary. They generate so much heat that they cause rapid drying of soil and air, and they burn out much sooner than fluorescent tubes.

Using Fluorescent Light Efficiently

The effectiveness of your fluorescent lights will be influenced by the way you use and care for them.

Length of Tubes. If you are preparing a fluorescent-light setup for your plants, buy the longest tubes you can manage to fit into the space you have. Why? Light at the ends of the tubes is weaker than that in the center and falls off more as the tube ages. If my experience is any guide, you never have enough light space under the tubes. The more you have, the more plants you're tempted to start, and when they're transplanted, you'll need all the space you can muster.

Tubes are available in 12-, 18-, 24-, 36-, 48-, 72-, and 96-inch lengths. Each foot of length uses 10 watts of power. If at all possible, avoid using tubes under three feet in length; they simply don't put out as much light for the power they use. Forty-eight-inch tubes are long enough to be efficient but short enough to fit conveniently into most household arrangements.

Amount of Light. For growing seedlings, your fluorescent-light setup should provide 15 to 20 watts per square foot of growing area. A single tube is, in most cases, both insufficient and inefficient, unless you have a long, skinny tray of seedlings under it. If you must use a single tube, construct a simple frame to hold foot-square mirrored tiles on both sides of the flats to reflect more light. A double row of tubes will give enough light for a flat up to about 16 inches wide, and two parallel double rows, like those attached to plant-growing carts, are even more efficient.

Types of Tube. If you are buying components and putting together your own light center or centers, you have a choice of the channel tube—a single- or double-mounted tube on a slim metal base, without a reflector—or the more common two-tube industrial-type fixture with a bent metal reflector. Channel tubes work well on shelves and undersides of cabinets, especially if surrounding surfaces are painted white to reflect more light. Industrial reflector fixtures may sometimes be obtained secondhand, but they are also widely available in lighting supply stores and from household mail-order catalogs.

High-Intensity Discharge Lamps. If price is no object, the most efficient lights you can buy today are the high-intensity discharge lamps, which—along with their fixtures—are costly to purchase but less expensive to run. There are two kinds of high-intensity discharge lamps: metal halide and high-pressure sodium. The metal halide HID lamps produce 94 lumens per watt (including ballast); cool-white fluorescent tubes produce 66 lumens per watt; and special plant-growth lights give off even less light: 37 lumens per watt for Agrilite, only 20 for Gro-Lux. The high-pressure sodium type is even more efficient (132 lumens per watt), but because its light has an unnatural yellow cast, some gardeners prefer the more natural, slightly less efficient white light of the metal halide lamps; these lose power more rapidly than sodium lamps. HID lamps cannot be mounted as close to the plant as fluorescent lights; the current limit is two feet for lights with wide-angle reflectors and six feet for those with standard mounting.

According to an industry spokesman, metal halide lamps of less than 400 watts, and those mounted in other than vertical positions, constitute a radiation hazard. Cracked bulbs can emit eye-damaging rays, much like those released by welding (according to a letter to the editor from Agrilite Company official, *Horticulture,* March 1987, p. 7).

Efficient Use. My first grow light was a 20-watt tabletop stand—a toy that helped me to get through a long northern Midwest winter when snow covered the ground until April. I raised tomato, pepper, lettuce, and pansy seedlings under that little light, and spring came after all. By then, of course, I was hooked. The following year, after moving back to Pennsylvania, to an old house full of nooks and crannies, I had a decentralized system—plant lights on every floor, from the basement to the kitchen to the bathroom, and I began raising all the seedlings we needed, racing the season to bring the earliest possible lettuce and cabbage to the table.

When we moved to our farm, we accepted a plainer, simpler house at first because we were hungry for land. In an old house that boasted not a single closet, we needed our shelves for books and canned garden produce, so I splurged on a four-shelf plant cart with four 40-watt tubes attached to three of the shelves.

I was delighted with the way my plants grew under fluorescent lights. They were stocky and green, with a special bloom to them. My only regret was that the lights consumed electric power. I tried to use them as efficiently as possible. The following tips can help you to get the most out of your lights.

1. Keep the tubes clean. Dust on the tubes decreases their efficiency.

2. Add reflecting surfaces to your setup. Use of reflecting surfaces like mirrors or aluminum foil under and around the lights gives the plants more light for the same power output.

3. Use flat white paint on shelves and reflecting boards. Flat white paint reflects more light than glossy paint.

4. Don't let the temperature drop too low. Fluorescent lamps seem to function best if the temperature doesn't fall below 50°F (10°C). Lights operated at around 40°F (4°C) may not perform as well.

5. Keep fluorescent lights turned on. To get more usable time out of the tubes, avoid turning them on and off more than absolutely necessary. A long burning time after each start is conducive to more economical operation and longer tube life. Most lamps last one to two years (10,000 to 20,000 hours) if turned on only one or two times a day. Efficiency decreases by 10 percent after a few months of use.

6. Get double use out of your lights. It is also possible, I've found, to save electric power by installing fluorescent lights for plants in spots where illumination is needed anyway. For example, we kept a fluorescent light stand on top of the refrigerator where the light it shed helped to illuminate a dark corner. We also installed a fixture on the underside of a shelf in the bathroom of our old house— light for the room *and* the plants. An imaginative look at your own home surroundings will no doubt suggest other possibilities.

7. Make use of the warmth of the lights. The tubes themselves give off little heat, but the ballast—the step-down transformer that makes it possible for the lights to use household current—does become warm. Most fluorescent lights have a ballast at the end of the fixture. Some recent arrangements have a remote ballast. Judicious planning of flat placement in relation to the warmer end zones of the tubes can make it possible to utilize this extra warmth to advantage, for example, in germinating seeds or starting sweet potato slips.

8. Reuse old tubes. Since light brightness decreases with the age of the tube, many gardeners routinely replace their seed-starting fluorescent tubes each year. You can still use the old tubes for general lighting purposes.

Grow-Light Setups

How you arrange your grow-light setup depends on the amount of space you have and the type of light you will be using.

Let There Be Light

Rapid-start fluorescent fixtures may occasionally develop starting problems if humidity around the light center is unusually high. Ventilation to promote better air circulation usually solves this problem. If you notice that a tube flickers, the starter—a small metal cylinder—probably needs to be replaced.

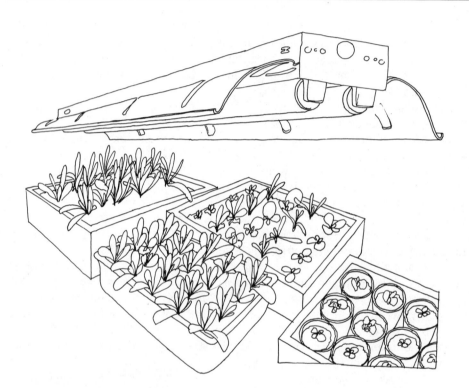

A typical grow-light setup.

When Space Is Limited. Not many of us have an entire room to devote to plants. A basement is often the most spacious area available, and that usually works very well unless furnace heat affects the plants or water is not readily at hand. Sometimes a bit of shoe horning is required to fit lights into an apartment or small home. Once, when plant space really got tight around here, I even tied a 36-inch two-tube light fixture to the underside of a piano bench so that it was suspended a few inches above a flat of plants. A crazy-looking arrangement, but it kept my tomato seedlings going until the cabbage and lettuce on the big cart were ready to graduate to the cold frame and leave room for the next wave of plants. If you're looking for ways to sneak in another light fixture or two or three, perhaps the following list will suggest some possibilities:

- Use a fluorescent study lamp you may already have.
- Install fixtures on bookshelves or storage shelves.
- Make a closet into a fluorescent light center with several tiers of lights and storage space for plant supplies.
- Install fixtures on the bottom surface of kitchen wall cabinets.
- Fit old buffet tables, radio cabinets, and other furniture with fluorescent tubes after removing interior partitions.

If your indoor seed-starting space is severely limited, you'd be wise to consider some of the following when deciding which plants to start early:

- Plants that need a long period of growth to prepare them for setting out.
- Those that are most costly or difficult to find commercially.
- Those that produce well over a long period.
- Fine-seeded plants that might get lost in the garden row.
- Vegetables you like best.

Distance from Light. Seedlings, in general, need more intense light than mature plants. If they are not getting enough light, they will develop long stems before their first leaves appear. Either they're overcrowded or they are too far from the light. I keep my seedlings extremely close to the light tubes, as close as possible short of touching the leaves to the glass, and never more than three to four inches away for the first three or four weeks after germination. Then, if they are stocky and growing well, I lower them by an inch or two and continue to lower the flat placement gradually as plant height increases. Light spreads more when plants are farther from the tubes. It's also a good idea, since light at the ends of the tubes is relatively weak, to trade positions of the various flats of plants from the ends to the more fully lit center every week or so.

The shelves on my plant cart are adjustable, but I find it easier to jack up the seedling flats by using egg cartons, shallow cardboard cartons, piles of magazines, and other improvised supports that may be removed gradually as the plants grow.

Industrial reflector fixtures, the kind we once used in a basement light center, may be suspended by chains. If you leave enough slack, about 12 to 15 inches, in the chain to raise and lower the light so that it can be set at the right height for either small seedlings or taller plants, you'll have a good adjustable setup.

The Hours of Light

In my experience, seedlings do very well with 16 hours of light a day, the usual recommended time. I have made do with 12-hour light exposures under limited light space. The lights burned day and night, and the seedlings took turns basking under them for 12 hours at a time. The tomatoes got a bit spindly, but most plants grew surprisingly well on this schedule. Sixteen hours is much better, though. Even 18 hours is not too long, but it's longer than necessary for most seedlings. Lighting

---

Cooler Temperatures Help in Low Light

Plants can better weather a period of light deprivation if they are kept cool, especially at night. According to studies conducted by a group of Russian scientists, the consequent lowering of the plant's metabolism that occurs at cooler temperatures apparently forestalls damage to the reproductive system, which would otherwise prevent flowering in plants subjected to more than three days of darkness. In general, plant metabolism slows at temperatures under 50°F (10°C).

---

time should not be increased above 18 hours. Too much light will disturb a plant, just as too much water or fertilizer will.

Providing Plants with Darkness

The light a plant receives, you remember, makes it possible for the leaves to manufacture starches and sugars, the components of growth. But a period of darkness is necessary for the plants to put these new compounds to use. Plants don't "rest" at night. They digest and grow. Both processes go more smoothly when night temperatures are 5°F to 10°F (3°C to 5.5° C) lower than daytime levels. The ideal ranges are 70°F to 75°F (21°C to 24°C) during the day and 55°F to 65°F (13°C to 18°C) at night.

Scientists have concluded that plants are actually attuned to the dark period rather than to daylight. Darkness is indispensable to normal growth. Some plants, in fact, blossom only when nights are long (chrysanthemums and poinsettia). Others, like onions, need short nights to develop fully. Most vegetable plants are day-neutral, meaning their full development isn't dependent on a light-dark rhythm of any certain pattern.

Turning off the lights at night, then, is as important as turning them on in the morning. When I shifted two batches of plants under lights that were on continuously, I was careful to provide "night" for the plants that received illumination from 8 P.M. to 8 A.M. by covering them with cartons or newspaper tents during the day.

11 Problems

Fortunately, seedling plants seem to thrive within a rather generous range of conditions. If your seedlings show compact growth, plenty of good green leaves, short internodes between leaves, and slow but steady growth, all is well.

Occasionally, though, excess or deficiency in one or another of the young plant's life requirements will cause trouble. If you interpret the problem correctly and treat it promptly, you have a good chance of saving the seedlings.

Signs of Trouble

Troubled plants give clues. What should you look for? Check for the following signs.

Leaf Curl. A plant whose leaves curl under, especially in bright light, may be suffering from overfertilization.

Solution. Naturally, you'll decrease the amount of fertilizer your plants receive if you notice symptoms of overfeeding. If the problem is severe, you might need to replant the seedlings in another flat of fresh potting soil. Be sure to leach out the extra fertilizer salts from the soil in the old flat before reusing it.

Yellowing of Lower Leaves. This is sometimes a sign of overfeeding, although it can also indicate magnesium deficiency.

Solution. Follow the same procedures as for leaf curl above.

Dropping of Leaves. Along with plant stunting, the loss of leaves may be caused by exposure of the plant to leakage of partly consumed gas from the stove or water heater. You can test for gas leaks by buying yourself a bouquet of carnations. If the flower petals curl upward, gas has probably escaped into the air. Tomatoes will droop in the presence of gas, too. Natural gases like butane and propane don't seem to hurt plants, but manufactured gas does.

Solution. Repair the leak.

Leggy Plants. Plants with long, often weak stems that have large spaces between the leaves growing from the stem may be suffering from any one of the following conditions: insufficient light, excessively high temperature, and crowding of plants.

Solution. A leggy plant can't be reshaped to conform to the stocky, well-grown ideal form, but there are steps you can take to promote more normal growth from this point on. First, of course, remedy the cause: Supply more light, lower the temperature, or thin the plants. Plants kept at a window might need some auxiliary evening light on cloudy days. Then, if at all possible, transplant leggy seedlings. The root pruning and general mild trauma of being moved will set back their growth a bit. Set the plant deeper in the new container.

Bud Drop. Particularly with pepper seedlings, bud drop may occur if the air is excessively dry.

Solution. Try setting the flat in a tray containing pebbles and water, or mist the blossoms with a fine spray of water at least once a day.

Leaf Discoloration. Discolored leaves usually indicate a nutrient deficiency.

Pale Leaves. In seedlings that receive sufficient light, pale leaves are a sign of nitrogen deficiency. Tomato seedlings that are severely deficient in nitrogen develop a deep purple veining, especially prominent on the undersides of the leaves.

Reddish Purple Undersides. A plant deficient in phosphorus will have a reddish purple color on the undersides of its leaves. In addition, the plant will often be stunted, with thin, fibrous stems. Soil that is too acid may contribute to the unavailability of phosphorus.

Bronze or Brown Leaf Edges. Leaf edges that have turned bronze or brown may reveal a plant in need of potassium. Brown leaf edges may also appear on plants that are overwatered. (See table 5, page 71.)

Solution. The differences among the various symptoms of deficiency are not always clear cut, especially in young plants. Your best bet, if you do notice leaf discoloration that can't be accounted for otherwise, is—in the short run—to give the plant a dose of fertilizer that contains trace minerals. In the long run, include some compost in your next soil mixture, or transplant the ailing plants to a medium that contains compost.

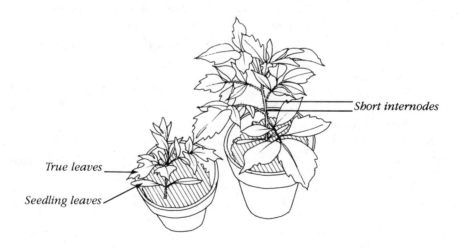

True leaves

Seedling leaves

Short internodes

Healthy seedlings show dense, compact growth and short internodes.

Discolored Roots. This is often the result of a buildup of excess fertilizer salts in the soil. Either the plant has been overfed, or overheating the soil in an attempt to sterilize it has released a high concentration of soluble salts into the soil. Water-logged roots often turn dark and may have an unpleasant odor.

Solution. Replant the seedlings in fresh soil if possible. Your second choice would be to flood and drain the flats several times in an attempt to leach out the toxic salts. If the flat is waterlogged, either provide better drainage or replant seedlings in better aerated and drier soil.

Mold. If you see mold on the surface of the soil, it indicates poor drainage, insufficient soil aeration, possible overfertilization, and/or a lack of air circulation.

Solution. Remedy the cause and treat the symptom by scratching some powdered charcoal into the soil surface.

Insect Damage. Damage on seedling plants grown indoors most probably means that conditions are not ideal for the plant.

Solution. Information on controlling pest damage on plants grown indoors can be found in chapter 23.

Damping-off. When damping-off occurs, you don't get much warning. Your first sign that the problem has hit your new plants is the total collapse of a few seedlings; the green leaves are still intact, but the stem has characteristically withered away right at soil level. Young seedlings are the most vulnerable.

Solution. Once a seedling has been attacked by the damping-off fungus (actually, there is a complex of microorganisms, any of which may cause the trouble) it can't be revived. The lifeline between root and stem has been cut off. However, you can try to prevent the problem by following these practices:

- Maintain good air circulation around seedlings by keeping the soil level high in the growing containers and thinning seedlings to avoid over-crowding.
- Avoid overwatering.
- Sow seeds in a sterile medium such as finely milled sphagnum moss or vermiculite.
- Presoak seeds in a small amount of water containing one or two crushed garlic cloves—an ancient practice that still makes sense, now that we know more about garlic's fungicidal properties. You can also treat seedlings with a garlic spray. (Blend one clove of garlic with one quart of water and strain.)
- Sprays of chamomile or nettle tea are sometimes used on seedlings to help prevent damping-off.

It is often possible to save the remainder of a flat of seedlings if only a few have died. Immediately move the flat to a more open area; make sure that it is well drained; and remove the affected seedlings. There is no sense in transplanting the others—you'd just transfer the disease to new soil. You can save the flats though. Dry them out and ventilate them. I've salvaged several such flats and used them to produce early crops after all.

Skimpy Root Growth. This won't be evident, of course, until you transplant the seedlings. It may be caused by any one or more of the following factors:

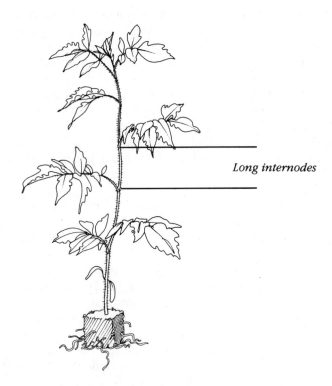

Long internodes

A leggy plant has long stems and large spaces between leaves.

- poor drainage
- low soil fertility
- concentration of excess fertilizer salts
- temperature too low
- insufficient air space in soil mixture

Failure to Sprout. If your seeds don't even sprout, the cause may be any one or more of the following factors:

- temperature too low or too high
- soil that was allowed to dry out
- seeds too deeply planted
- top watering that floated seeds off
- seeds that were old and poorly stored
- insufficient contact between seeds and soil
- toxic substances in soil
- damping-off disease
- lack of light for those seeds that need light in order to germinate, or lack of darkness for certain seeds, mostly flowers but also a few vegetables, that need this condition (see chapter 6)

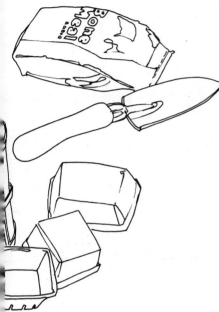

Section Two

Moving Plants Outdoors

It is surprising in how many fields one can step over the frontiers of science beside one's garden path.

Lawrence Hills

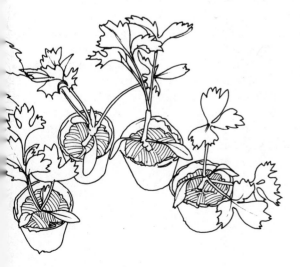

12 Soil Preparation

Gradually the late-winter sun rises higher in the sky each day, and its rays are warmer, more intense, even through the brisk March wind. Soon we're able to eat our evening dinner by daylight. As the last winter snow melts into the earth, I begin to shuffle seed packets and round up the garden tools. I know it's still too early to dig, but there's no harm in being ready, is there?

When to Work the Soil

Working the soil too early is a mistake. When the earth is still saturated with melting snow or spring rain, it is easily compacted by treading across it, or even worse, driving heavy equipment on it. In addition, large clumps of wet soil turned over at this time will only bake into impervious clods that will be very difficult to break up later, and as you remember, plant roots grow best when there are some air spaces between soil particles. Heavy, wet soil doesn't break up into the loose, air-retaining texture that is best for plants. Its clumpy texture is also likely to trap pockets of air around plant roots, and that is just as bad as no air.

How can you tell whether your garden has dried out enough to be worked? The truest test of soil condition is that age-old gesture of the gardener—fingering a handful of soil. Pick up about half a cup of earth in your hand. Now squeeze the soil together so that it forms a ball. If the ball of earth can readily be shattered by pressing with your fingers or dropping it from a height of three feet or so, it is dry enough to dig. If the ball keeps its shape or breaks only with difficulty into solid sections rather than loose soil, it still contains too much water. Clay soil that is too wet will feel slick when rubbed between thumb and forefinger. If it is very wet (75 to 100 percent moisture), the mass will be pliable, and a ribbon of earth can be drawn out and pressed with your finger. Working soil that wet can spoil its texture for the whole season.

Heavy clay soil will form a ball even when moisture content is less than 50 percent. Soil that is somewhat coarser, a sandy loam or silt loam, tends to crumble when moisture content is low but will probably form a ball at about 50 percent. At 75 to 100 percent moisture, it will be dark, pliable, and may feel slick between the

66

fingers. Coarse-textured sandy soil will not form a ball if moisture content is below 50 percent. At 75 to 100 percent moisture, it can be pressed into a weak ball, but even then it shatters easily. Coarser soil, of course, may be worked at a higher moisture content than fine-particled clay.

Fall Tillage

By preparing our garden in the fall, we've been able to move our spring digging date up by at least two weeks. In November, when frost has laid low all but the root vegetables that we've purposely planted at one edge of the garden, Mike plows up the garden. As the one-bottom plow carves up thick slices of earth, I stand ready with a hayfork to stuff coarse mulch into the newly opened trench. Where mulch has worn thin, I toss in all the leaves we've been able to collect. The next pass of the plow blade roughly covers the mulch-filled trench and opens up a new furrow into which I rake another humus-building line of hay and leaves.

The patch remains rough all winter, catching all the poor-man's fertilizer (snow) that falls. By spring, frost action has mellowed and pulverized the rows of raised soil. The garden dries out a good two weeks earlier than it did when left flat under a season's-end accumulation of mulch and weeds.

Preparing the Garden in Spring

With fall tillage, our garden does not need spring plowing, just a going-over with the rotary cultivator. I often hand-dig an extra-early row for peas.

We often manure the garden in the fall now, but when our barn was full of animals, we did our manure spreading in spring. Fall cleaning of the animals' pens is fine for the garden, but leaves too thin a layer of bedding for winter. Chicken manure, which the hens have finely pulverized and mixed with their well-shredded bedding, is loose and fine enough to blend in easily with just a few passes of the tiller. The heavy sheep, with their sharp hooves, do the next best job of chopping up their bedding. Manure from the large loafing pen of milking goats tends to be coarser because it has a lot of waste hay in it.

When we farmed an acre of ground, on our first homestead, we tilled our vegetable garden. Two or three passes with the rotary cultivator, each slightly overlapping the previous path of the tines, fines the soil up quite well. It's a good idea to till in both directions (north-south and east-west). Depending on the size and shape of the patch we're working, we'll sometimes till in a grid pattern, making the furrows at right angles to each other. At other times, we'll till in a spiral, perhaps finishing off with a pass across the garden.

During previous years, in city and small-town gardens, we practiced hand digging entirely. In a small, closely worked garden it is possible to dig a few rows at a time as higher land drains or as time becomes available. Dig as deeply as possible so that the soil is loose to a depth of eight to ten inches. As each shovelful of earth is turned over, shatter it with your fork or spade. If it doesn't fall apart readily, you're probably digging too early.

It's a good idea to let the soil settle for a day or two after tilling or digging and raking, especially before planting fine seeds like carrots and parsley that might get washed too deeply into the furrow if a heavy rain follows planting. I seldom wait, though, to plant large seeds like peas and beans.

Breaking New Ground

When turning a sodded area into a garden, get a head start, if you can. Dig the patch in the fall and turn it grass side down so that all that green manure can break down over winter. If digging is impossible, cover the area with a layer of old boards or plastic to kill the grass over winter.

Often, though, you'll be digging in the spring. That's when expansion fever seems to hit most of us gardeners. To avoid having all that grass come back at you all summer long, shake out each clump of sod as you dig it up and toss the grass, roots and all, on the compost heap. Then you'll be able to rake the ground into a fairly smooth seedbed. If the size of the sod plot you're digging makes this kind of operation impractical, then I'd let the plot weather for a few weeks after digging, smooth it as well as you can, and use it either for crops with large seeds, like corn and beans, or for seedlings that you will plant into the garden. Mulch heavily to suppress the grass that will still try to come up.

When to Dig. The very best way to develop a sense of the right time is to have the experience of forking up a section of loose, crumbly soil that rakes out so fine you can practically comb it. You know, then, that you've chosen the right time, and you'll find when the next growing season starts that your impatience to begin is tempered by your memory of that moment of perfection. You won't want to lose it by starting too soon.

Improving Problem Soils

The ideal soil for gardens contains half solid matter and half pore space. Half of the pore space should contain air, and the remainder should be filled with water. Roots and the microorganisms that break down soil components into forms they can use live in the spaces between soil particles.

Clay Soil. This type of soil retains more nutrients than sandy soil, but its flat, slippery particles pack so tightly together that they exclude air and impede root growth. The best way to loosen heavy clay soil is to add organic matter—leaves, spoiled hay, green manure crops, and kitchen scraps. Lime, which causes the tiny soil particles to adhere to each other in larger clumps, also improves the tilth of clay soil. Humic acids in rotting manure act in the same way.

Sandy Soil. To begin with, fewer nutrients accumulate in sandy soil because it is not as active chemically as clay. Those nutrients that it does contain are more easily lost from sandy soil by leaching. Once again, the solution is a generous helping of organic matter—this time to bind the loose, independent soil particles together and to act as a reservoir of nutrients and moisture.

Mycorrhiza. In addition to improving soil structure and providing nutrients, organic matter helps to nourish plants by supporting mycorrhiza—the beneficial association of plant roots and soil fungi, in which the filaments of the fungi extend into the plant roots and also into the soil. The fungi, which can extend surprisingly far into the soil, absorb nutrients from the soil and channel them back to the plant roots. They also protect plants from some disease organisms, in some cases forming a sheath around each tiny root. If possible, try to rotate vegetables that do not form mycorrhizal associations—beets, spinach, cauliflower, cabbage, brussels sprouts, and radishes—with other garden plants, most of which *do* enter into this mutually helpful arrangement.

Quick Compost

I'd like to give you my recipe for quick compost. I worked this out one year when, at the beginning of the spring gardening season, I had nothing but coarse, partly decomposed compost in my two bins. I needed some fine, crumbly compost to put in the hole with my spring plantings of squash, cucumbers, and cantaloupes, as well as for tomato transplants and other special seedlings.

Lacking a shredder, and not really wanting one either, I needed to find ingredients that were already in small pieces or well chewed up. A little scavenging around the farm produced the following ingredients:

- Sawdust, well aged, into which I had poured manure tea several times during the previous year.
- Locust tree leaves—tiny, fine, curled up, and crisp dry, these legume products broke down quickly. It was an easy, pleasant task to scrape up bags of the leaves from the ground in the locust grove.
- Wood ashes.
- Goat manure and hay bedding, both finely shredded by the prancing of our buck as he played with a chain-hung tire.
- Manure tea, which I poured into the pile several times a week for the first two weeks after building it.

To enclose the pile I simply stacked cement blocks in two parallel lines, using the solid block wall of my large compost pile as the back end. The total volume of the pile was not large—about one cubic yard—but it made enough to get my greedy feeders off to a good start.

To build the pile, I layered the ingredients, using roughly equal parts of sawdust and leaves, with a three-gallon bucket of manure sifted over each bushel or so of leaves and sawdust. A handful of wood ashes and a heaping shovelful of soil salted each layer. Sawdust, even old sawdust, contains a high proportion of slow-to-rot carbon, but the addition of nitrogen in both manure and manure tea and the legume leaves helped to hasten its breakdown.

When all the ingredients were piled up, I moistened the pile thoroughly so that it was wet without being soggy. Then, over the next 14 days, I poured about ten gallons of manure tea into it. The weather was cold in March, when I built the pile, and the microorganisms sleepy. It took a good two weeks to heat. After four weeks, I turned the pile, and it heated again. Eight weeks after I'd made it, it was ready to use—dark brown, crumbly, spongy, and rich. It was so easy and so satisfying that I started another pile right away.

Spreading Rock Powders

Every few years, we add rock minerals to each of our three garden areas. We apply them just before spreading the manure so that it is all tilled in together. Acids produced by the decaying manure help to release important nutrients in the ground rock. Later, when crops have been planted, the growing roots produce carbon dioxide, which also aids in breaking down the powdered minerals.

The more finely ground a rock powder is, the more quickly it will become available to growing plants. All natural rock minerals work slowly in the soil, breaking down gradually over a period of several years.

Rock Phosphate. In addition to phosphorus, which is in the form of calcium phosphate, this powder contains a good many trace elements. Superphosphate has been chemically treated to make it highly soluble and therefore quickly available. The gain in rapid absorption, though, is more than offset by damage to the population of cellulose-digesting fungi in the soil. Repeated use of this product has been shown to impair the soil's natural ability to build humus out of buried organic matter.

Application. Although, because soils differ in their needs, no general prescription can be given for application rates of rock phosphate, the usual practice is to spread a pound of the powder on each ten square feet of garden surface (or two tons to the acre) every three to five years.

Lime. It not only tempers soil acidity but also helps to minimize toxic effects of such plant-inhibiting elements as aluminum, iron, manganese, and others. Its effect on the soil is physical as well as chemical. By encouraging flocculation, or clumping, of soil granules, it counteracts the tendency of clay soils to form a solid mass of fine sticky particles.

Application. The generally recommended rate of application is a pound to each 20 square feet (or one ton to the acre) every three or four years. Less may be needed in areas of limestone soil, and heavier applications may be indicated where soil is strongly acidic. Consider your crop, too. Blueberries, potatoes, flax, and watermelon, for example, do not like lime. Most other garden vegetables are benefited, or at least not harmed, by soil with a pH closer to neutral (pH 7).

Greensand. This rock powder is an ocean mineral. It is an excellent source of potash and many trace elements. It not only enriches the soil, but also improves tilth and water retention.

Table 5

Soil Deficiency: Symptoms and Treatments

Primary Nutrients	Signs of Deficiency	Natural Sources of Elements
Nitrogen (N)	Pale leaves, turning to yellow, especially lower leaves; stunted growth	Manure tea Bloodmeal Manure Feathers
Phosphorus (P)	Reddish purple color on leaves, especially undersides, veins, and stems; thin stems; stunted growth.	Rock phosphate Bone meal Fish emulsion Waste wool
Potassium (K)	Bronze coloring, curling, and drying of leaf margins; slow growth; low vigor; poor resistance to disease, heat, and cold.	Wood ashes Greensand Tobacco stems Granite dust
Secondary Nutrients		
Boron (B)	Slow growth; specific symptoms vary according to vegetable: beets and turnips—brown corky spots on roots; cauliflower—hollow stem, brown curl; celery—cracking; chard—dark stripe with cracking; tomatoes—stunted stems and curling, yellowing, and drying of terminal shoots.	Granite dust Vetch or clover
Calcium (Ca)	Curling of young leaf tips; wavy, irregular leaf margins; weak stems; poor growth; yellow spots on upper leaves.	Limestone
Copper (Cu)	Slow growth; faded color; flabby leaves and stems; lettuce—leaf elongation; onions—thin, pale outer skin; tomatoes—curled leaves and blue-green foliage.	Agricultural frit
Iron (Fe)	Spotted, pale areas on new leaves; yellow leaves; yellow leaves on upper part of plant. (Iron is more often unavailable because of insolubility rather than actual soil lack.)	Manure Dried blood Tankage (Humus helps make iron more available.)
Magnesium (Mg)	Plant may be brittle; yellow mottling on older leaves; margins and tips may be brown; corn shows yellow stripes on older leaves. (Deficiencies ordinarily show up late, near seeding time, when the element is most needed.)	Dolomite Limestone

(continued)

Table 5 —*Continued*

Secondary Nutrients	Signs of Deficiency	Natural Sources of Elements
Manganese (Mn)	Slow growth; beets—leaf turns deep red, then yellow and brown; spots develop between veins; onions and corn—narrow yellow stripes on leaves; spinach—pale growing tips.	Manure Compost
Sulfur (S)	Yellowing of lower leaves; slender, hard stems.	Compost
Molybdenum (Mo)	Drying of leaf margins; stunting; yellow leaf tissue between veins.	Agricultural frit Rock phosphate
Zinc (Zn)	Leaves unusually long and narrow; also yellow and spotted with dead tissue; beans—cotyledon leaves have reddish brown spots; beets—dry margins, yellow areas between veins; corn—wide stripes at leaf bases.	Manure

Sources: Editors of *Organic Gardening, The Encyclopedia of Organic Gardening* (Emmaus, Pa.: Rodale Press, 1978).
J. E. Knott, *Handbook for Vegetable Growers* (New York: John Wiley and Sons, 1962).
United States Department of Agriculture, *Soil, the Yearbook of Agriculture* (Washington, D.C.: U.S. Government Printing Office, 1957).

Application. The usual rate of application is about two pounds to each ten square feet of garden surface (or four tons to the acre) about every three to five years. I often sprinkle it in the furrow when planting, especially with potatoes.

Granite Dust. This is another excellent source of potash, along with small quantities of trace minerals.

Application. It should be spread at the rate of about a pound to each ten square feet of garden (or two tons to the acre) every three to five years. If your garden is large and you intend to spread a lot of granite dust, use a dust mask; long exposure to the fine dust can damage the lungs.

Relative Tolerance of Vegetable Crops to Soil Acidity

The vegetables in the slightly tolerant group can be grown successfully in soils that are on the alkaline side of neutrality. They do well in soils with a pH up to 7.6 if there is no deficiency of essential nutrients. Calcium, phosphorus, and magnesium are the nutrients most likely to be deficient in more acid soils with a pH less than 6.0.

Slightly Tolerant *(pH 6.8 to 6.0)*	*Moderately Tolerant* *(pH 6.8 to 5.5)*	*Very Tolerant* *(pH 6.8 to 5.0)*
Asparagus	Beans, Lima	Chicory
Beets	Beans, Snap	Dandelions
Broccoli	Brussels Sprouts	Endive
Cabbage	Carrots	Fennel
Cabbage, Chinese	Collards	Potatoes
Cauliflower	Corn	Rhubarb
Celery	Cucumbers	Shallots
Cress, Garden and	Eggplant	Sorrel
Upland	Garlic	Sweet Potatoes
Leeks	Gherkins	Watermelons
Lettuce	Horseradish	
Muskmelons	Kale	
Okra	Kohlrabi	
Onions	Mustard	
Orache	Parsley	
Parsnips	Peas	
Salsify	Peppers	
Soybeans	Pumpkins	
Spinach	Radishes	
Spinach, New Zealand	Rutabagas	
Swiss Chard	Squash	
Watercress	Tomatoes	
	Turnips	

13 Mapping Out the Garden

Once the soil has been prepared, it's time to find the planting plan we doodled over by the late-winter fire. Already much erased and revised, it will no doubt be changed once again as we confront the reality of freshly raked soil and extra last-minute seed purchases. Subject to revision though it may be, the planting plan is a valuable gardening tool.

Although I wouldn't presume to prescribe what form your plan should take, I *would* like to suggest several considerations to keep in mind while poring over the graph paper with seed orders and pencil in hand.

Planning Rows

If you are planning a row garden, consider the lay of the garden and the spacing of the rows.

Orientation of Rows. Your first decision, as you face that blank piece of paper, is to determine in which direction your rows will run. If possible, choose a north-south direction so that sun striking the garden from the east and west will cast a plant's shadow onto the space between the rows rather than onto the next plant.

On a sloping plot, where rain tends to carry loose soil downhill, contouring the rows so that they run parallel to the hill (and at right angles to the direction of the slope) will help to conserve much valuable topsoil. Mulching will help, too, as will terracing on a very steep slope.

Row Spacing. The space you leave between rows will be determined by the kind of plants you grow, the amount of mulch you can get, and the sort of equipment you plan to use to cultivate the garden. Rows must be separated far enough to give the plant sufficient room to grow and the gardener space to walk and often space to till or hoe.

Single rows of nonspreading plants like beets, lettuce, peas, carrots, and such are giving way, in more and more gardens, to 4- to 36-inch-wide bands of these vegetables. We've had good results with garlic, lettuce, carrots, peas, herbs, and beans planted in wide rows. Careful hand weeding between the vegetables is necessary in the early stages, but later the plants help to take care of each other by shading

74

On a sloping plot, rows should run parallel to the hill to prevent soil from being carried down the slope by rain.

out weeds and keeping soil moist. Wide rows use space more efficiently than narrow drills.

Row middles that will be tilled should be about six to eight inches wider than the tine-to-tine measurement of the tiller. (You can adapt some front-end tillers to work row spaces as narrow as eight inches by removing the two outer tines and reversing the center tines so that their blades face inward rather than outward.)

Mulched rows may be closely spaced, but if you have a lot of coarse mulch, as we do, you will want to plan your garden so that the loose, shaggy stuff will be used to mulch wide rows like tomatoes. The finer mulches like sawdust or old leaves, or neat bundles like hay, can be saved for narrow rows.

Garden Beds

Rows are customary, but not necessarily traditional. There are many good arguments for planting vegetables, herbs, or flowers, closely spaced, in small plots or blocks. Soil may be intensively improved. The beds may be raised to improve drainage. Weeding and harvesting are convenient. When well cared for, small garden beds are delightfully attractive. Many arrangements are possible. You can divide your garden area into blocks separated by paths. Each block will be solidly planted to a single vegetable or an especially chosen combination of vegetables. You might want to keep a grassy path between a double row of narrow vegetable beds. Vegetable beds may be located on the lawn, next to the house, or along a walkway.

(continued on page 78)

Table 6

Plant and Row Spacing for Vegetables

Crops	Space between Plants in Rows (inches)	Space between Rows (inches)
Artichokes	48–72	84–96
Asparagus	9–15	48–72
Beans, Broad	8–10	20–48
Beans, Bush	2–4	18–36
Beans, Lima, Bush	3–6	18–36
Beans, Lima, Pole	8–12	36–48
Beans, Pole	6–9	36–48
Beets	2–4	12–30
Broccoli, Raab	3–4	24–36
Broccoli, Sprouting	12–24	18–36
Brussels Sprouts	18–24	24–40
Cabbage, Early	12–24	24–36
Cabbage, Late	16–30	24–40
Cardoon	12–18	36–42
Carrots	1–3	16–30
Cauliflower	14–24	24–36
Celeriac	4–6	24–36
Celery	6–12	18–40
Chervil	6–10	12–18
Chicory	4–10	18–24
Chinese Cabbage	10–18	18–36
Chives	12–18	24–36
Collards	12–24	24–36
Corn	8–12	30–42
Corn-salad	2–4	12–18
Cowpeas (Southern Peas)	5–6	35–48
Cress, Garden and Upland	2–4	12–18
Cucumbers	8–12	36–72
	24–36, hills	
Dandelions	3–6	14–24
Dasheen (Taro)	24–30	42–48
Eggplant	18–36	24–48
Endive	8–12	18–24
Florence Fennel	4–12	24–42
Garlic	1–3	12–24
Horseradish	12–18	30–36
Jerusalem Artichokes	15–18	42–48
Kale	18–24	24–36
Kohlrabi	3–6	12–36
Leeks	2–6	12–36
Lettuce, Cos	10–14	16–24
Lettuce, Head	10–15	16–24

Crops	Space between Plants in Rows (inches)	Space between Rows (inches)
Lettuce, Leaf	8–12	12–24
Muskmelons and Other Melons	12 24–48, hills	60–84
Mustard	5–10	12–36
Okra	8–24	24–60
Onions	2–4	16–24
Parsley	4–12	12–36
Parsnips	2–4	18–36
Peas	1–3	24–48
Peas, Southern (See Cowpeas)	...	...
Peppers	12–24	18–36
Potatoes	10	30–42
Pumpkins	36–60	72–96
Radishes	½–1	8–18
Radishes, Storage Type	4–6	18–36
Rhubarb	24–48	36–60
Roselle	24–46	60–72
Rutabagas	5–8	18–36
Salsify	2–4	18–36
Scolymus	2–4	18–36
Scorzonera	2–4	18–36
Shallots	4–8	36–48
Sorrel	½–1	12–18
Spinach	2–6	12–36
Spinach, New Zealand	10–20	36–60
Squash, Bush	24–48	36–60
Squash, Vining	36–96	72–96
Sweet Potatoes	10–18	36–48
Swiss Chard	12–15	24–36
Tomatoes, Flat	18–48	36–60
Tomatoes, Staked	12–24	36–48
Turnips	2–6	12–36
Turnip Greens	1–4	6–12
Watercress	1–3	6–12
Watermelons	24–36 72–96, hills	72–96

Source: Oscar A. Lorenz and Donald N. Maynard, *Knott's Handbook for Vegetable Growers,* 2d ed. (New York: John Wiley and Sons, 1980).

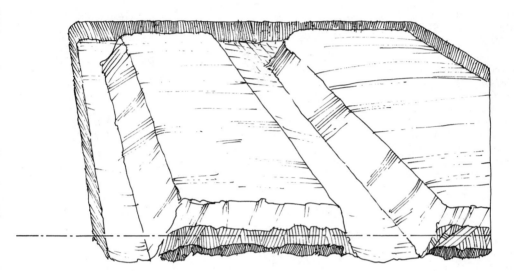

Raised beds.

One practical problem with garden beds bordered by grass is the encroachment of the grass into the bed. This may often be solved by sinking a thin metal edging strip between vegetable bed and grass or by building a raised vegetable bed. Some gardeners support the deeper soil with wooden boards or railroad ties, but such enclosures are not really necessary. Plant roots will hold the soil in place. When you form the beds, slope them slightly from the base to the top surface.

Raised Beds. If I were starting a new garden today, I'd plant intensively in raised beds. Here's why:

- Raised beds are easy to work (once established).
- The deep, loose soil encourages excellent root growth.
- The absence of foot traffic prevents soil compaction.
- Compost and other soil-improving additions can be concentrated where they will be effective, not wasted on footpaths.
- Raised beds warm earlier in spring.
- The solid cover of plants helps to shade out weeds and conserve soil moisture.

When starting a raised bed, remove weeds and sod and add them to the compost pile. To add humus, dig in about one bushel of manure or compost to each 25 square feet of garden. Toss on a few shovelfuls of wood ashes. Rake up all loose soil from paths and, if possible, haul in some extra soil from elsewhere on your place.

Double digging. If you double-dig your beds, you probably won't be sorry, because deeply worked soil gives plant roots more of what they need. To double-dig, remove the soil from a 1-foot-deep trench at the edge of the bed and save the dug-up soil to fill the last trench. Then, using a spading fork, loosen the packed earth in the bottom of the trench. Fill the first trench with soil dug from trench number two and continue to dig your way down the bed, trench by trench. Beds should be 3 to 3½ feet wide and any convenient length. Wider beds are hard to reach into. Leave at least one footpath between beds.

I've talked to a fair number of gardeners who chose not to double-dig their raised beds, and they seem to be getting good results, too. One couple first tilled their ground and then shoveled all the loose soil aside and went over the same ground again with their tiller to loosen the soil more deeply. Others have simply piled up as much soil and organic matter as they could collect to raise the beds 8 to

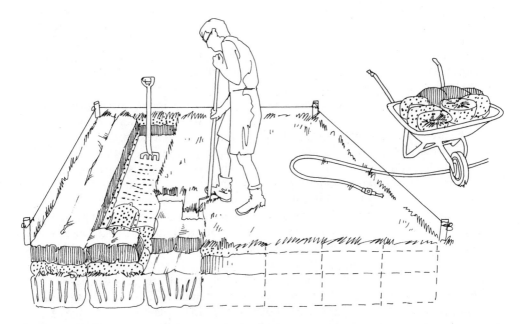

In double digging, soil is removed from one end of the bed to be placed at the other end. Successive trenches are dug, and soil is moved up the bed. During this process, the soil can be worked deeply to benefit root growth.

12 inches above ground level. All are reaping the many benefits of the raised bed. It is a trade-off of more intensive preparation and careful planning for easier care and less tilling later on.

Where to Plant What

Once you've determined the basic design and layout of your garden, you'll want to decide where to place your plants. In general, tall plants like corn or sunflowers should be planted on the north side of the garden so they don't shade adjacent vegetables. If it is necessary to put them on the south end of the garden, plant parsley, lettuce, or other midsummer shade-tolerant plants next to the corn. Placement of your crops in the garden will be partly determined by their growing habits.

Rambling Vine Crops. Squash, pumpkins, melons, and cucumbers may be planted in rows or hills. There's a certain amount of confusion surrounding the term "hill." Generally, a hill is simply a small designated area, not necessarily raised, in which a group of seeds, usually those of spreading plants, is sown. Except in very small gardens, where vines can be trained to climb fences, hills use space more efficiently than rows when growing rambling plants. Planting vining crops in a hill rather than in rows allows for easier placement of compost and manure and simpler, quicker insect control early in the season.

Some gardeners do make hills, which are actually small elevated mounds an inch or two above the normal soil level. These small raised beds drain well and warm quickly.

Climbers. You can save space in the garden by planting cucumbers, pole beans, certain melons, Malabar spinach, and others if you put them at the edge, next to a fence or trellis, so save an end row for some of these space-takers.

Bush Vegetables. In recent years, breeders have developed space-saving bush varieties of popular long-vined vegetables like squash, pumpkins, and melons. Those shorter-vined plants also have shorter internodes—less space between the leaves. Some of them bear earlier, and some have a more determinate habit than their roaming cousins—that is, they may stop bearing and growing earlier in the season rather than continuing until frost. Most bush varieties produce full-sized fruit. Some previously developed bush-type vegetables were inferior in flavor to those grown on full-sized plants. This situation seems to be improving, though. Bush-type cucumbers seem to suffer the least loss of flavor, watermelons the most. Some of the muskmelons are so-so, but Musketeer has a fine flavor. A plant with more limited leaf area can be stressed by a large fruit load, so pinching off excess flowers and providing plenty of organic matter in the soil should help to encourage better flavor.

Overwintering Crops. Parsnips, salsify, and carrots, and late-fall bearers like brussels sprouts, escarole, parsley, and collards, should be planted at the edge of the garden where they will not be disturbed if you intend to till or plow the rest of the garden in the fall.

Perennial Vegetables. Crops like asparagus, rhubarb, comfrey, and Jerusalem artichokes should, of course, be planted either at the edge of the garden where they will not be plowed up or in a separate bed.

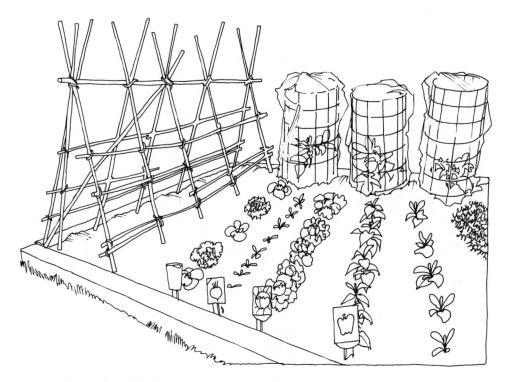

When planning a garden, try to place tall plants on the north side so they don't shade other crops. Plant climbers near an edge where they can grow up a trellis, and if you grow any perennial vegetables, give them their own bed, if you can, so they won't be disturbed.

Crop Rotation

Keep your vegetables moving! Alternating the kinds of crops you grow in a given space is just about the least expensive and least time-consuming method you can use to maintain and even improve the quality of your soil and the health of your vegetables. Each kind of vegetable that you might grow absorbs soil nutrients in different amounts. Corn and leafy vegetables need a lot of nitrogen; sweet potatoes get along with small doses of nitrogen; root vegetables need potash; legumes (with the help of nitrogen-fixing bacteria on their roots) actually add nitrogen to the soil. By alternating vegetable families, you give the soil a chance to replenish nutrients, and you discourage the proliferation of insect pests. If you plant vulnerable food crops in a different location each year, the bean beetles won't have a ready feast waiting for them the moment they hatch. In addition, plant diseases caused by fungi and bacteria are less likely to threaten successive crops if their host plants keep hopping around. The rotation unit can be a row, half-row, a raised bed, or even a separate garden plot.

What constitutes a vegetable group? Some folks simply alternate those vegetables that are generally considered more demanding on the soil—corn, cabbage, squash, melons and their relatives, and tomatoes—with those that are satisfied with somewhat leaner soil—legumes, root vegetables, herbs, and onions. A third-year planting of a soil-enriching cover crop like buckwheat, clover, oats, or rye would be a wise addition to such a simple rotation. At the very least, you'd want to rotate corn, which is an especially heavy feeder, as well as any vegetables with which you've had serious disease and insect problems in the past.

For a more complete and effective rotation, divide your garden vegetables into seven categories: legumes; leafy vegetables; root vegetables; cucurbits; onions and their relatives; tomatoes and related peppers, potatoes, and eggplant; and corn and other grains. If you plant green manure crops, make that your eighth plant group. A sample rotation, then, might be a planting of spring peas followed by fall kale, with parsnips occupying the row the second year, squash the third year, onions next, tomatoes or related vegetables in year five, and corn the following year, after which you'd either put in a cover crop of oats or buckwheat, or begin again with legumes— perhaps soybeans—then cabbage, carrots, cucumbers, garlic, eggplant, and so on.

The most impressive garden rotation scheme I've ever seen was worked out by Vertis Bream, an accomplished Pennsylvania gardener who divided his garden into six raised beds, each 3½ feet wide, 30 feet long, and 8 to 12 inches high. The Breams plant a different group of crops in each bed every six years. For example, a bed that's growing beets, potatoes, carrots, and onions this year will be planted to wheat after the root vegetables are pulled up. After the wheat is harvested in the second year, they plant a clover cover crop, followed by corn in the fourth year, then beans, and finally, in year six, a bed of lettuce, tomatoes, and cole crops. They keep garden diagrams and move the potato row from row 1 to row 2, then 3, 4, 5, 6, and then back to row 1 as the years go by.

As these plans suggest, you will sometimes grow two different groups of plants in a certain row in a given year. Garden rotations can look complicated because we're dealing with limited areas and a large variety of plant types, and often with in-season plant successions. Don't worry about making it all come out straight, down to the letter and inch. Just do the best you can to give each kind of vegetable a different spot in the garden each year—a good reason for saving each year's mud-spattered garden plan.

Green Manure. With careful planning, you can end the season with a section of garden that is free of crops, a perfect chance to put in a soil-building crop of winter rye, or start in spring with oats to be plowed under before they head. Either way you'll add humus and nutrients to your garden. In order to have a solid block of land ready for a fall cover crop, it's necessary to group a bunch of early-maturing spring plantings together so that when they are harvested in August or September, the rye can be planted right away. (Be sure to get coarse-seeded winter rye, not the fine-seeded grass that may repeat on you in next year's garden.)

Succession Crops. To keep your garden continuously productive, plant crops in succession. They can be dovetailed in a very intricate way. All kinds of variations are possible. Here are a few examples, from my garden:

- An early planting of peas followed by a late corn planting (but use a fairly early-maturing kind of corn).
- Early cabbage followed by late beans.
- Spring lettuce giving way to fall beets.
- Early onions succeeded by fall lettuce.

Once I planted bean seeds along the row as I harvested leaf lettuce. By the time the lettuce row was used up, all the beans were up and growing. More often, I start seedlings in flats for transplanting into the garden when an early crop is finished. As I gather peas, the kale that will take their place is hardening off on the east patio. When I pull onions, I have Chinese cabbage seedlings ready to take their place.

Interplanting

When planning your garden, you might want to consider interplanting your vegetables. Interplanting saves space. It's seldom necessary, for example, to devote a whole row to spring radishes. Plant them, instead, along with your lettuce and carrots and use them as thinnings when they're ready. One year I grew a long row of bush beans between the widely spaced, just-planted rows of tomatoes and squash. By the time the vines closed over the gap, I'd harvested many meals of good green beans. Pumpkins do well at the edge of the corn patch where they have space to ramble. In studies at the University of Maryland, sweet corn and soybeans planted in the same row produced satisfactory harvests. Although individual vegetable yields were lower and corn ripened three days later, the total harvest from the row was larger.

Companion Planting

Companion planting, the pairing of plants that benefit each other in close proximity, adds yet another dimension to the juggling of rows and beds. Reports on certain vegetables sometimes vary widely; some authorities say that onions and beans do well together, while others warn that beans don't like onions. There *is* a firm scientific basis for the study of plant relationships, though: Plants are known to produce root exudates that do, in many cases, affect soil life and roots of other plants around them.

A whole body of companion planting lore exists that is difficult to dismiss, but difficult also, so far, to prove conclusively. Some of these protective interplanting arrangements seem to work for some gardeners but not for others. A few sound to me like myths that have simply been repeated from one writer to another without serious trial. Others have been proven effective in published studies—for example, the nematode-suppressing effect of French marigolds. Catnip oil, too, has been shown to repel 17 species of insects, but unfortunately, like the mints, it is an invasive plant that can choke out flowers and vegetables.

In my garden, I make only a few intentional companion plantings, usually

---

Marigolds Are Ancient Companions

On vases and grave furniture unearthed in Equador and Peru, archaeologists have noted depictions of the marigold *Tagetes minuta* painted right next to pictures of crops on which the people depended for food. Farmers in these pre-Inca civilizations were, in fact, able to grow corn, tomatoes, beans, and potatoes on the same land, without any crop rotation, for more than 1,000 years by fertilizing with bird guano and fish waste and interplanting their crops with marigolds.

---

radishes with cucurbits and marigolds for nematode control, but I do try to keep different plant varieties well mixed. Although I don't follow the lore closely in my garden, partly out of busyness and preoccupation with other matters, I'm no more ready to dismiss it entirely than I am to endorse without reservation the various companion-planting combinations I've heard of. One thing is certain: Growing things relate and interact in amazingly complicated and subtle ways. Much study remains to be done, and close observation by gardeners continues to be a valid source of information about the success of companion planting.

Helpful Companions. According to traditional lore, certain herbs, weeds, flowers, and vegetables have the effect of deterring insects and encouraging plant growth when planted near compatible plants. Borage, chamomile, and lovage are supposed to enhance growth and flavor in nearby vegetables. Garlic planted near roses and raspberries should deter aphids. Horseradish has the reputation of repelling potato bugs. Marjoram is said to improve the flavor of nearby food plants. Mint, sage, and rosemary are the traditional enemies of cabbage moths. Nasturtiums, which actually attract aphids and thus are useful as a trap crop, are sometimes erroneously listed as an aphid repellent. Catnip is often planted to deter flea beetles. Yarrow planted near aromatic herbs is thought to enhance their production of essential oils.

Companions for Garden Space. Another branch of companion planting pairs shallow-rooted plants like onions and celery with others, like chard and carrots, whose roots delve deeply into the subsoil. This seems to me a sensible practice that makes good use of that third dimension of garden space—depth.

Interplanting. There seems to be no question that monoculture (planting large areas to a single plant species) encourages heavier insect infestation. In repeated studies, diversified plantings have suffered less insect damage than monocultured fields. Therefore, the practice of mixing aromatic herbs and flowers with vegetables, alternating rows of different species, and interplanting various species within the row should help, at least, to confuse the insects and, at best, to promote positive plant health.

Harmful Companions. Some plants, in the companion-planting tradition, harm rather than help each other. Fennel, the most notorious outcast, is said to discourage growth of most garden plants. Dill and carrots, basil and rue, sage and cucumbers, cabbage and grapes, and chives and peas or beans are all plants that reputedly make poor companions because one or both are badly affected by close proximity.

Allelopathy. The inhibition of seed germination and plant growth by certain plant-produced natural compounds is sometimes responsible for otherwise unexplained poor plant growth. Plant toxins are usually exuded by roots, but they are also generally present, in varying amounts, in stems, leaves, and fruits.

Walnut Trees and Juglone. Walnut trees release juglone, which retards many plants. I found that out the hard way in my own garden when I planted a row of beets near an English walnut tree at the edge of the garden. Beets, I have since discovered, are particularly sensitive to toxins in the soil. The beet seeds didn't even germinate. Other plants—onions and marigolds—have grown fairly well in the same spot, but most things I plant near that tree produce rather halfheartedly. Shade and competing tree roots could also be a factor, but 200 species of plants are known to be susceptible to juglone. Black walnut trees produce more of the toxin than English walnuts or butternuts.

Like beets, some crops are more vulnerable to the effect of toxins. Soybeans, tomatoes, okra, asparagus, alfalfa, lilacs, apples, peonies, and chrysanthemums, among others, have been shown to be seriously inhibited by juglone in the soil. Those plants that are resistant to juglone include red cedars, redbuds, quinces, black raspberries, corn, beans, carrots, and zinnias.

Other Allelopathic Plants. Other notorious allelopathic plants are ailanthus, artemisia, absinthium, eucalyptus, sunflowers, and sometimes sycamores.

Some crop plants inhibit weeds. For example, Kentucky-31 fescue grass retards growth of black mustard and trefoil. Weeds can suppress crop plants, too. In some experiments, residues of lamb's-quarters, pigweed, velvet lead, and yellow foxtail have inhibited the growth of corn and soybeans.

Although published observations of the effect of plants on the growth of other plants date back hundreds of years, scientific study of the phenomenon has only

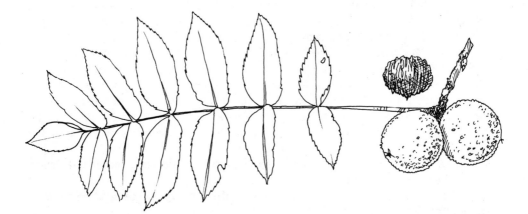

Black walnut.

begun to scratch the surface. So many plants and variables are involved that few sweeping generalizations can be made. Allelopathy can throw a wild card into some of your planting schemes, but keep an open mind and don't let it worry you. Just try to plant your sunflowers separately and keep your garden away from walnut trees. (A distance of 1½ times the height of the tree should be safe.) If the subject interests you, conduct your own experiments, and jot down what you planted and what you observed.

Taking Pollination into Account

When laying out rows and beds, remember that corn is wind pollinated and should therefore be planted at least four rows deep, preferably six to eight. If you plant only a small amount of corn, make a block of four or more short rows rather than planting a single long row.

If you are planning to save seeds of wind-pollinated garden vegetables like corn and spinach, you will want to follow recommended spacing requirements to avoid crossing (see table 6).

Special Problems

Perhaps you have a spot in your garden that is poorly drained, a corner that is shady, a section with hardpan, or a rocky area. Each of these special situations may be met and sometimes even partially solved by your choice of plants.

Partial Shade. Plant summer lettuce, parsley, raspberries, or rhubarb.

Hardpan. Treat the problem by planting deep-rooted vegetables like comfrey or Swiss chard to break up the impervious layer of subsoil.

Poor Drainage. Avoid planting globe artichokes, sweet potatoes, or other lovers of warm, loose soil in that spot, until you've corrected the problem by trenching to divert water and/or digging in more humus or making raised beds.

Rocks. Potatoes will do well, especially under mulch, but any crop that needs frequent hoeing and cultivating will be a challenge. If you use a tiller, get pointed tines for working around rocks. When your whole garden is rocky, you accept it and work around it (and often have as fringe benefits a lovely rock garden and several fine rock walls), but when only one part is rocky, you might as well minimize the wear and tear on your hoe by selecting crops to plant there that can be mulched early.

Short of Space. If you're short of space, make some good management decisions to get the most out of the available ground.

The Vegetable Garden. Consider some of the following criteria when deciding what to plant in your vegetable garden and how to plant it:

- Plant vegetables with the highest cash value.
- Grow vegetables that lose quality in shipping.

- Plant earlier, making use of hardy vegetables, early varieties, and plant protectors such as tunnels and cold frames.
- Stretch the harvest with continuous succession plantings.
- Grow high-yielding vegetables such as tomatoes, lead lettuce, turnips, summer squash, and edible-pod peas.

The Flower Garden. Efficient use of space allows you to grow more flowers, too. Consider the following suggestions:

- For more flowers in a small space, grow climbers like morning-glories.
- Use container plantings on patios and porches.
- Plant flowers in windowboxes.

Planting Flowers with Vegetables

There's no good reason, other than staid custom, for keeping vegetable and flower plantings segregated. More gardeners now feel free to mix flowers, herbs,

Create lovely garden displays by growing flowers and vegetables together.

and vegetables in the same plot with delightful results. Flowers brighten the patch in the backyard, and ornamental vegetables grow proudly in the front border. Not every vegetable plant assumes a pleasing form, but those that do—rhubarb chard, head lettuce, peppers, and many more—deserve their place in decorative planting.

In a gorgeous display planting at Longwood Gardens in Kennett Square, Pennsylvania, vegetable and flower plants grow together in borders similar to conventional perennial beds. The plot is laid out as if it were someone's backyard, with one-third of the area in lawn and a path separating a border bed along the fence from the main vegetable bed, which curves gracefully in a long sweep down and across the yard. Where you'd expect to see masses of delphiniums and lilies, escarole, beets, and beans thrive instead, interspersed with marigolds, nasturtiums, and portulaca. A low border of alyssum sets off the arrangement. The surprising effect is that the true beauty of well-grown vegetables stands out. Boundaries between vegetables and flowers fade even more when you remember that blossoms of some flowering plants—calendula, nasturtiums, and chives, for example—are edible. When planning an ornamental edible garden, keep in mind Kate Gessert's advice in her book *The Beautiful Food Garden:* "Be playful, be flexible, take risks."

14 Hardening Off

Your seedlings have been growing in the protected environment inside your home. The garden outdoors exists under a new and variable set of conditions: sunlight, wind, and cooler temperatures. Seedlings must be gradually introduced to their new environment, rather than plunged directly into the ground. This process is called hardening off, and consists of two parts.

Toughening Plants Indoors

The first part is to toughen the plants, quite literally to harden them, so that they will be less vulnerable to outside weather extremes. Sappy, succulent growth is easily damaged by sun, wind, and cold temperatures. What you want to do, then, is slow down the growth of your plants for about a week before you introduce them to the more extreme outside weather. To do this, water them less often and don't fertilize them during that final week indoors. Keep temperatures on the cool side— if possible a few degrees lower than the temperature that prevailed up until now. The result will be a shorter, more fibrous plant that will suffer less from the transition to the outdoors.

At about the same time you begin hardening off plants indoors, block out any plants that are growing in flats. To do this, cut between the seedlings, across the flat from left to right and from top to bottom so that each plant will be centered in a cube of soil. Blocking out severs roots that would be broken in transplanting anyway. Thus, the roots can, in effect, begin their recovery before transplanting. In addition, blocking out stimulates the plant roots within each block of soil.

Acclimating Plants to the Outdoors

When your seedlings have been toughened by enduring a cool, dry week without fertilizer, they are ready to begin to become accustomed to outdoor conditions. Even the best-prepared seedlings will fold up if subjected immediately to a full day of direct sun and even gentle breeze, though, so proceed gradually, eager as you may be to get those flats planted out in the row.

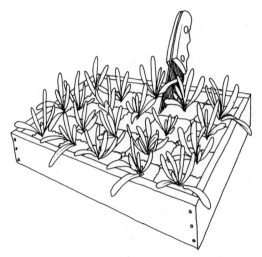

Blocking out seedlings.

Sunlight. Begin with a few hours' exposure to filtered sun—in the shade of a bush, porch railing, or improvised shelter. Gradually increase the amount of direct sun the plant receives until, at the end of a week or ten days, the seedling is able to take a full day of sunlight. Sunlight dries the plant and the soil, and in shallow flats it is important to keep the soil moist enough to avoid wilting. Most flats will need watering daily or at least every other day. In addition, the ultraviolet light given off by the sun, formerly filtered out by window glass or not present in significant amounts in fluorescent light, constitutes a stress for the plant. Only by gradual exposure can the seedling acclimate itself to new, higher levels of ultraviolet light.

Wind. Wind, which is often strong and gusty in early spring, can damage plants in three ways. Its drying effect can increase transpiration in the plant to the wilting point. A badly wilted plant is often permanently impaired. Wind also breaks stems of taller, unstaked seedlings. In addition, it may whip even small seedlings about to the extent that their roots are loosened or torn from the growing medium. It is important, then, to choose a sheltered corner for the plant's first week outdoors, one where you either have, or can arrange, protection from prevailing winds.

Cold. Cold outdoor temperatures pose a less serious threat to your earliest seedlings—the cabbage, lettuce, endive, and so forth—than to the later tomatoes and peppers. Celery, eggplant, and melons are especially sensitive to cold temperatures. Have cartons ready to put over outdoor flats when frost threatens.

The use of cold frames and hotbeds can make possible an even earlier transition to the outdoors. For more information on building and using these structures see chapter 16.

15 The Art of Planting Out

Your plants have come a long way in only a few weeks, from seeds to tiny seedlings to hopeful transplants to flourishing leafy green promises of bright bloom and early good eating. With every stage you've managed successfully—germination, transplanting, maintaining fertility and water balance, hardening off—your young plants have become more valuable to you. With each week that passes, they become more difficult to replace.

All the attention given to choosing the best seeds and starting the plants off right, then, has built up an investment in your plants. It's important now, at planting-out time, to keep that care consistent when setting seedlings in the garden. It is easy to lose the work of months overnight, from overeagerness, poor timing, or insufficient protection. There are, it seems to me, three skills involved in helping young plants to make the transition between pampered confinement inside and life on their own in the big garden outside: good judgment, good planting technique, and knowing when and how to protect your plants from harsh conditions.

Judging the Signs of Spring

Setting out plants at the right time, in the right way, is the mark of a seasoned gardener. I've made plenty of mistakes in putting out my plants. That's how I've learned what's important. Not that I've arrived at any completed state of gardening expertise—far from it. There are too many variables in gardening—weather, soil, plant vigor, insects, wind, timing—for any of us ever to claim complete enlightenment or certainty about the whole process. All I can really say is that I'm learning as I go. It's a path, not a destination, and there's a lot of ground yet to be covered.

Good judgment about moving seedlings outdoors began way back in late winter when you decided when and how to start your plants. If you've been able to resist the winter-weary impulse to start your seedlings too early, you'll now have strong, compact plants to set outside, rather than spindly retarded ones that have been kept on hold for too long. If the plants aren't overgrown, you won't be so tempted to put them out before conditions are favorable.

Judgment of outdoor conditions doesn't come overnight. It's the result, rather, of patient observation, experimentation, and record keeping. If you're new to gardening or to the particular piece of soil you're working with this year, allow yourself a few years to develop a sense of earth and weather signs and how they interact.

Record the last (spring) and first (fall) frosts each year. Take note of the prevailing wind direction. Find several protected, lightly shaded, and sun-dappled corners for hardening plants. Discover which areas in your garden thaw first and which drain more rapidly. In many gardens there is a small corner or strip where, with a little hand digging after early warming, an early chance planting can be made. If it takes, you start eating fresh from the garden a few weeks early. If it doesn't, there's still time and space to make the certain main-crop plantings that must wait a bit for better odds.

Microclimate. What this means is that you must study your microclimate. It helps to know average frost dates for your area, but it helps even more to know how cold air moves on your terrain and which patches of ground become workable first. Cold air sinks. Warm air rises. Land in a valley, where there is no lower point to which frosty air may drain, can be a frost pocket. Here on our farm, on the south side of a mountain, our open land lies on a long gradual slope running from north to southwest. The upper gardens may escape frost damage that hits the lower patch. Even though there is still lower land below this lower spot to which the cold air can continue to move, it is more vulnerable to frost than the more elevated plot.

City buildings hold heat and block wind. Large trees shelter plants from wind but not from frost. Large bodies of water and even small ones, to a lesser extent, can moderate the temperature of the surrounding land.

Observe and experiment, then, and keep notes on what happens. When did you set out the first tomatoes? Where? Did frost hit after you planted them out? How did you protect them? When did they start bearing?

After a few years of gardening in Philadelphia, we found that it was generally safe to set out tomatoes by May 9 in the sheltered block of old houses where we lived. When we moved to Wisconsin, we soon discovered that it was foolish to plant tender plants out before Memorial Day. Now, on our south central Pennsylvania farm, we count on setting plants out around the middle of May, up to a week later than the prevailing safe time in the southern part of our county.

Signs of Spring. You can run into trouble, though, if you work strictly by the calendar. Not every year is the same. Keep the calendar at hand, certainly; it will help you to know what to expect. But try also to listen and look for the actual signs of spring—the return of certain birds, the budding of key trees, the blooming of flowers—that let you know what the situation actually is right around you. In trying to learn this approach myself, I've found that much of the traditional country lore, like the advice to plant soybeans when oak leaves are the size of mouse ears, is surprisingly accurate. (For more information on specific signs to watch for, see the individual plant entries in the encyclopedia at the back of this book.)

Spring peepers.

Spring peepers, although a cheering early sound of spring, don't tell you much about planting out safely. Those little tree frogs get spring fever early, and country wisdom has it that they'll see through glass (ice) three times before spring weather settles in. For me, the cheeping peepers are a reminder, though, to start looking for the first wild greens of the season. Sometimes I can till the garden within a week after they start their trilling, but I don't rely on them for planting information.

The more you become attuned to what is going on around you outdoors, the sharper your sixth sense about the weather will become. Soon you will find it possible to sniff the air, watch the moon, feel the breeze, and make at least an educated guess about the possibility of frost at night.

Phenology

There is actually a branch of science that deals with the interrelationships that occur in the natural world. Phenology, the study of the cycle of development of living things as it unfolds throughout the year, is based on observations dating back to early Chinese civilization. Observation networks in many countries record blooming and fruiting dates of certain key plants, called "indicator plants." Honeysuckle and purple lilac, which are widely distributed, are often used to gauge the progress of the season. Old as it is, the science of phenology hasn't yet been thoroughly researched. Your own recorded observations made over a period of years could reveal dependable patterns that would help you to plan your plantings.

Signs of Frost. When we notice any of the following signs, we get ready to protect our tender seedlings:

- A calm air, no breeze blowing, and a clear sky often indicate imminent spring or fall frost, especially if the air is dry.
- An early evening drop in temperature to around 40° F (4° C) can be a frost warning.
- The last week before a full moon is traditionally a time of frost danger. Observation does seem to be supported by theory here, too. It seems reasonable to assume that the moon, which influences ocean tides, also affects tides of air.

Learning from Neighbors. Take a cue, too, from your experienced neighbors and relatives. In most neighborhoods, there's a certain amount of friendly rivalry to see who can plant the first peas and onions and harvest the earliest tomato. There's also a large store of gardening expertise and often a little plant trading. Your neighbors may have much to teach you. It is also possible that, after a few years of learning and experimenting, you'll have a lot to share with them. I'm thinking of the farm-raised professional man who hesitated to plant zucchini for fear that they would cross-pollinate and affect this year's watermelon harvest. (See chapter 28 to find out why he needn't have worried.) Then there's the neighbor who liked hot peppers but didn't know you could raise them from seed. Another acquaintance planted spring radishes but never tried winter radishes or planted a fall garden.

Developing a Good Planting Technique

Judgment, important as it is, may sound elusive, up in the air. The second skill you need is quite down-to-earth: good planting technique. It can be achieved by following these steps.

1. First, choose a good day for setting out your transplants—not a clear, sunny, breezy day, but rather a damp, drizzly, warmish day. You may not always be able to wait for such a day, but if one does come along at the time when other conditions are favorable, then drop everything and devote the day to planting.

Sun and wind, you see, remove large quantities of water from the plant by transpiration, and the roots, disturbed in transplanting, are hard put to make up for this water loss. You can get an idea of the extent of water loss, even under ideal conditions, from the following figures published in the book *Botany for Gardeners,* by Harold Rickett. In one season, a tomato plant loses 35 gallons of water by transpiration; a potato plant, 25 gallons; a giant ragweed, 140 gallons. The less setback a plant suffers in being put out in the garden, the sooner it will be off and growing and putting on melons or tomatoes or good green leaves.

2. Use a shovel or an extra-large trowel to dig a generous hole—one that will provide ample space for the plant's root ball.

3. If you possibly can, put a handful of compost into each planting hole,

especially for melons, cucumbers, and squash. When adding compost, dig the hole a few inches deeper, mix in the compost, and toss on a handful of fine loose soil before setting the plant in place.

4. A tomato plant placed flat on the ground in a shallow trench with just the top one or two sets of leaves showing above ground will form roots at each leaf node and escape drying winds. It will also be easier to cover when frost threatens. Cabbage and lettuce plants may be set more deeply than they grew in the flat if they have developed a one- to two-inch stem; otherwise, cover them with soil only up to the bottom leaves. Soil in the crown of the plant will encourage rot.

Plants that have been blocked off in advance will often have a sufficiently dense network of roots to hold the soil around them until you settle them in the hole. A bare root plant suffers more. Some gardeners dip bare roots in a slurry of thick, muddy water to coat the roots, especially if the day is dry and sunny. As you position the plant in the hole, try to arrange the roots so that they do not double back on each other or fold back toward the top of the hole.

Some gardeners like to soak the planting flat with liquid fertilizer just before removing the plants to set them in the garden. It is true that this one-step treatment saves the trouble of individual plant feeding to a certain extent. I prefer to leave the plant a bit on the hungry side, reasoning that the roots will then be more likely to put out extensive new growth in search of nourishment.

5. Water each plant *as you plant it,* rather than setting a group of plants and returning later to water the whole row. After the seedling has been positioned in the hole, pour at least a quart of water on it to settle it in the hole.

6. Next, fill in the hole with loose soil—never with clods, which would leave root-drying air spaces.

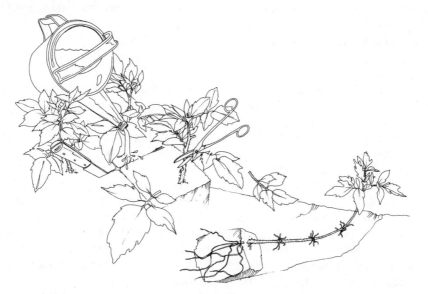

When planting tomato seedlings, block out plants and harden them off. Prune lower leaves and plant each seedling so that only the top leaves are above ground. Roots will grow from each node along the stem.

7. Finally, press the loose soil gently but firmly around the plant so that no air pockets are left. At the same time, form the soil around the stem into a shallow, saucerlike depression which will catch rain and funnel it toward the roots.

Continue to plant the rest of the seedlings in the same way. If you're making an early planting in hopes of moving your harvest day up a week or two, it's a good idea to save a few seedlings of each kind of vegetable you're gambling on as insurance in case anything would happen to the first planting.

When planting seedlings started in peat pots or pellets, keep the following points in mind: Peat pots that stick out above soil level are amazingly effective wicks, drawing moisture away from the plant's roots. For that reason, it is important to tear off at least the top inch or so of the fiber pot so that only the stem and leaves are above ground. Since I have less faith than I once did in the permeability of peat pots to plant roots, I sometimes peel off the whole pot—gently, to avoid tearing roots—when setting out a potted seedling. (If the roots have already penetrated the pot, I don't remove it.)

The plastic netting surrounding those compressed peat-pellet containers is also suspect in my eyes, but I've noticed that roots usually seem to penetrate it more readily than they do the peat pots. The netting is easy to tear off, though, in a case where root penetration might be doubtful.

Protecting the Plants

The final trick you need for weathering your seedlings successfully through their big move to the outdoors is protection. Protection from sun, wind, and frost. Leaving new transplants to their own devices usually leads to disappointment. Sun and wind work against their survival for the first week that they have their roots in earth, and even after that, sneak frosts can wither a whole planting in just an hour or so of early morning chill.

Protecting Plants from Sun and Wind. Sun and wind damage to a newly set out plant are usually confined to its first three or four days in the ground, but may occur up to a week after planting out tomatoes and eggplant, or bare-root seedlings.

Pale leaves indicate sunscald. Cover them promptly and increase exposure gradually over the next week or two. I often cover these plants with protectors used against frost, such as glass cloches, plastic sheeting, or mulch for most of the first two or three days—no longer than that, though; they need ventilation and light.

Sometimes I use the pint-sized plastic, open-grid berry boxes to filter out some sun on early lettuce and broccoli transplants. When I must set out young plants in sunny weather, I scatter a light covering of dry hay stems, borrowed from the mulch pile, over their crowns. Just a handful for each plant will suffice. The hay shades the plants while admitting some air and sun. By the time the leaves grow above it, they are strong enough to make good use of full sun.

To shield early cabbage from strong winds, many country gardeners push an old shingle into the ground on the north or west side of their plants. (To determine the direction of prevailing winds in your garden, tie cloth strips to several stakes or poles

Baskets protect young plants from frost.

and note in which direction they blow.) An opened, bottomless half-gallon milk carton set around the plant makes a good wind and sun shield. Even brush and twigs pushed into the soft earth around the plant will serve to cast dappled shade and deflect wind.

One year, when I put out an early planting of head lettuce in a sheltered corner near the back step, I covered the young plants lightly with brush left from staking the pea patch. For the first few days, I draped feed bags over the brush to cast dappled shade on the tender plants. It looked perfectly awful, but the combination of varying shade and plenty of air, followed by the lighter shade from the twiggy branches after

I'd removed the burlap, brought the plants on gradually without any sunscald or windburn.

The first week in the ground is a critical time. The extremes of early-spring weather can prove overwhelming to a tender young plant. A bit of hovering and preventive protection at this time can make the difference between a setback and a head start.

Protecting Plants from Frost. I've had to learn not to be vain about the garden in early spring, but rather to stand ready to cover the young plants with all the odd things I can muster if frost seems likely. Frost watch must sometimes continue for as long as two weeks after you put tender plants inside. Even head lettuce and endive must be defended from severe frost.

Improvised Protectors. I use half-bushel baskets, flowerpots, coffee cans, cottage cheese cartons, berry baskets, and gallon plastic cider jugs with one end cut off. The lightweight plastic protectors were sometimes blown off by the wind until we started the practice of routinely topping them with a small stone or clod of earth to hold them down.

Two precautions:

- Leaves that touch the inside of a metal protector can get zapped by the frost, conducted right through the metal. I use the cans over the smaller, lower plants.
- A flowerpot has a hole through which cold air can pass, and in a severe frost that can be enough to kill the plant. It's happened in my garden, so I now put a little stone or plug of moss over the pot's drainage hole. Berry baskets have side holes, but since they permit air circulation, this problem doesn't seem to arise.

There's usually about a month between the setting out of lettuce and cabbage in April and tomatoes and peppers in May, when I keep the garden cart loaded with my motley plant caps, ready to wheel out to the field when needed. Heavier protectors such as flowerpots, pans, or cans may be left right beside the plant, but the light-weight plastic containers tend to blow away, so I usually collect them and bring them out when needed.

In addition to the items you might find around the house or yard, there are other types of covers and there are certain methods that ward off early frost damage.

Glass Cloches. These can be made by cutting off the bottom of glass jugs. They will not only keep out frost, but also act as miniature greenhouses for risky early plantings. Ventilate the top during the day (remove the cap), or the plant may cook in the sun's heat.

Tents of Plastic Sheeting. Support sheeting by draping it over wire U-forms, and weight down the edges of the plastic with soil or stones. These tents let in sun and keep out frost and wind. Be sure to ventilate them at the ends and at the bottom rather than the top to prevent the escape of the rising warm air.

Kelp. Plants sprayed with a dilute solution of this seaweed (see chapter 19) have

Various items found around the house and yard can be used to protect seedlings from too much sun, wind, and frost.

shown greater resistance to frost. In one study, treated tomato plants survived a temperature of 29°F (–2°C).

Mulch. Loose straw or hay drawn closely around plants and lightly over their crowns provides enough trapped-air insulation to ward off a light frost.

Water. Water can insulate, too. In experiments reported by Professor Frederick Fay of the University of Toronto, plants continued to grow in a cold frame topped by water-filled plastic bags (set right on the glass), even though air temperatures outside averaged 20°F (–7°C). A space of at least four inches must be kept between the plant and the outside air.

Freezing Water. You can even take advantage of the warming effect of freezing water. That's right—water gives off heat when frost hits it. One pound of water releases 144 Btu's when it freezes. In an enclosed space like a cold frame or a plastic tent, several wide, shallow pans of water will help to take the edge off the frost as they freeze. It's important to use shallow containers that will chill rapidly. Deep bowls or pans will not cool quickly enough to do your plants any good. Letting the sprinkler run directly onto vulnerable plants also protects against frost kill.

Rescue. Then there's the morning-after rescue. If, despite your best efforts or because of an extra-sneaky frost, you wake up to see the characteristic darkening of frost-touched leaves in your brave new garden rows, keep calm and reach for the hose.

Frost damages plants by forming sharp-edged ice crystals, which puncture cell walls. If you can hose the plants with a fine spray of water *before the sun hits them,* you can often prevent the cell damage that kills the plant. If a hose isn't handy, use the watering can with the sprinkling head on it as I've done more than once, or use your trombone sprayer. If the sun is coming up fast and you have many plants to spray, toss an old sheet or blanket over the plants that must wait to prevent their thawing—and consequent collapsing—before you get to them.

16 Cold Frames, Hotbeds, and Cloches

Trying to extend the growing season a little in either direction can be a gamble. However, using structures like cold frames, hotbeds, and cloches or row covers often pays off in more garden food and earlier flowers by protecting your freshly sown seeds or just transplanted seedlings from harsh conditions.

Cold Frames

Once you have a cold frame, you'll wonder how you ever got along without one. My own simple cold frame adds several weeks to the productivity of my garden, in both spring and fall. Here's how I use it:

- To start seeds in early spring.
- To harden off spring transplants.
- For late-summer plantings (or transplants) of easy-to-cover fall vegetables like endive and lettuce—and even to winter-over parsley.
- For late-fall sowings of vegetable seeds, such as asparagus, sorrel, or corn-salad, that will come up in early spring but would get plowed over if sown in the big garden.

My small three- by six-foot frame can't serve all those functions at once, of course. I really need several more cold frames. (Are you reading this, Mike?) However, by careful planning and prompt rotation of plant populations, I keep the frame busy from March to December.

Location and Design. Your first consideration in planning a cold frame should be location. A southern exposure is a must, and some protection from cold winds will help to maintain the sun-generated warmth in the frame. If the site slopes slightly to the south, so much the better. The spot should be well drained, too. Don't place your cold frame in a hollow where water will collect.

100

A cold frame lets you begin your growing season earlier in the spring and end it later in the fall.

Your aim should be to build the frame so that the sun's rays will strike the glass at as close as possible to a 90 degree angle. Since the sun's position in the sky changes with the seasons, from its low, weak point in winter to its commanding summer position high in the sky, you obviously can't ensure a 90 degree angle at all times. Because March and April are the critical months for seedling protection for many gardeners, the greatest good for the most plants can probably be obtained by making the angle of the cold frame cover in such a way that early-spring sun will hit it

The Angle of Your Cold Frame Lid

Most gardeners will be content simply to build a cold frame with a sloping lid, but if you want to fine-tune yours, here is a formula that you can use.

The angle of the sun's rays from the horizon will be equal to 90 degrees plus the amount of the sun's declination (its angular distance from the equator at noon—this figure, which is different each day, is published in almanacs) minus the latitude where you live. Now to determine the angle of your cold frame lid, use this formula:

$$\text{lid angle} = 90 \text{ degrees} - \text{sun's angle}$$

head-on, perpendicularly. (See the box, The Angle of Your Cold Frame Lid, for specific details on how to determine this angle.)

Plants that are in the cold frame over summer are often being protected from strong sun by slatted covers or screening, as in the case of summer lettuce, so the summer angle isn't important. Fall plants need frost protection and intensification of the gradually waning sun, which is somewhat lower in the sky than it was in early spring.

You can build a cold frame that will be a remarkably effective sun-catcher during both spring and fall if you make it in sections that can be bolted together, and make a special fall insert to catch the pale, weak, late-year sun. In a *Farmstead* magazine article, Rocky Roughgarden describes the modular cold frame he built for his Massachusetts garden. The glass side of the basic spring section slopes at a 35 degree angle from the earth. A wedge-shaped fall insert, bolted on in early autumn, raises the angle of the glass to 55 degrees from the earth. The Roughgarden cold frame boasts, in addition, an extra cover insulated with scrap Styrofoam. The exact angle of the sun will vary, of course, in different latitudes, and at different times. Unless you want to build several inserts, you'll need to decide when you want to capture the most extra light and plan the angle of the insert accordingly.

Dimensions of the cold frame can be suited to your site and the materials at hand, but in most cases a maximum width (front to back) of three feet is ideal. A frame that is too wide—over four feet—will be difficult to reach into when you tend the plants. Length of the frame is not critical, but make it as long as possible; soon after it's built, you'll wish it were longer. If you will be creating the top of the cold frame from windows or storm sashes, consider the number of windows you will use and their length when deciding exactly how long to make the cold frame. Practical dimensions range from 6 to 12 inches high for the front board and 12 to 18 inches high for the rear.

Construction. The construction of the cold frame, for most of us, is determined by the materials we have at hand. We made ours out of discarded wood-framed storm sashes and two-inch locust boards milled from trees cut in our woods. Locust has the advantage of lasting a long time in the ground. Cedar, cypress, and redwood are also decay-resistant woods that should outlast the more easily replaced sash. When using pine, oak, or other less rot-resistant woods, brush on several coats of linseed oil or

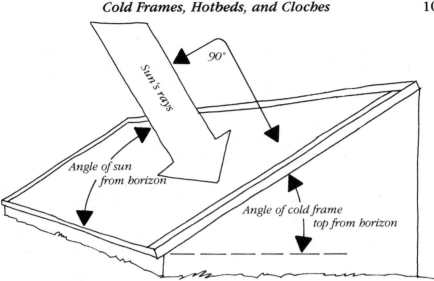

Ideally, the sun's rays should hit a cold frame at a 90-degree angle during spring.

copper naphthenate to retard rotting. Avoid using creosote, pentachlorophenol, or mercury compounds to treat wood intended for use around plants.

Some gardeners make a deep frame of especially rot-resistant wood and bury it eight to ten inches in the ground. Others simply set the frame on level ground. You can even attach handles to the ends of the cold frame to make it portable so it can be used to cover special row crops in the garden. Frame sides may be screwed, bolted, nailed, or hooked together.

You'll want to reinforce the corner unless the frame is designed to be taken apart for moving or storage. We nailed two- by two-inch stakes of scrap wood in each corner of our frame. Purists object that interior supports interfere with placement of seed flats; some who wish to make very efficient use of every square inch of the frame may prefer to nail one- by three-inch boards to the exterior corners for bracing. It is often a good idea to drive supporting stakes into the ground around the cold frame to keep it well anchored. We did not add struts to support our windows, but in a long cold frame or one using heavy windows, or where snow cover is heavy, you might want to nail one or two one-inch-wide strips of one-inch-thick lumber across the frame to support the sash. These cross ties may be countersunk in the front and rear boards of the cold frame; however, it's easier to work in the cold frame when there are no cross ties.

Wood is a good insulator, far better than metal, and so the boards should be a good inch thick, if possible. Two-inch-thick lumber will protect your plants even more, although it will be more cumbersome to handle. Either 10-inch or 12-inch-wide boards would work well. Dimensional lumber would be my first choice, but you could also use ½-inch, ⅝-inch, or ¾-inch exterior plywood.

Construction of the cold frame is simple. Just follow these steps:

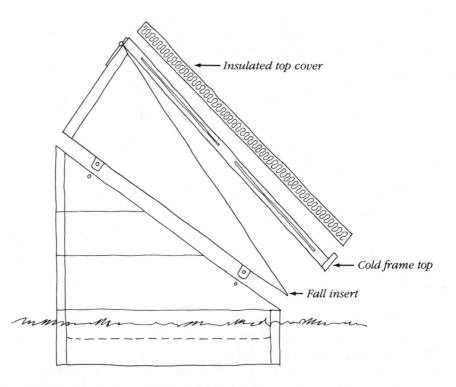

The Rocky Roughgarden cold frame. (Redrawn from "The Modular Cold Frame," by Rocky Roughgarden, Farmstead, *Spring 1977, pp. 32-33.)*

   1. Dig a hole about a foot deep and three to four inches larger in width and length than the dimensions of your projected frame.
   2. Cut the side pieces, the front, and the back from sturdy wood. If you're using plywood, nail a length of scrap lumber about two to three inches wide and three-fourth inch to one inch thick to each end of the long back and front boards. Then use nails (or, for stronger construction, screws) to attach the side boards to the front and back boards. To further brace the frame, nail a length of one- by two-inch scrap lumber to the upper edge of the back board (see the diagram on the facing page). The two short side pieces of the frame should taper from back to front to give you the desired angle, and the long back board, of course, will be higher than the long front board. Check to determine whether the rectangles are true by measuring from corner to corner in an X. Both lines should be the same length. Nail or screw the pieces together.
   3. Reinforce the corners with metal braces or short two-by-fours, unless you have used the scrap lumber described above for plywood.
   4. Position the cold frame in the hole, taking care to level it well.
   5. For the sun-catching top of the frame, most people use secondhand windows or storm sashes hinged to the back board. Here again, wood-framed windows will retain much more warmth than those that are metal-rimmed.

On frigid nights, insulating the cold frame from the outside can help to keep internal temperatures above that critical frost point. You can pile old blankets, leaf-stuffed bags, boards, or rugs over the sash and also bank the frame with hay or bags of leaves. An extra layer of glazing, even a simple sheet of clear plastic, will trap heat by blocking convection and conduction—the often-cited greenhouse effect—and this can reduce heat loss by 34 percent.

A Few Variations. Variations on the basic box-frame and glass-sash construction of the cold frame can run from the substantial to the makeshift. For a more durable structure than wood, you could build a brick, stone, concrete block, or poured concrete frame. For any such frame that is held together by mortar, you'd need to dig deeply enough to lay the bottom layer several inches below the frost line, so the cement wouldn't be cracked by the inevitable rocking action of alternate freezing and thawing.

For a collapsible frame, you could attach hinges to the back corners of a wood box-frame and then attach the front panel with screws or hooks. You could also make a conventional box-frame portable by attaching handles to its sides.

For a quick cold frame, if you have enough hay bales and plenty of space, arrange the bales in a rectangle, open at the top, to support a storm window. For easier access, use a scrap board for the front side. Be sure to position the bales so that

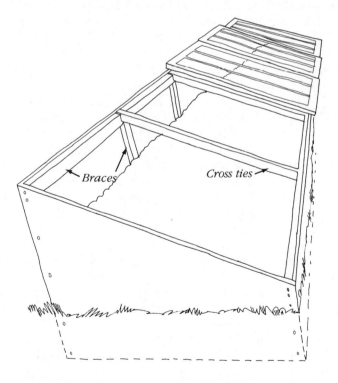

Cross ties and braces reinforce the structure of a cold frame.

the string bindings are on the sides rather than on the ground where they would rot quickly.

Monitoring Temperature. Warm spring sun can make cold frame interiors surprisingly hot. You might want to put a thermometer in your frame so you'll know when internal temperatures climb into the danger zone, much above 80°F (27°C). Different plants, of course, thrive under different temperature ranges:

- Most tender plants—60°F to 75°F (16°C to 24°C)
- Half-hardy plants—50°F to 65°F (10°C to 18°C)
- Hardy plants—45°F to 60°F (7°C to 16°C)

James Crockett, former host of "The Victory Garden" television program, offered helpful advice: When the outdoor temperature rises to 40°F (4°C), Crockett says, you can open the cold frame cover six inches. At 60°F (16°C), swing the lid wide open. Then, in midafternoon, close the lid to retain the midday heat.

Automatic temperature-sensing vents are available from greenhouse suppliers, but they're not practical for most improvised frames, which usually have lids too heavy for the lightweight mechanism to lift.

Accessories. Once the basic structure is made, you'll need a few accessories to make the use of your cold frame easier and more efficient.

To Anchor an Open Lid. On good days, when winds aren't high, you will want to swing back the lid of the cold frame to expose the plants directly to the sun. Sudden wind gusts, though, can be surprisingly forceful. If your cold frame stands up against another structure, or you can't rest the lid on the ground, you will want to devise a way to anchor the glass lid so that the wind won't bang and break it. Our cold frame is built on the south side of the cement-block-enclosed compost pile. When I raise the lid, it leans back and rests on the block wall, but a strong wind can blow it down with disastrous results. (If you think I'm warning you about this because I slipped up on it the first time around, you're right!) After Mike replaced the broken glass, he put a screw hook into each of the two top windows and fastened wire around and through a block on each end of the compost-pile wall. Now when I open the glass top, I catch the wire loop in each hook so the wind can't blow it down.

You'll avoid all this trouble if you simply lay the windows on top of the cold frame. Although not quite as airtight as the hinged construction, this simple setup makes it easy for you to work in the cold frame: You simply lift the window off.

To Prop the Lid Open. You'll need a prop of some kind to elevate the cold frame top a few inches when intense sun warms up the interior. This can be as simple as a handy stick or stone, but an adjustable stop, made of a notched two-by-four, will give you more leeway in arranging ventilation. Cut three steps into the prop stick at different heights. To use it, simply set the prop stick on the ground and rest the lid of the cold frame on one of the three steps.

To Provide Some Shade. In early spring, I use the cold frame to bring on the early lettuce and cabbage under glass. By the end of April it's usually safe to start hardening off the peppers and tomatoes I'll plant in the garden in mid-May. When I put out flats of these vegetables, I leave the glass windows open during the day, unless the wind is strong, and close them at night when temperatures dip. For the first week or so, I use improvised slatted covers to shade out about 60 percent of the sun's rays so the plants don't suffer from sunscald as they would if suddenly exposed to the full impact of direct sunlight.

Watering the Plants. Plants in the open cold frame dry more quickly than those kept under lights. For some, daily watering is necessary when the sun is intense. Most flats, unless shallow and lacking moisture-retaining moss, do well for me when watered about every other day. Plants that are being hardened off, you'll recall, should be kept on the dry side but not allowed to wilt.

Hotbeds

A hotbed is simply a cold frame that is supplied with extra warmth. The heat, usually from the bottom, helps to promote seed germination and steady growth of seedlings that are cold sensitive.

Using a Soil-heating Cable. One effective hotbed design uses a soil-heating cable imbedded between vermiculite and soil. A 30-foot cable will heat a 3- by 6-foot cold frame. To build such an arrangement follow these steps:

1. Spread a two-inch layer of vermiculite on the bottom of the cold frame for insulation.

2. Lay the cable on the vermiculite in long loops, keeping wires eight inches apart from each other and three inches away from the edge of the bed.

3. Cover the cable with an inch of soil or sand.

4. On top of the soil, place a layer of wire screening to prevent the cable from accidentally being dug up.

5. Cover with four inches or more of good fine soil. This will give steady bottom heat.

A Temporary Hotbed. If you already have your cold frame set up and don't want to install a permanent heating arrangement, you could warm the frame temporarily with a chicken brooder lamp. Or, for very gentle warmth, take a tip from one of our neighboring master gardeners and tack a string of outdoor Christmas tree lights around the inside of the frame.

A Manure-heated Hotbed. Conservation-conscious gardeners are turning increasingly to the manure-heated hotbed, since it's possible to achieve the same end without drawing on electric power. In addition, you improve your soil and eliminate equipment purchase and care. Most old gardening books offer elaborate plans for the construction of hotbeds of heroic proportions. Our grandfathers, who farmed with horsepower, had an abundance of this quick-heating manure at their disposal. Even today, though, the common keeping of pleasure horses by people who don't always have a use for the by-products puts horse manure within the reach of city and suburban as well as country gardeners.

The following instructions will take you through the simple process of making a manure-heated hotbed:

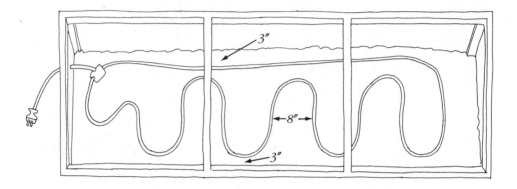

When placing a soil-heating cable in a hotbed, keep wires eight inches apart and three inches from the edge of the bed.

1. Dig a rectangular hole three to four inches larger all around than the dimensions of your cold frame and 1½ to 2 feet deep.

2. A week to ten days before planting, pack horse or chicken manure into the hole up to six inches from soil level.

3. Hose this layer down to start decomposition.

4. Spread six inches of fine soil on top of the manure.

The manure will reach its peak of heat within three to six days. Wait to plant, though, until the soil temperature falls to 85°F (29°C), or you'll cook your seedlings. Ventilate the bed as necessary to keep the temperature below 90°F (32°C).

Improvisations. If you like to improvise, perhaps some of the following ideas, used successfully by other experienced gardeners, will work for you:

- Make a mini-hotbed from a discarded refrigerator crisper drawer set in a hole over packed, moistened manure as described above and topped with its original glass.
- Back a portable cold frame up to a cellar window, which can be opened to admit heat from the house.
- Make a long, low tent from two screen doors covered on each side with a layer of tacked-on plastic sheeting. Attach triangular end frames of outdoor plywood in which ventilation holes have been drilled. Use this as a cold frame, or bury manure underneath it to make it a hotbed.
- Make frames from scrap wood covered with poultry netting (or use old screen doors). Tack on plastic for an inexpensive portable cold frame.
- Line sturdy cardboard cartons with black plastic that is stapled on. Top

with a flap of clear plastic that is stapled to the back end and clipped to the front end with clothespins. The waxed cartons used for chicken parts work especially well; you can get them from grocery stores.

As you've no doubt noticed by now, part of the satisfaction of rigging up a cold frame or hotbed—for me at least—lies in using materials at hand to make a useful thing. See what *you* can come up with.

Cloches and Other Season-Extenders

A bit of extra protection in early spring can make a significant difference in the growth of your plants. Probably the most effective use of temporary row or plant covers is to moderate harsh spring weather for seedlings planted at or near their normal planting time. You can also use spot covers to enable you to get away with planting up to three weeks early, but don't expect much more than that. Although they certainly do reduce heat lost by convection and radiation, small temporary covers can't retain enough of the sun's heat to tide a tender plant over a severe freeze.

Cloches. For many years, French market gardeners used cloches—bottomless clear glass bell jars—to protect their early plantings. More recently developed plant protectors are easier to use and store than the heavy breakable glass. Whether or not you still call them cloches, as many people do, these covers are used by many cost-conscious present-day market gardeners, who wouldn't bother with them if they didn't confer a useful amount of protection.

You can make your own cloches, of course, by removing the bottoms from plastic gallon jugs. These are easy to stack, although they do get a bit misshapen after a few seasons of use, and they have the extra advantage of being free. Press them well into the soil when you set them over your plants so the wind won't topple them.

Tunnels and Row Covers. Tunnels may be purchased or homemade. You can use clear plastic to arch over your early rows, supported by supple bent twigs or wire hoops stuck in the ground. The sides of the plastic should be thoroughly battened down by banking with soil or covering with boards, and the top surface should be slit at intervals to allow excess heat to escape on a sunny day. Sheets of perforated plastic for tunnels are available from mail-order suppliers. A sheet of corrugated fiberglass, bent into a U shape and wrapped with wire on each end to retain its tunnel form, will last longer than plastic sheeting and needs no additional support except, perhaps, a few stakes to keep it anchored at the sides.

Two recently developed commercial products confer additional protection.

Wallo' Water. The Wallo' Water is a clear vinyl ring of tubes, 18 inches high and 18 inches in diameter. When filled, the enclosure will raise the air temperature around a plant by at least 5°F (3°C) and the soil temperature by about the same amount. The way it works is that the water—25 pounds of it—soaks up the sun's warmth and then gradually releases it to the plants at night. For more support and warmth, you can set a tire around the Wallo' Water.

It's an interesting idea, and it certainly does help to encourage plants that are set out moderately early—say, seven to ten days ahead of the usual time, but it can't work miracles. I wouldn't use it with tender plants like tomatoes and squash in freezing March weather.

It takes about five minutes to fill the tubes with water (a process that can be a comedy of errors if you don't read the directions). Put each ring in a five-gallon bucket for support and fill the tubes on alternate sides, rather than all on the same side, so the device will stand straight sooner. Although stakes are not supposed to be necessary, the rings do occasionally slump, so you might want to slide two smooth stakes inside the rings right from the start to help keep them upright. If you fill the tubes only two-thirds full of water, you can bend the top flap over to form a closed teepee. To prolong their useful life, be sure to thoroughly drain and dry the tube rings at the end of the season.

Reemay. This lightweight spun polyester fabric is easier to use than the Wallo' Water, but short-lived. To use it, you just drape it over your rows immediately after planting seeds or transplanting seedlings and anchor it well on both sides with loose soil. Allow 1 to 2 feet of slack, depending on the size of your plants. This provides leeway for the plants as they grow. The 67-inch-wide material is available in lengths of 20, 50, and 100 feet. A sheet of Reemay is wide enough to cover a bed 3½ feet wide or two average garden rows. The fabric is so light in weight that plants can easily shoulder it upward as they grow.

Reemay protects seeds and young seedlings in several ways. It reduces light transmission by about 20 percent. Young plants still receive plenty of light, and on a sunny day the space under the cover is less likely to overheat than it would be with a clear cover. The temperature under a sheet of Reemay rises by 8°F to 10°F (4.5°C to 5.5°C) on a sunny day, and although heat is lost again at night, the bonded cover provides a good 2°F to 3°F (1°C to 1.5°C) of frost protection. It also retains moisture, which helps to encourage germination, and when thoroughly anchored without any gaps, it's an effective insect barrier. (If you use it for cucumbers or squash, remove the Reemay when blossoms form so pollinating insects can do their job.) Reemay is permeable, so rain can soak in and hot air can escape. In warm weather, you might need to loosen it to permit more ventilation; when outside air temperature is 85°F (29°C), the air under Reemay heats to 95°F to 97°F (35°C to 36°C).

Reemay offers a few disadvantages as well. It encourages weeds as much as it does vegetables, so you'll need to do some weeding. Because sunlight degrades the material, it often starts to shred after a three-month stint in the garden. If you pack it away out of the light as soon as it's no longer needed, you might find that it will last for another growing season.

Other Row Covers. Other brands of porous row covers include Agronet, Agryl, and Vispore. Agronet is available in a 42-foot width. A newer row-cover material, Tufbell, is made of polyvinyl alcohol and is three to ten times more stable in the presence of ultraviolet light than other row covers. It is also, so far, more expensive. Tufbell is not yet as widely available as Reemay. No doubt these products will continue to be improved in years to come. The best use for these fabrics seems to be to ease plants through those marginal days of early-spring weather, when an ounce of prevention can enhance their growth.

17 Care of
the Young Transplant

Extra measures of support given to new plantings—mulching, watering, fertilizing, and staking—should be carried out with regard for variations in weather, plant growth habits, and row spacing. Although differences in climates and individual gardens make rigid prescriptions unworkable, perhaps a few loosely scheduled suggestions would help you to time these practices correctly to encourage rather than impede the plant's growth.

Mulching

Mulch spread between rows smothers weeds and helps to slow the evaporation of soil moisture. Biodegradable mulches like straw and leaves have the additional advantage of improving soil tilth and fertility.

Kinds of Mulch. Generally, you have the choice of mulching with organic materials like leaves and straw or with synthetic plastic products. Mulches of organic materials tend to keep the soil cooler than if the ground were bare, whereas polyvinyl sheets, both black and clear, warm the ground under them. Use any fine mulch you may have for narrow rows and save the coarse stuff for wider aisles.

Newspaper. According to recent reports, most newspapers are now using low-lead inks on their newsprint, but not necessarily on their glazed-paper supplements. Both unprinted newspaper and black ink contain about 5 ppm of lead (which is a pretty low figure compared to the 1,000 to 2,000 ppm found in city carpet dust); therefore, newsprint should be safe. We use it on our garden, but we do not mulch with magazines or with newspaper supplements, some of which have a higher lead content.

Maple Leaves. Fresh maple leaves contain phenol compounds that retard root development of plants in the cabbage family and other vegetable crops and possibly some flowers, too. It's best to let the leaves age for a season so that rain can leach out the offending phenols before using the leaves for mulch.

When to Mulch. Begin mulching the cool-weather crops—lettuce, peas, cabbage, beets, and carrots—as soon as their seedlings look as though they might

111

amount to something. Wait until the soil warms up before mulching heat-loving crops like tomatoes, peppers, eggplant, cucumbers, and melons. My mulching program progresses gradually as the season wears on. Each day, starting in May, I spread a few more armfuls of discarded barn hay around the lettuce and peas. By mid- or late June the soil has warmed, and my husband has cut several wagonloads of weeds for me to use in mulching the large spaces between the tomatoes, squash, and other heat lovers. By early July we have all of our mulch spread for the season.

If you use black plastic to mulch melons, as I do when I accumulate enough black plastic bags that I had used for holding leaves, put it around the plants soon after they've germinated to help warm the soil. Clear polyethylene sheets promote the highest soil temperatures but also encourage weeds.

How Much Mulch. How much mulch is enough to keep weeds down? It depends on the mulch you use. Here are some guidelines:

- Discarded carpet: a single layer.
- Grass clippings: about four inches when first spread (they'll pack down).
- Hay, baled: about a five- to six-inch thickness flaked off from the packed bale; loose: eight to ten inches.
- Leaves: eight to ten inches.
- Newspaper: three to six sheets, held down by stones or other mulch.
- Sawdust: two to three inches.
- Wood chips: at least six inches without a liner. Two to three inches is enough if a layer of newspaper or other light-impervious material is put down first.

Fertilizing

Newly transplanted seedlings have little need for extra plant food. They're busy forming more feeder roots and adjusting to their new environment. If you've spread fertilizer before digging the garden and have put some compost in the holes at planting time, most of your plants should be well supplied for the season.

However, some heavy feeders, like cauliflower, cabbage, melons, and rhubarb, often produce better if given extra feedings, starting a week or ten days after setting out. Some gardeners give their tomato plants extra fertilizer to keep them producing well through a long growing season. An extra side-dressing, if you can manage it, may improve corn. In a large garden, it's usually difficult to get around to doing these extra things. We put extra manure on our corn plot one year when we had not been able to plow in enough fertilizer when preparing the land, but for the most part we depend on the manuring we did before planting to carry most plants through the growing season. Exceptions, in our garden, are lettuce, cabbage, cauliflower, and melons, to which we try to give several booster feedings of either diluted fish emulsion or manure tea. These nitrogenous boosters, since they promote leafy growth, should not be given to your root vegetables, or you may have great leafy carrots with spindly roots.

To prepare manure tea for your garden or to pour on the compost pile, follow these steps:

1. Fill a burlap sack with manure and place it in a barrel or trash can.
2. Fill the can or barrel with water.
3. Cover (to keep out flies), and let the brew steep for several days.
4. Dip off some of the enriched water into a bucket or watering can, dilute it to the color of tea, and pour it around the base of the plant.

For smaller amounts of manure tea, use this method. Put manure directly in old buckets, fill with water, and dip off as needed. Leaving the original manure in the bucket, refill with water. After three to four batches have been made from the same manure, pour the slurry on the garden and start again with fresh manure.

Staking

Vining or climbing plants that are well supported usually produce more fruit or flowers. Because air circulation is better, they're less likely to suffer from fungus diseases. Crop-supporting arrangements are limited only by your imagination and available resources. The following are some suggestions to get you started.

Cylinders. Concrete-reinforcing wire can be formed into cylinders that make excellent plant supports, especially for tomatoes. Clip off the horizontal wire on one end so that you can poke the resulting prongs into the soil to anchor the cylinder.

Teepees. Eight-foot poles pushed into the ground and tied together at the top support climbing beans. Although soil cultivation is not as convenient around broad-based teepees, they are more stable in the wind and easier to set up than single poles, which must be buried more deeply. You can form teepees using last year's sunflower stalks, too.

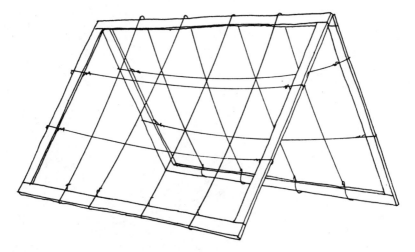

An A-frame for staking vegetables.

Box Frames. You can make box frames from scrap lumber to set right over seedling plants. The plants will then lean on them as they grow.

A-Frames. Two rectangular frames strung with wire or twine and hinged or tied together work well for both cucumbers and tomatoes.

Pole Tents. These can support an impressive wall of flowers or beans. To make one, drive two eight-foot-long 2- by 4-inch posts 18 inches into the ground, spacing them ten feet apart. Nail a ten-foot-long 2- by 2-inch board between the two posts at the top. Then nail three stakes to each of two 1- by 2-inch ten-foot-long boards. Drive the stakes (with boards attached) into the ground on either side of the posts, four feet apart. Tie a length of twine to one long bottom board, run it up and over the top of the support, and tie it to the opposite bottom board. Continue to tie lengths of twine to the boards, spacing them 6 inches apart. Nails may be driven in the top and bottom boards to stabilize the twine.

Maypoles. These work well not only in the garden but also, surprisingly, in container plantings. In National Garden Bureau trial plantings, 12 bean plants growing in a ten-gallon tub with a tall central pole climbed strings that were stretched between the top of the pole and pegs set in the soil at the edge of the tub.

Hay Bales. Try placing bales of hay between rows of tomato plants and stretching wire wide-mesh fencing across the tops of the bales. The hay mulches the plants and the wire supports them. Leave a path between every two to three plants for easy access.

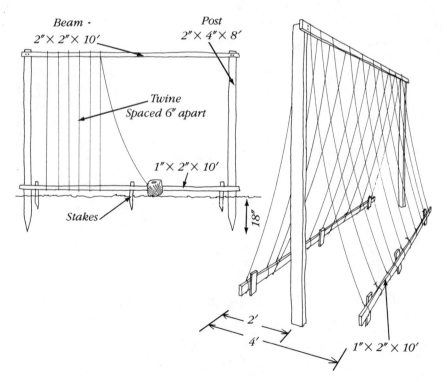

A side view and end view of a pole tent used to support flowers or beans.

Staking Variations. You can even make a few simple variations on the basic garden stake for more effective support structures.

Gather stakes for your tomatoes and drill several holes in each one, spacing them evenly apart. The holes should be positioned at the same level on all of the stakes. Set the stakes between tomato plants in a row and thread twine through the holes, forming loops to enclose each plant. You needn't tie the string to the plant.

Set strong stakes at a 30 degree angle, pointing west, in rows running north to south. Weave twine between the stakes at eight- to ten-inch intervals as the plants grow. Prune off tomato suckers. The plants benefit from excellent exposure to the morning sun, and the fruits, which tend to hang down on the shadier side of the leaning stakes, are shielded from the heat of the afternoon sun. Weeding is easier with this method than with cages or frames.

When to Set Out Stakes. Stakes for plants that need support—tomatoes, pole beans, peas, morning-glories, and others—should be driven into the ground while the plants are still too small to need the stakes. This avoids damage to the more extensive root system of the older plant. It's also a good idea to take note of the prevailing winds that blow over your garden, and position the stake so that the plant will lean toward it, rather than straining away from it when the wind blows.

Weed Control

The old adage "one year's seeding makes seven years' weeding" has its basis in fact. Although few of them have impermeable seed coats, weed seeds have an incredibly long life span, often 10 years and even longer. In one experiment, 11 species of wild seeds buried in 1902 were still viable 20 years later; 8 of these were alive after 40 years; and 2 still germinated 60 years later. Moral: Try to avoid letting weeds go to seed.

Mulch, hoe, or pull weeds when plants are small. The first month or so after transplanting or seedling emergence is the most critical time for weed control. According to Mort Mather, a successful market gardener, timing is the secret of weed control. Mather tills the ground ten days before planting and then rakes out any sprouted weeds just before sowing seeds. He cultivates a second time ten days after planting, even if only a few weeds are visible (more are sprouting underground). Mather does his third (and for some fast-growing crops final) weeding ten days after the second. Later, when plants are bearing, weed competition may be less damaging, and in some cases, it even helps the plant by shading, keeping roots cool, or bringing up trace minerals. I've been following Mather's advice for the past two years and I've found that the good early start really does make weed control easier for the rest of the season.

Watering

When should you water your garden? Consider the condition of the soil and the plants, and consult your rain gauge. Young seedlings need water when the top inch of soil is dry. Later in the season, you can let the soil dry to a depth of three to four inches before watering. Wilted plants tell you that the soil needs more water, but it's best to supply water *before* the plants reach that stage. If a week goes by without an inch of rainfall, you'll want to water at least your young plants and those that are especially susceptible to drying, if not the whole garden.

Special Cases. Mulched gardens can endure a drought that would parch unprotected vegetable patches, but even a mulched garden will begin to suffer after two to three weeks without rain. Soil that is rich in organic matter will retain more water. Although dense clay soil loses less water than sandy soil, the water it retains is seldom fully available to the plants. The remedy for both clay, which is too compact, and sandy soil, which is too loose, is to dig in plenty of compost, manure, and cover crops. If water is so scarce that you must severely limit your garden waterings, the most effective time for a good soaking (after getting the seedlings established initially) is, for most plants, 10 to 20 days before the plant is expected to mature. For ornamentals, this means 10 to 20 days before flowering; for other plants, before the fruit ripens or the seed head forms.

Watering Devices. In a small garden, you can use buckets and sprinkling cans to pour water directly on the base of the plant—not on the leaves. Sprinkler heads that screw onto the end of a garden hose are inexpensive and easy to use, but about one-third of the sprayed water is lost to evaporation. If you are using expensive metered water, a more efficient system might pay for itself. Less water evaporates from soaker hoses, which leak water directly onto the soil through canvas or perforated plastic, but they can be used for only one bed or row at a time.

Drip Irrigation. Drip irrigation, a more efficient but also more complicated system, delivers water directly to multiple rows through branch lines that are fed by a main line leading from the water tap. You leave the lateral lines in place all season, usually under soil or mulch. Drip lines can also be attached to a rain barrel or other water-storage tank, in which case the water level in the barrel should be five to seven feet above ground so the water will flow by gravity. Drip hoses that are permeable along their entire length are easier to install than those with emitters (small valves that release the water), and they usually don't require a water filter to prevent clogging. You can buy automatic control devices to program watering even when you are away.

How to Water. If you must water, follow these guidelines:

1. Water as infrequently as possible. Constant moistening of the upper layer of soil encourages plant roots to remain close to the surface where they are vulnerable to prolonged drought and also to hoeing, if the ground is bare.

2. Water thoroughly when you *do* water. Give the plants a steady soaking, enough to make the soil wet four to six inches down. Allow 2 to 4 gallons of water per square yard of growing surface (640 gallons delivered to 1,000 square feet amounts to one inch of water).

3. Get the water to the roots.

4. Increase the humus content of your soil so as to improve its water-retaining capacity.

18 Direct Seeding

Planting seeds right in the ground is a ritual that links us with people of generations past who sowed the ancestors of these same seeds in even more precarious times. It links us with other people around the globe, many of whom depend desperately and directly on the fruit of the seeds they sow. For the gardener, seed sowing is essentially a hand operation. True, there are mechanical seed-sowing devices, and preplanted tapes of seeds may be had for a price, but somehow these have never tempted me. At a sacrifice, perhaps, of some precision and seed economy, I prefer the immediate contact between hand, seed, and soil.

When to Plant

Every spring we gardeners are usually ready with our seed packets long before the soil is in condition to work. After we've started the early cool-weather crops indoors and planted flats of the later tropicals like tomatoes and peppers, we go over the seeds for outdoor planting and divide them into groups:

- First plantings: dill, leaf lettuce, onion sets, peas, radishes, turnips.
- Second early plantings: beets, cabbage, carrots, fennel.
- Mid-season early plantings: hardy annual flowers; purple green beans; later carrot, lettuce, and radish plantings; celeriac; early corn; leeks; sweet onions; parsnips; salsify; New Zealand spinach.
- After-frost plantings: tender annual flowers, green beans, corn, cucumbers, summer lettuce, soybeans, tampala.
- Warm-weather plantings (early summer): brussels sprouts, Chinese cabbage, cauliflower, endive, kale, head lettuce, winter radishes, rutabagas.
- Mid- to late-summer plantings for fall: beets, more dill for pickles, leaf lettuce, pansies, radishes, spinach.

Then, when we've dug or plowed or tilled the garden, and the soil has settled for a day or two, we start with the earliest, hardiest seeds on our list. The soil should be in good tilth, as finely worked and raked as possible. It may be cold, but it should not be wet and lumpy. Peas and lettuce will germinate in soil as cold as 40°F (4°C), followed by spinach and cole crops (cabbage, radishes, and so forth) at around 45°F (7°C). Onions, parsley, chard, and beets wait for 50°F (10°C), soil and warmth-

117

loving squash and okra for no less than 70°F (21°C), while eggplant and muskmelons hold out for minimum temperatures of 75°F (24°C). Turnips are the most obliging; they'll germinate at temperatures as low as 60°F (16°C) or as high as 105°F (41°C). In almost all cases, emergence is much more rapid at warm temperatures. Some cool-weather crops, like spinach, germinate slowly but more completely at the lower temperatures they prefer, and rapidly but spottily at higher temperatures.

As the soil warms and the weather moderates, we work our way through the increasingly tender vegetables, all the while watching the signs in the world around us as well as our schedules and records from previous years. Some old herbals, in fact, advise the gardener to go out naked to sow his spring crops. At first hoot, this suggestion seems ridiculous as well as impractical. For most of us, it remains impractical, but we must admit that the gardener who goes out unclothed will surely be sensitive to air and soil temperature and therefore less likely to plant seeds too early.

Years ago, and even today in cultures that acknowledge our absolute dependence on plants, the planting of seeds would be a ritual, carried out with a sober and hopeful sense of its significance. In his book *Indian Corn in Old America,* for example, Paul Weatherwax describes the many ceremonies and taboos associated with the planting of corn, including a period of abstinence from sexual activity preceding the time of planting, culminating in ritual intercourse at or just before seed sowing.

Planting by the Moon

Planting by the moon is another example of applied folklore that is still widely practiced. Now that the moon's influence over bodies of water is universally accepted, the assumption that living things are also affected by changes in the moon's relationship to the earth seems reasonable. If the moon really does exert a force that raises groundwater as it does the ocean tides, perhaps the rising of nutrient-carrying water from the roots through the stem to the leaves is stimulated too. Research studies at Tulane University in New Orleans have shown that seeds absorb water and germinate on a regular cycle that coincides with the lunar month, and other studies on trees and potatoes have demonstrated a similar periodicity.

I have not experimented in any serious way with moon planting. It makes sense, in many ways, even though details are sometimes fuzzy and occasionally contradictory. I'd really like to plan more of my plantings by the phases of the moon. So far, though, I've had too many other variables to dovetail at spring planting time.

Don't let me discourage you, though. Let's just say that I haven't yet given the system a fair trial. Actually, some experienced practitioners of the art of moon-phase planting maintain that the method *helps* them to organize their garden chores by establishing a schedule for planting, weeding, and harvesting chores. You even get a period of rest. The influence of the moon, by the way, is thought by many people to affect not only plants but also animals, and so you can find directions in some country almanacs for propitious times to castrate, dehorn, and wean livestock.

The lunar month is divided into four quarters. During the first two quarters,

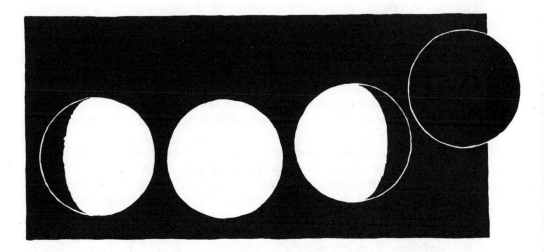

Some gardeners plant by the cycles of the moon.

beginning with the appearance of the new moon, the moon is waxing, or growing. This is the time to plant or transplant aboveground crops—leafy vegetables, grains, flowers, fruits, the cole family, parsley, peppers, and cucumbers. The third quarter, when the moon has begun to wane, is the time to plant root crops.

"Don't plant anything during the last quarter," advise the moon-planting sages. That's a barren time, best reserved for weeding and cultivating. Furthermore, the day the moon changes quarters and the four days when the moon is almost full and just past full are not propitious for important projects, according to some moon-sign authorities.

Consult your almanac or feed-store calendar for exact timing of the moon's phases and Louise Riotte's book *Planetary Planting* for more information on using this practice to guide your garden work. Your calendar may provide extra information that could be confusing, unless you realize that in each lunar month, according to those who map out such things, the 12 signs of the zodiac rule in the same succession they follow in the yearly calendar (Capricorn, Aquarius, Pisces, and so forth). If you *really* get into moon planting, you'll avoid the barren signs (Gemini, Leo, and Virgo) and plan important plant work during the fruitful signs (Cancer, Scorpio, and Pisces). Personally, I have enough to contend with in just getting the ground prepared between snow melt and spring rains, but perhaps I'm missing something by not making a consistent effort to plant by the moon.

Presoaking Seeds

Presoaking the seeds before planting may hasten germination of some kinds of seeds. Parsley and carrot and other slow-germinating seeds, or seeds with hard coats like morning-glory or New Zealand spinach, are often soaked in warm water overnight before planting. When ready to plant, drain the seeds. Tiny seeds should then be mixed with a dry substance like sand or coffee grounds so they don't clump together. Long soaking can reduce the seed's viability by cutting its oxygen supply. To aerate the water, shake the container from time to time or add an aquarium aerator stone.

I often presoak pea and snap bean seeds for a few hours before planting. Lima beans and soybeans contain so much protein that they may swell too much and then split, if soaked too long. When legume seed coats split, the seeds may lose vital nutrients and fall prey to disease fungi when planted. When I do presoak these seeds I drain them after they've been submerged for an hour.

Inoculating Seeds

Inoculation improves plant vigor and yield by coating the seeds with helpful bacteria at the time of planting. Most inoculants come in powdered form, and the seeds should be moistened to help the powder to adhere. The kinds of inoculants available to home gardeners will no doubt be more numerous in the near future. At present, specific strains are available for garden legumes, potatoes, and various grains, and there is a special mixture for use on a variety of garden vegetables.

An inoculant containing the nitrogen-fixing bacteria in the rhizobium group, commercially available as Nitragin, is familiar to most gardeners. It is applied to pea and bean seeds at planting time to ensure that a good supply of soil- and crop-improving rhizobia are present in the soil. If you are planting seeds in a spot where rhizobium-inoculated crops have grown before, you may not need to reinoculate. I usually treat the seeds anyway, just to be sure. In the days before inoculants were commercially available, farmers would mix soil from a previous stand of high-yielding peas or beans into the row for the next crop.

Specific strains of the bacterium are available for different legume crops like alfalfa or clover, as well as for soybeans. If you are planting soybeans in your garden, either the field variety or a kind bred especially for table use, you need a special soybean inoculant. (See the source list at the back of this book.)

Sowing Seeds

Now it's time to plant those seeds.

Mark the Row. First, you'll want to mark the row so that you have a guide for sowing seeds in a straight line. An old farmer of our acquaintance maintains, with a twinkle in his eye, that you can grow more plants in a crooked than a straight row. But since the placement of each row affects the one next to it, straight rows (unless you are contouring around a hillside) use space more efficiently and prevent a lot of headaches if you use a Rototiller. There's nothing like getting to the middle of a row with the tiller and finding you can't get through without destroying some plants. I know—I've done it. Marking rows with a string tied to stakes at each end of the row is a method very familiar to most gardeners. What is not so generally recognized, though, is that your row will be marked more accurately if you tie the string near the bottom of the stakes, rather than at the top, so that it doesn't waver around in the wind. Then hoe or rake to one side of the string.

Prepare the Row. When making a furrow, common practice is to use the pointed edge of a hoe to form a V-shaped trough. This is all right for large, sparsely planted seeds, but a wider, shallower depression in the soil allows seeds to spread out rather than tumble together in a clump.

When planting root crops, especially carrots and parsnips, I like to make a mounded row by raking up loose soil from both sides of the row. After smoothing the top of the mound to a flat 12-inch-wide surface, I broadcast the seeds, tamp them gently with a rake, and sprinkle a light covering of fine soil over them.

Sow the Seeds. Planting depth affects both aeration and temperature surrounding the seeds. Soil at lower levels is cooler and more tightly packed. In most cases, shallow planting is preferred in spring. Cover the seeds with no more than three times their diameter in soil. Firming of the seedbed is important to ensure close contact between seeds and soil. Ground moisture in spring is usually sufficient for germination. Occasionally I'll water carrot and parsley seeds into the row to get them well settled.

To help avoid overcrowding, fine seeds may be mixed with three or four parts of sand, fine dry soil, or coffee grounds. Too much sand or other diluent, though, will cause an uneven distribution of the seeds.

Some gardeners like to make spot plantings. They place groups of two to six seeds in the row, leaving a few inches between groups. Later, the plants are thinned.

Early Weeding. Vegetable specialist Victor Tiedjens recommends a method of eliminating early weed competition that should work well if the soil has a fine texture. Rake an inch of fine soil over the row where you've just sown and covered seeds. Then, in about ten days, go along the row with the flat edge of your rake, knocking off the small ridge of soil and with it thousands of potential weeds.

Identify Rows. A last reminder from an absentminded gardener: Mark the row you've just planted with a stake at each end and label it. Seed packets pulled over the stakes look cheerful and hopeful, but the wind usually blows them off. Some gardeners staple the packets to the stakes. Record, indoors, what you've planted and where. After a few rains, the evidence of your having worked the ground will be erased, and it's embarrassingly easy to find yourself replanting the same row.

When Conditions Are Less than Ideal

You know, and I know, that there'll occasionally be times when we'll be putting seeds in the ground when by rights we probably shouldn't. If the soil is still rough and full of clods but you *must* plant corn, sow it thickly—six seeds to the foot—and try to rake away the clods before pulling dry soil over the seeds.

If the soil is still colder than it ought to be, but you want to gamble an early planting, sow thickly then, too. If it is not a large planting, spread a layer of compost in the furrow before sowing the seeds. Cold soil slows down the uptake of nutrients by the plant. Studies have shown, though, that putting fertilizer right in the row during a cold spring boosted the eventual yield of corn. Thus, making nutrients immediately available to the plant may well help to counteract unfavorable conditions.

More strategies for planting seeds in cold soil:

- Presprout the seeds. Once germinated, they'll accept lower temperatures for continued growth.
- Plant cold-resistant varieties of vegetables, such as the Sub-Arctic tomatoes and Royal Burgundy snap beans.
- If possible, make early plantings on a south slope or in ground that receives reflected warmth from a building or stone wall.
- Plant early crops in humus-rich soil, which warms more quickly than heavy clay.
- Use tunnels and row covers for protection (see chapter 16).

Thinning

When I first started growing vegetables, thinning seemed wasteful to me. I even tried to transplant the radish seedlings I had pulled out. When I found out that

Although you might not want to pull healthy seedlings out of the garden, thinning seed-lings provides more space and results in healthier mature plants.

unthinned lettuce rarely makes a decent rosette of leaves, carrots entwine and stay small, and fennel shoots to seed with thin stalks, I began to accept what I'd read, that each vegetable needs a certain amount of space in order to grow to its genetic potential. Thinning may be painful, but growing crowded, second-rate plants is wasteful. Charles Dudley Warner writes, "The Scotch say that no man ought to be allowed to thin his own turnips, because he will not sacrifice enough to leave room for the remainder to grow; he should get his neighbor, who does not care for the plants, to do it."

Most plants you've seeded directly in the ground will need to be thinned, unless you have better control of your planting hand than many gardeners do. Squash, cucumbers, and pole beans are usually intentionally overplanted so that the strongest plants can then be chosen and retained. Thinning should begin early, as soon as the seedlings can be easily grasped, to minimize damage to roots of neighboring plants. You'll soon notice that the remaining plants are growing better with more space around them.

Some edible-leaved plants like Swiss chard, lettuce, spinach, and turnips may be thinned gradually, with the ever-larger thinnings contributing to the soup pot or the salad bowl. Rather than thin only part of the row of these plants thoroughly, thin the whole row somewhat sparingly and then return in a few days to pick thinnings again for the supper table. Firm the soil back around the plants if pulling out thinnings has loosened it.

Spacing varies according to the plant, of course. Check the listings for the individual plants in the encyclopedia section of this book to find out how many inches or feet each plant needs, and then gradually thin your way to that ideal spacing as the seedlings grow. If you live in a damp or foggy area where mildew and other fungus diseases thrive, you'll want to be especially careful not to crowd plants.

19 Treating Plants with Seaweed Extract

Some remarkable results have been obtained both in the laboratory and in the field by researchers using a spray of diluted liquified seaweed extract to treat plants. Gardeners are beginning to apply the results of these studies in their backyard patches, often with surprisingly noticeable benefits, and perhaps most telling of all, commercial growers of fruits and vegetables have begun to purchase seaweed products to spray on their crops. Some report a significant increase in yield from treated plants. Others notice that beneficial ladybugs are attracted to kelp-sprayed plants.

Let's take a look at some of the beneficial results that have been obtained under controlled conditions. The following results have been measured and recorded by different scientists working with various crops in several nations:

- increased crop yields
- longer storage life of fruits and vegetables
- improved seed germination
- frost resistance in growing plants
- resistance to attack by insects and fungi

Studies, beginning in 1959 and continuing today, have repeatedly demonstrated that something does indeed happen when plants are sprayed with a seaweed extract solution. Why have we gardeners been so slow to apply these findings to our family food crops? Kelp solution is nontoxic, easy to use, and moderate in cost. However, the idea is new to us; skepticism has too often squelched curiosity.

Actually, the practice of using seaweed to improve plant growth is very old. All along the shores of northern Europe, since at least the twelfth century, tillers of the land have collected seaweed for cattle fodder and soil improvement. Portuguese fishermen who kept their catches fresh by piling wet seaweed over the fish tossed the slimy stuff aside, often on their gardens, after it had served its primary purpose. Gradually, as people recognized that plants grew better in plots treated with seaweed, they gathered and spread it intentionally. Even earlier, centuries earlier, in fact, coastal Oriental gardeners had discovered the value of seaweed as a soil improver, and they systematically used it to help maintain the fertility of their land.

During the seventeenth, eighteenth, and nineteenth centuries, the ash of burnt seaweed, known then as kelp, was used as a fertilizer. In 1912, a patent on a liquid preparation of dried seaweed was issued in England. A few studies on the use of liquid seaweed preparations were done during the 1930s, but by far the bulk of the scientific evidence that tends to bear out the effectiveness of this plant treatment has been printed since 1960.

The Benefits of Kelp

You're still shaking your head, perhaps, wondering whether any one substance can affect a plant from seed germination through flower formation, maturity, and fruit production. Perhaps you'd like to hear about some of the measured effects that have been reported. Hundreds of studies have been done. The following examples are representative:

Crop Yields. Celery plants sprayed at 20 days of age with a 1:100 solution of seaweed extract yielded larger stalks than the untreated control plants. Snap beans treated one week before bloom showed a 10 percent increase in yield on the first picking. Weekly spray applied to cucumber plants during the fruiting season resulted in a yield increase of 41.8 percent, in a three-year study. Dr. T. L. Senn observed earlier fruiting of pepper plants to which medium and low concentrations of the solution had been applied. In one test, corn yield (number of ears) was actually lower in the treated plants, but the ears were 15 to 20 percent larger than the controls.

Longer Storage Life. In studies reported by Kingman and Senn at Clemson University in South Carolina, peaches from trees sprayed with seaweed solution had a longer than usual shelf life, and the growth of decay organisms seemed to be inhibited. Those cucumbers mentioned above that were more prolific after seaweed treatment also kept longer than cucumbers from control plants.

Improved Seed Germination. Seeds of some species of plants germinate more

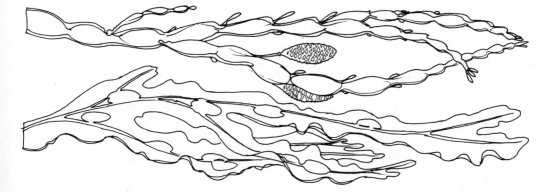

Two types of kelp, knotted wrack (top) and bladder wrack (bottom).

rapidly and completely when treated with seaweed extract, according to Dr. Senn, who also noted that the treatment caused accelerated respiratory activity in the seeds (see chapter 6 for more on seed respiration). Too strong a solution inhibited germination, though. In an experiment exploring the effect of seaweed extract on the germination of seeds of creeping red fescue grass, Button and Noyes reported improved germination of seeds treated with concentrations of 0.5 percent and 1 percent extract. A 5 percent solution reduced germination sharply. Seeds that were allowed to dry between soaking and planting did not perform dependably.

Frost Resistance. Seaweed-treated tomato plants studied at Clemson University were able to survive temperatures as low as 29°F (−2°C). Untreated control plants were killed by this frost. Eskimo have long used seaweed to protect vegetables from frost damage, although the first recorded instance of this practice occurred during the 1940s.

Resistance to Insect and Fungus Damage. Studies done at the Virginia Truck Experiment Station demonstrated that a seaweed preparation protected tomatoes from fungus damage as effectively as a fungicide. Some early work done in France and England suggests that nematodes tend to avoid fields that have been treated with seaweed.

Dr. Byrley Diggers of Rutgers University recorded an appreciable drop in the red spider mite population of seaweed-sprayed apple trees. Although studies done by another researcher did not corroborate this evidence under greenhouse conditions, this whole area is still largely unexplored, and it is possible that controlled experiments will suggest further applications. E. Booth of the Institute of Seaweed Research has reported that the reproductive rate of the red spider mite is known to fall when the carbohydrate and phosphorus content of the leaf increases. If more studies find that seaweed extract does indeed reduce red spider mite populations, research might continue on to determine if the extract affects the carbohydrate and phosphorus content of plant tissues.

Although it is too soon to say with any certainty that seaweed extract sprays will definitely protect a plant from insect and fungus attack, it is not a bit too soon for gardeners to try the method for themselves. It can't hurt, as long as low concentrations (about 1:100) are used, and it might very well help.

The Secrets of Kelp's Effectiveness

Test results have given us an inkling of what can be done by fortifying plants with seaweed extract solutions, all the way from improving germination to lengthening fruit storage life. Although some results are not yet conclusive, evidence continues to mount. The more curious among us are probably wondering, "How does it work? What does seaweed contain that would account for its effectiveness in boosting plant yield, vigor, and resistance?"

Some definite facts are known about the properties of seaweeds and their effect on plant tissues. New theories are being tested right now, and speculation continues about the many facts of this interaction that are still not understood. Even what we do know is often best expressed as what we know doesn't happen.

Elements and Minerals. Because nutrients are never in short supply for sea plants, they don't develop deficiencies as land plants can do. Seaweed contains all of the important elements necessary for plant growth, including trace minerals. It decays readily because it contains very little cellulose.

The nutritional value of the spray, which has been established by test results, seems to be unrelated to the seaweed's content of the major elements in plant growth—nitrogen, phosphorus, and potassium. Studies have shown that treated plants made better use of boron, copper, iron, manganese, and zinc. In fact, tomatoes that had been sprayed were found to have more manganese than the seaweed itself had contained. The conclusion seems logical that some active agent in the seaweed helped to release previously unavailable manganese from the soil.

In trying to account for the peculiarly wide-ranging effectiveness of seaweed, earlier speculation centered around the trace elements. A plant subjected to cold temperatures, for example, has been found to lack the ability to synthesize certain vital components of metabolism. The theory that trace elements in the seaweed make up for this deficiency and thus confer a degree of frost protection has not yet been discredited.

Auxins and Gibberellins. Substances other than trace elements have entered the picture, also. Auxins, plant-produced chemicals that control growth, have been thought to be responsible for the growth-promoting effect of the seaweed, and yet tests so far have not borne out this theory. Solutions of auxins are known to be unstable. Although auxins were detected in fresh, whole seaweed, none could be identified in the solution of dried, powdered seaweed. Seaweed samples have shown gibberellin-like activity, but only when the solution was freshly prepared; after four months the gibberellin content was not large enough to be effective. (Gibberellins are plant hormones that stimulate growth.)

Cytokinins. The most recent theories on the effects of seaweed center around cytokinins, growth-promoting hormones that have been discovered in seaweed. Cytokinins have been reported in the seaweed *Laminaria digitata* (as far as I can determine, studies have not been completed on other species of seaweed). According to Dr. Gerald Blunden of the Portsmouth Polytechnic College in England (speaking at an *Acres, U.S.A.* conference), cytokinins are powerful, safe, and nontoxic to people. They are quickly absorbed by the leaves of plants, and they remain there. Their beneficial effect on the plant's ability to make carbohydrates and chlorophyll helps to stimulate longer plant life, increases production, and makes for higher protein content. In Dr. Blunden's opinion, it is the cytokinins that account for the special effects credited to seaweed.

Cytokinins are apparently easily lost when seaweed solutions are applied to the soil. They are known to be readily absorbed by leaves, though, especially young leaves, so foliar spraying has come to be the recommended method of treating plants with seaweed extract.

Commercial seaweed preparations may differ in the level of their cytokinin activity, depending on the method of extraction (high or low heat or hydrolysis), the species of seaweed used, the time of year it was collected, and conditions affecting the storage life of either the unprocessed seaweed or the extract. A seaweed solution with standardized cytokinin content is available. Other unstandardized brands have, nevertheless, been shown to give good results in university and experiment station tests.

Regardless of the brand of commercially available seaweed extract used in the experiments, researchers agree that the treatment is effective in very low concentrations, concentrations so low that the amount of trace minerals being applied must surely be very small. Heavy concentrations, above 5 percent, seem to have a negative effect. Better results are often obtained by spraying twice with a weak solution diluted to 1:400 than once with a solution diluted to 1:200.

Using Kelp

What does all this mean for you, the seed-starter? Just this: that kelp might well be useful to you in germinating seeds, in protecting plants from frost, in increasing blossom set and yields, in protecting your plants from insects and destructive microorganisms, and in increasing storage life of fruits and vegetables.

If you remember nothing else from this chapter, make note of the following nine points:

1. You can use the solution to do the following:

   • Presoak seeds before planting.
   • Soak fiber pots before planting—not a proven technique, but one you might like to experiment with.
   • Spray tomatoes, beans, and cucumbers in the flower stage, just before bloom.
   • Protect plants before frost.

2. Kelp sprays are more readily absorbed by plants growing in phosphorus-rich soil.

3. Low concentrations of seaweed extracts are effective; high concentrations are toxic.

4. The spray is most effective when applied while the leaves are still young.

5. As reported by commercial growers, kelp solution sprays seem to be more effective when applied early in the day, when the plant's leaf pores are still open.

6. Adding a tiny amount of a spreader-sticker substance (available at garden supply stores) will help the solution adhere to the leaves.

7. You should use the first setting on your sprayer; tiny droplets are more easily absorbed than large ones.

8. You should spray not only the upper surfaces, but also the undersides of leaves.

9. For vegetable plants, one application per season is often enough. Use no more than three. Remember that cytokinins are absorbed by the leaf.

Concentration. Proper concentrations for different uses are as follows:

- For soaking seeds and bulbs before planting, or for presoaking peat pots, add one to two tablespoons of concentrate to two gallons of water (a 1:400 to 1:200 dilution).
- To spray vegetable plants, add one tablespoon of seaweed concentrate to 1 gallon of water or, for larger quantities, one-half pint of concentrate to 25 gallons (a 1:400 dilution).
- For watering seedbeds and rooting cuttings, add one tablespoon of seaweed concentrate to one quart of water or, for larger quantities, one-half pint to 7½ gallons (approximately a 1:120 dilution). To control mildew, use eight tablespoons to 1 gallon of water (a 1:50 dilution).
- For potted plants and houseplants, add four teaspoons of seaweed concentrate to two quarts of water or, for larger amounts, one-half pint to ten gallons (approximately a 1:160 dilution).

We are just beginning to get a more complete picture of what kelp can do for seeds and for garden and farm plants. Many questions remain unanswered, but the message I get from the studies that have been done is this: Seaweed extract appears to be beneficial. Let's use it, and take careful note of our results. Mike and I have just begun to use kelp spray for frost protection, watering seedlings, and presoaking seeds. This year we're spraying more young vegetable plants. Although it is too soon for us to draw any definite conclusions, we are sufficiently convinced of the merit of the method to recommend that other gardeners try it, too.

Kelp or Seaweed?

The word kelp originally referred to the ash of burned brown seaweeds. Today the word kelp is used as a general term for the several varieties of large, many-celled ocean algae that are used to make extracts. Species most commonly harvested include *Ascophyllum nodosum* (knotted wrack), *Fucus vesiculosus* (bladder wrack), and various species of Laminaria. Seaweed extracts sold in stores are sometimes called kelp concentrates, sometimes seaweed extracts or concentrates. Use them interchangeably.

20 The Fall Garden

One of the surest signs of gardening expertise, to my way of thinking, is the fall garden, still producing good food all the way to the first frost and even beyond. Something about fall gardens pleases me so much, in fact, that my husband and I have begun to seek out good fall gardens in our travels. One of my favorites had rows of silvery cabbage; red tomatoes; blue-green brussels sprouts; ribbony leeks; fat, complacent rutabagas; and ruffly green escarole, all edged by a row of marigolds blooming with the deep, rich, almost glowing oranges and golds of those last few weeks before frost.

Such a garden doesn't just happen. It is very easy to let the rows peter out and the hills grow scraggly when the first planting of beans finishes and tomato picking takes over. Keeping a productive garden right up to the edge of winter takes a little planning and sometimes a little shoehorning, but the slight extra trouble will repay you well. Succession planting can double your garden's yield. Most cole crops, root vegetables, and types of lettuce will stay in good eating condition longer in cool fall weather than in spring.

The trick is to remember, in late spring and early summer, how good it was to have fresh young beets and turnips, crisp new greens, and delicate Chinese cabbage to put on the table last fall. After the spring rush, when both the early lettuce and cabbage and the later tomatoes and peppers have graduated from grow light to cold frame to garden, and the heat-loving melons have just been set out, I turn my attention to starting the fall crops.

A child's discarded wading pool makes a good starting bed for fall seedlings. If it's cracked on the bottom, all the better; you need to provide for drainage anyway. Some vegetables, like carrots, Chinese cabbage, winter radishes, and turnips, are planted right in the garden row, usually where an earlier crop—peas, early cabbage, lettuce, or radishes—has finished.

For summer sowings, when weather is likely to be dry, I hoe out a slightly deeper furrow than I did in spring, and I take the watering can and run a stream of water all along the furrow before planting the seeds. Then I rake and firm dry, loose soil over the seeds. Often, especially for carrots, I scatter fine dry grass clippings over the planted row, an additional ounce of prevention against crusty soil.

For most other midsummer plantings, I start seeds in flats, and transplant them to the row after a week of hardening off. Brussels sprouts, escarole, cabbage, lettuce, broccoli, and cauliflower can be seeded right in the row, of course, but I find I can raise better plants and make more efficient use of garden space if I start the plants indoors or in flats in the open cold frame, and set them out in their appointed rows or odd corners when they're three to four inches high and flourishing.

How to Figure the Last Planting Date

	Days to Maturity[1]	+ Days to Germination[2]	+ Days to Transplanting	+ Short-Day Factor[3]	+ Days before First Frost[4]	= Count Back from First Frost Date (days)
FROST TENDER						
Beans, Snap	50	7	direct seed	14	14	85
Corn	65	4	direct seed	14	14	97
Cucumbers	55	3	direct seed	14	14	86
Squash, Summer	50	3	direct seed	14	14	81
Tomatoes	55	6	21	14	14	110
SURVIVE LIGHT FROST						
Beets	55	5	direct seed	14	N/A	74
Cabbage, Chinese	45	4	21	14	N/A	84
Cauliflower	50	5	21	14	N/A	90
Endive	80	5	14	14	N/A	113
Kohlrabi	45	4	direct seed	14	N/A	63
Lettuce, Head	65	3	14	14	N/A	96
Lettuce, Loose-Leaf	45	3	14	14	N/A	76
Peas	50	6	direct seed	14	N/A	70
SURVIVE HEAVY FROST						
Broccoli	55	5	21	14	N/A	95
Brussels Sprouts	80	5	21	14	N/A	120
Cabbage	60	4	21	14	N/A	99
Carrots	65	6	direct seed	14	N/A	85
Collards	55	4	21	14	N/A	94
Kale	55	5	21	14	N/A	95
Radishes, Summer	25	3	direct seed	14	N/A	42
Radishes, Winter	55	3	direct seed	14	N/A	72
Spinach	45	5	direct seed	14	N/A	64
Swiss Chard	50	5	direct seed	14	N/A	69
Turnips	35	2	direct seed	14	N/A	51

(continued)

Table 7—*Continued*

Source: *Organic Gardening*, July 1980, pp. 50–51.

Note: N/A = Not applicable.

[1] These figures are for the fastest-maturing varieties I could find. Fast-maturing cultivars are best for fall crops. For the variety you have, get the correct number of days from your seed catalog.

[2] These figures assume a soil temperature of 80°F.

[3] The short-day factor is necessary because the time to maturity in seed catalogs always assumes the long days and warm temperatures of early summer. Crops always take longer to mature in late summer and fall.

[4] Frost-tender vegetables must mature at least two weeks before frost if they are to produce a substantial harvest.

My Timetable

Don't wait until fall to begin thinking about your fall garden. Actually, plans and seed starting for the fall garden begin in spring. As an example, look at my schedule.

May. I start seedlings of kale, brussels sprouts, and fall broccoli.

June. I make second plantings of carrot, cucumber, and zucchini seeds directly in the garden.

July. This is a busy month. By July I'm transplanting the kale and its cousins into rows where I've pulled onions or peas. July 25 is traditional turnip-seed planting day. A July planting of snap beans will bear in September. Around mid-month I plant Chinese cabbage seeds in flats. At the end of July, I plant head lettuce seeds.

August. I plant leaf lettuce the first week in August. Also in early August, I plant spinach, usually pre-germinating the seeds in damp towels in the refrigerator.

To time your plantings for local conditions, consult table 7 on page 131.

Later, in the fall, you can plant peas for spring in well-prepared furrows, and sow seeds of corn-salad for early spring eating. In the South and parts of the West, fall sowings of turnips, parsnips, lettuce, beets, and other vegetables that flourish in cool weather will live over winter.

Flowers

Most annual flowers are at their best in early fall. Calendula and snapdragons will survive light frost. I often scatter calendula seeds in the garden in early summer for fall blooms and cut back snapdragon plants after their first blossoming so they'll bloom again.

Picking beans, corn, and marigolds in September, lettuce, spinach, and chrysanthemums in October, beets and turnips in November, and kale and collards in December sure beats letting the patch run to weeds!

Vegetables and flowers for the fall garden.

Vegetables for the Fall Garden

These vegetables can be started from seed in the summer to plant in the fall garden:

beans, snap	collards	kohlrabi
beets	corn	lettuce
broccoli	corn-salad	radishes
brussels sprouts	cucumbers	spinach
cabbage, Chinese	dill	turnips
carrots	escarole	zucchini
cauliflower	fennel	
celtuce	kale	

21 Insect and Animal Pests

Although most of us are prepared to accept a certain amount of insect damage and even animal predation later in the season, when young plants are struck they may never have a chance to reach harvest size or to produce their fruits. If you've ever lost a plant to a cutworm or rabbit I'm sure you can remember how indignant you felt, and how helpless to reverse the damage. It's one thing to be willing to share the harvest with other life-forms that may have come on the scene first; it's quite another to lose seedlings before they have a chance to get off the ground.

Insect Pests

Protecting young plants from insect attack is largely a matter of prevention, plus a few defensive strategies you can use if bug trouble is serious. Here's a list of insect pests that commonly feed on young plants, along with some suggestions for saving your crop from the marauders.

Cabbage Moth Larvae. These well-camouflaged, pale green caterpillars eat holes in cabbage-family plants.

Control. They may be controlled by spraying plants with a solution of *Bacillus thuringiensis,* a disease that affects the digestive systems of the larvae. You can buy it as Dipel or Thuricide in your garden supply store.

Carrot Rust Fly Larvae. A yellowish maggot, one-third inch long, that feeds on roots in the soil for a month before hatching to produce another generation of flies. These pests are responsible for eating those unsightly tunnels in carrot roots. The first generation of larvae hits plantings made before early June, and the second generation is not usually active before September.

Control. They may be foiled by timing and rotating your plantings. Planting carrots in different areas and harvesting with an eye on the calendar can prevent a damaging buildup of these garden pests.

Cucumber Beetles. Small yellow bugs just under one-fourth inch long, with black stripes down their backs, these beetles prey on the whole cucurbit family (melons, cukes, squash, and pumpkins). Spotted cucumber beetles, with black spots on a yellow body, also attack cucurbits.

The cucumber beetle threatens seedlings in three ways. Its small white larvae, about one-third inch long, feed on the underground parts of plants. The adult beetle eats the delicate young stems and cotyledons of newly emerged seedlings and later

134

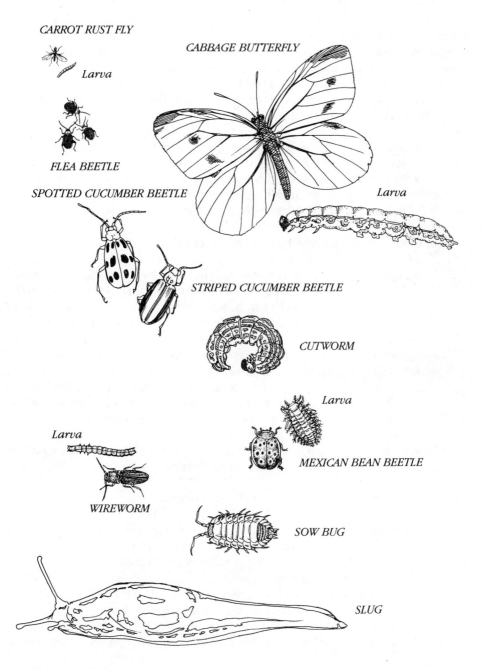

CARROT RUST FLY

Larva

CABBAGE BUTTERFLY

FLEA BEETLE

SPOTTED CUCUMBER BEETLE

Larva

STRIPED CUCUMBER BEETLE

CUTWORM

Larva

Larva

MEXICAN BEAN BEETLE

WIREWORM

SOW BUG

SLUG

Common insect pests of the garden.

the blossom and tender leaves. Even worse is the disease-carrying habit of both the striped and spotted cucumber beetle. They spread bacterial wilt and mosaic, two murderous diseases that affect all cucurbits. When a planting is badly infested, it may never bear a single fruit.

Control. Every measure you take to defend your seedlings against this small but powerful pest will increase your chances of a good yield when summer comes. The suggestions that follow will not eradicate every beetle from your patch, but the use of one or several of these practices should at least provide you with the makings for a few more jars of dill pickles or for a batch of frozen melon balls.

Prevention of cucumber beetle damage starts early. The following measures should be taken before planting:

- Order disease-resistant varieties of the plants that can be affected. A good seed catalog will give you this information.
- Try starting some seedlings indoors so they will be tougher and larger when you set them out—a defense against the direct action of the beetles but not, unfortunately, against the diseases they carry.

Then, at planting time, try the following methods:

- Plant a ring of radishes around each hill of cucumbers, squash, or pumpkins. This can be done up to a week before planting the main crop in the hills, or the radish seeds can be scattered as you plant. I've done it both ways. Advance planting of radishes ensures a ready and waiting beetle deterrent when the cucumber seeds sprout, a ploy that is especially useful when you plant presoaked cucumber (or other cucurbit) seeds.
- Bury onion skins in the hole when you plant your cucurbits. According to an experienced gardener who reported her success in following this advice, which she had read in an early issue of *Rodale's Organic Gardening,* the method really works. It doesn't seem to make much difference in my garden, but perhaps it will in yours.
- Try interplanting beans with your cucumbers. Some gardeners report that the beans seem to repel cucumber beetles.

Here are some controls you can apply after planting:

- In some cases, mulching around the plant apparently impedes progress of the larvae from underground to their adult feeding stations above ground. You'll need to do more than mulch, though, if your cucumber beetles are as numerous as mine!
- To trap the beetles, set out saucers of water with a thin film of oil on top.
- Dust with rotenone early in the season, before the first generation of beetles has a chance to reproduce. Rotenone, which breaks down in three to seven days, should be used cautiously near ponds since it is toxic

to fish even when greatly diluted by rain and pond water. Derived from plants, it will not accumulate in your body as chemical insecticides will, but it is not entirely innocuous. Wear a mask when dusting with the powder and a long-sleeved shirt when spraying a solution.

• Scientists from the United States Department of Agriculture's Beneficial Insect Introduction Laboratory have found a nematode in the high Andes that parasitizes cucumber beetles there. Research is just beginning, but if the nematode proves to be a useful cucumber beetle control here, gardeners from coast to coast will be cheering, and none more loudly than I!

Cutworms. These grayish pests are 1 to 1½ inches long and curl up when exposed. The damage they do is large compared to the amount of plant tissue they eat. The trouble is that they operate at soil level, and so when they cut through a stem, the top of the plant dies. Well-rooted tomato plants that have been decapitated by cutworms will sometimes sprout again, but since you can't be sure that they will, you must still replant.

Control. Be sure to protect your plants from cutworms when you set them out. The very next day may be too late.

Two natural controls for cutworms include a tiny wasp, *Trichogramma pretiosum,* which parasitizes cutworm eggs, and the beneficial nematode *Neoplectana carpocapsae,* offered as Seek, which preys on cutworms, wireworms, cabbage root maggots, and other soil-dwelling caterpillars. Birds will also pluck cutworms from tilled soil, but you probably shouldn't rely on them to protect your plants from this pest.

The cutworm must wrap itself around the stem of a plant in order to "bite" through it. Consequently, any device that will make this encircling maneuver difficult or impossible for the cutworm will protect the plant. We have used each of the following three methods successfully:

1. Cut slits in an index card or any stiff paper and wrap the card snugly around each plant stem to form a cylinder around the plant. Half the card should be above soil level and half below. This works, but it's a nuisance to do when cold spring winds are stiffening your fingers.

2. Stick a small twig into the ground three inches below soil level and protruding two to three inches above the soil near each seedling. The twig should be right next to the plant stem, touching it, so that it forms a tough barrier that foils the soft-bodied, bristly-mouthed larvae. If you've dug wood chips into your garden soil, you'll probably have plenty of twigs within reach as you work your way down the row with the flats and trowel. Otherwise, collect a supply in advance. Dry asparagus stalks are often handy and work quite well.

3. Tear scrap aluminum foil into pieces about three inches by two inches. Wrap the foil around the plant stem as you get the seedlings in the ground, again making certain that half will extend below ground level and half above.

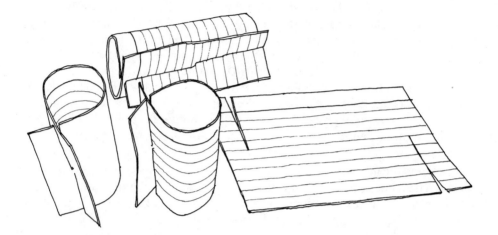

One way to block cutworms from getting to your seedlings is to cut slits in the ends of an index card and form a cylinder that can be placed around the seedlings and into the soil.

Flea Beetles. These tiny, very active, black insects are the size of a pinhead. They appear early and riddle certain plant leaves with small round holes. Eggplant, tomato, broccoli, and other small cabbage-family seedlings are especially vulnerable. Last year I lost a flat of cabbage seedlings to flea beetles, which devoured them in just a day. I had made the mistake of putting the flat out at the edge of the garden near some weeds infested with flea beetles.

Control. Flea beetles thrive in hot, dry places. They dislike moisture and shade. In small gardens they can be washed off the plants with a stream of water from the hose. It helps some to diversify your plantings, too. Flea beetle infestations are usually worse when large areas are planted to the same crops. Keep seedlings well supplied with fertilizer and water; rapidly growing plants can usually outlive an early-spring beetle outbreak. Also, keep the garden clear of boards and other hiding places over winter so that flea beetles won't have handy shelters to help them survive the winter.

Mexican Bean Beetles. The coppery-colored adult beetles resemble large ladybugs. Their destructive soft-bodied yellow larvae eat holes in plant leaves and scar and deform the beans, too. Affected plants die from loss of leaf surface.

Control. Bean beetles also often can be circumvented by making both an extra-early planting of a cold-resistant bean like Royalty and a late planting of a quick-maturing kind like Bountiful Stringless (matures in 47 days). The early beans will usually yield one good picking before the beetles have a chance to multiply, and late plantings, even though cold nights may slow them down a bit, often thrive in the absence of the heat-loving beetles during early fall.

For more certain control, try the recently released *Pediobius foveolatus* wasp, a tiny parasite that preys on Mexican bean beetle larvae. The Pediobius wasp is not native and can't be depended on to overwinter, so a new supply must be ordered each spring. The wasps should be placed in the garden when larvae first appear. Another natural ally, *Coccipolipus,* is a mite that attacks adult Mexican bean beetles. Research looks promising, and I hope the mites will soon be available to gardeners.

Nematodes. Nematodes exist everywhere, even in the polar tundra. Most of the thousands of different nematodes that have been identified are smaller than a printed period. The majority of nematodes are helpful, or at least innocuous, but the minority (one-tenth or so) that do damage plants are responsible for a vast amount of plant destruction. Tiny as they are, their sharp mouthparts, formed like a hollow spear, are capable of penetrating tough roots. Then a suction-forming bulb behind the nematode's esophagus draws out plant juices. All this violence goes on underground, unseen. Symptoms of nematode infestation are, in fact, vague and nonspecific. The plant may simply go into a slow decline. Often the nematode-weakened plant succumbs to a secondary bacterial or fungal invasion.

Control. Helpful organisms live in the soil, too. Predacious fungi, which thrive on organic matter in the soil, are capable of trapping and engulfing, even of hunting down nematodes. At least 50 different nematode-attacking fungi have been identified so far, and more are being found as studies continue. Once again, then, we find that good soil-building practices like composting, green manuring, and digging in mulch will defend plants on many levels.

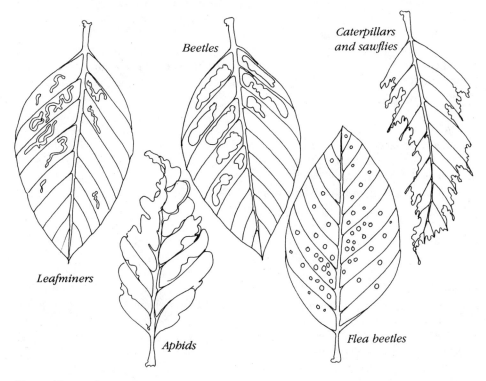

Types of insect damage.

Marigolds, especially the French varieties, produce a root exudate that makes the soil around their roots inhospitable to nematodes. It is sometimes necessary to plant the flowers in the same spot several years in a row to get the full benefit of their nematode-spooking activity.

Slugs. These soft-bodied, snaillike creatures without shells are on the prowl during the damp, early days of spring. They seem to be especially fond of lettuce and peas, and they seek acid soil.

Control. If you know what they like (beer or yeast solution in saucers or halves of citrus fruit) you can trap them. Or consider what they don't like—powdery, sharp, dry, or abrasive surfaces—and guard your garden accordingly by spreading lime, wood ashes, or diatomaceous earth around your plants. It also helps to spread ashes, sand, or cinders along garden paths.

Sowbugs. These are half-inch-long, gray crustaceans that roll up into balls when disturbed (hence their other common name, pillbugs). They feed on tender roots and shoots of young plants and can be a real problem in garden beds next to buildings, for they lurk in crevices and under boards and debris.

Control. I've never had any luck in getting rid of them once they were established. The best defense is prevention. Clear away all boards, boxes, piles of plant refuse, and other objects from the garden area, and try moving your vegetable patch to an open space in the yard if plantings next to your house are bitten hard by these pests. Woody and mature plants are not as attractive to sowbugs as young, tender seedlings. Watch out for sowbugs under the protective covers you put over new transplants, especially opaque covers.

Wireworms. These inch-long, brown, hard-shelled, segmented larvae of the click beetle eat the juicy interior tissue of roots and seeds. The damage they do to young seedlings can be severe in a badly infested plot. They often persist in gardens recently dug from sod.

Control. Baiting them with potatoes or other food is a short-term remedy. The best plan is to correct the conditions that favor wireworm population growth: poor drainage and lack of aeration. Dig up the sod patch early in the season and disk (cultivate with a harrow or plow) or till it. At the same time, incorporate as much compost, old mulch, rotted manure, or other organic material as possible into the soil to increase aeration deep down. The beneficial nematode *Neoplectana carpocapsae,* offered as Seek, preys on wireworms.

An Insect Solution as Control. A new insect-control method, that of spraying affected plants with a diluted solution made from the ground-up pests themselves, has been gaining favor among adventurous gardeners and experimenters. Although it's been suggested that the effectiveness of the spray is due to its ability to infect healthy insects with contagious diseases from infected insects, the whole idea is still very much in the experimental stage. The nice thing about it, though, is that it doesn't cost much to try.

Predatory Insects as Controls. The large wasps and yellow jackets that live near your garden are probably already at work feasting on insect pests. They kill leafhoppers, moth and potato beetle larvae, grasshoppers, and other undesirables. Blooming fennel, dill, and other plants of the Umbelliferae family often attract tiny beneficial wasps.

Lacewing and syrphid fly larvae consume moth eggs, aphids, thrips, and leaf-hoppers. Lacewing larvae, also called aphid lions, are more likely to remain in your garden than imported ladybugs. As with the Pediobius wasp, mail-order companies sell the eggs (see the source list in the back of this book for addresses).

With any biological control, try to time your release of helpful predators so that there are some insects for them to work on but before the infestation becomes overwhelming. Better yet, make several releases, which is still less trouble than spraying several times.

Animal Pests

Fending off raids by seedling-destroying wildlife has brought out the creative ingenuity of many a gardener. The following devices have worked for a good number of people. Perhaps one of them will help you. Ask elderly neighbors and relatives, too. Folklore is particularly strong and often effective in this department.

Deer. Deer are especially fond of legumes—soybeans, peanuts, snap beans, and peas. They can wipe out a whole bed or several rows in an evening of snacking.

Control. Spread wire-mesh fencing flat on the ground where the deer must cross it to get to the garden. Scatter lion manure around the garden; the essence of fierce predator that this conveys to noncarnivorous animals makes them uneasy. (This repellent, commercialized as Zoo-Doo, is produced by the Bronx Zoo in New York and the Portland Zoo in Oregon. If you don't live in either of these areas, try asking at the nearest zoo.) Sprinkle blood meal around the edge of the garden, or water around the plants with water in which you've soaked the liver of a butchered animal. In extreme cases, fence in the garden with six- to eight-foot fence.

Rabbits. Rabbits are most likely to eat beans and other legumes, lettuce, and sometimes beet greens—anything tender—but they also enjoy pansies. Both deer and rabbits will often chomp off the tender top leaves and leave the tougher bottom stems.

Control. Either the lion manure or the blood meal treatment used with deer will scare rabbits off until it gets rained into the ground. Then it must be repeated. I've read that a few dead fish spotted around will deter rabbits, but I haven't tried it.

Moles. Although their tunnels are unsightly on the lawn, the moles may not really be damaging your vegetable seedlings. They're after grubs, not your plant roots, when they burrow underground.

Control. If their search for food does disrupt your young plants, though, try one of the following controls:

- Place unwrapped sticks of Juicy Fruit chewing gum in the runways. It gums up the mole's digestive system. My gardening cousin Wally informs me that this method banished the moles from his garden.
- Plant scilla bulbs over the runways.
- Flood the runways with water.
- Poke into the tunnel a windmill or pinwheel on a stick or any other device that will send vibrations into the tunnel.

Common animal pests of the garden.

- Plant castor beans near the runs. (Be aware, though, that the seeds of this plant are poisonous.)

Once you start to ask around, you'll hear all kinds of mole cures. Some of them even work—sometimes. Try the gum first. It's the easiest.

Crows. Are crows eating your young corn?

Control. The crow is no dummy; any scare device you use to drive him off must be changed frequently. Try the following setups:

- Crisscross the patch with string about five inches above the ground. This makes takeoff difficult for the birds.

- Hang fur pieces—pelts and trimming from old coats—around the patch.
- Suspend shiny pie plates or other reflectors that blow in the wind and catch the light.
- Make a scarecrow, though this will be more picturesque than effective.
- String several kernels of corn on a long piece of horsehair and leave it in the field. Crows that get the hair stuck in their throats are said to give out alarm cries, warning other crows away.
- Plant thickly.

Red-winged Blackbirds. Red-winged blackbirds also eat corn.

Control. Try applying one teaspoon of turpentine to each pound of corn seeds several hours before planting.

The sight of a scarecrow creates delight more often than fear.

22 The Young Seed-Planter

Tending a small plot or a garden row of a certain crop can be a fine activity for the child, provided he or she really *wants* to have a garden. It is the garden assignment imposed from above, as a "worthwhile experience," that may lead to trouble, because the expectations of parent and child differ. I'm not talking here about the help that a child is expected to give in the main garden with weeding, picking, and shelling, but about the special plots that are sometimes unrealistically given to the child for his or her own use.

Above all, you want the child's first gardening efforts to be successful and satisfying. A child who experiences repeated gardening failures, such as poor germination or weed takeover, would be better off not having had a garden experience, but rather waiting until motivation and capability were better matched.

Help to Make It Work

Sooner or later, though, most children want to plant something. How will you guide them in their first seed-planting efforts? Whole books have been written on this subject (see Recommended Reading, at the back of this book). I'll just sketch here a few things we learned when helping our own children with their small gardens, which enjoyed varying degrees of success.

Choose Easily Grown Plants. Give the child something besides radishes to plant. I don't care what the gardening books say: Radishes can be tricky. Yes, they're fast and bright colored, but they can be strong tasting and maggot tunnelled just as easily as not. Beans take only a few weeks longer, and the large seeds are easy for a child to handle. Nasturtiums are good, and you can eat the leaves and the flowers. A small plot of corn attains spectacular growth, but be sure to protect it well from animal raiders. Cherry tomatoes grow fast and make good snacks. Sugar peas bear before the weeds get bad. They were one of our children's most dependable garden vegetables. Sunflowers are a natural for kids—fast growing and a good snack food. Onion sets shoot up quickly, but need pretty thorough weeding. Save the carrots and eggplant for later.

Use Good Soil. Choose a patch of good soil, not a rubble of builder's fill or hard clay ground where *you* wouldn't want a garden. Give the child every chance to succeed by providing the best possible conditions.

Start with a Small Plot. Begin simply and keep the plot small. Nothing is more discouraging than feeling overwhelmed by the garden, and no child needs the guilt feelings engendered by a weedy, unmanageable patch.

Plant Later in Spring. Help the child to plant late enough in the spring so that germination will be prompt and frost unlikely to damage the planting. Children like

Some children enjoy gardening as play.

fast results; it's better to delay planting for a week or two to help ensure steady growth.

Choose Easy Weed Control. Decide on the method of weed control before mapping out the child's garden. Rows should be spaced widely enough to permit easy hoeing or mulch placement. Children under 12 shouldn't operate a rotary cultivator without direct supervision. Plant a solid bed of something only if you are convinced that the child can handle the intensive hand weeding that will be necessary during the early weeks.

Work Together. Working along with the child, you can casually pass along the gardening lessons you've learned from hard experience—things like waiting until the ground dries before digging it, giving indoor seedlings a week or two to become accustomed to outdoor weather before putting them in the garden, watering deeply rather than sprinkling lightly every other day, and avoiding the temptation to sow seeds too thickly or to cover them too deeply. Recognize, too, that the child needs to learn some things from experience, and *will* learn, if you're not pushing too hard for perfection.

Eat the Produce. Whatever the child grows, use it respectfully and appreciatively. Even a few peas can make pea-potato soup. Zucchini may be picked young so as to use more of it. Perhaps the young gardener would like to eat it, stuffed with rice and meat and topped with cheese.

Allow Exploration. Accept the child's need to explore, which may take the form of pulling up half-grown beets or dismembering a flower to see what's inside. And if the child eats the beans when harvesting from his or her five plants, be glad. What else is a garden for?

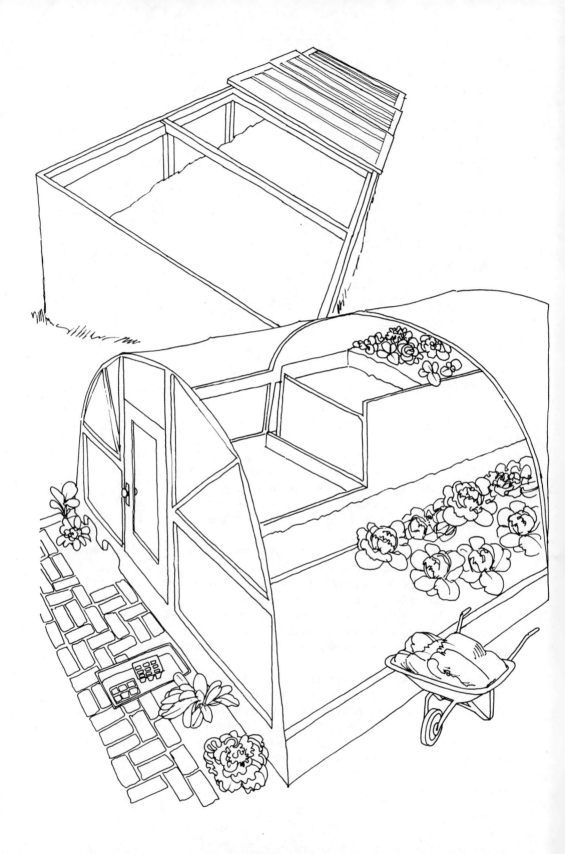

Section Three

Special Techniques and Situations

The world is so full of a number of things, I'm sure we should all be as happy as kings.

Robert Louis Stevenson

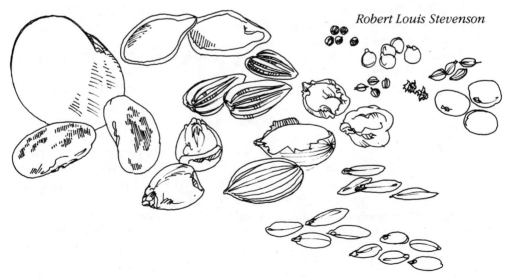

23 Starting Seeds in the Greenhouse

Remember how you longed for your first bike, or perhaps for your first car? That's how much I wanted a solar greenhouse! While we made plans for a new little house here on our homestead, I raised seedlings under fluorescent lights. Good seedlings they were, too, but I was always conscious of the abundance of sunlight that blessed our south-facing land. I knew I could raise many more seedlings in a greenhouse. Well, the dreaming and planning led to digging and building, and I now have a modest solar greenhouse attached to our home.

You know, of course, what happens when you get what you've longed for: You then have a new challenge, that of making the best use of your much-desired acquisition. I've learned a lot—and believe me, I'm still learning—about how to start seeds in my greenhouse and how to manage it, and I'd like to share here some hints for running a productive solar greenhouse.

When to Start Seeds

Most people who have solar greenhouses use them both to start spring seedlings for garden transplants and to grow some vegetable plants to perk up the winter menu. Seed starting can go on all year round.

For Planting Outdoors. My schedule for raising spring seedlings to be planted out in the garden goes something like this:

December—pansies (if I remember!)

January—chives, leeks, onions, and Siberia and other cold-resistant tomatoes

February—peppers

March—hardy annual flowers, broccoli, cabbage, cauliflower, lettuce, parsley, and some tomatoes

April—more annual flowers, eggplant, and more tomatoes

May—cucumbers, melons, okra, and squash

For Growing in the Greenhouse. To produce food in the greenhouse, it's best to plant seeds for fall and winter crops in August and September, while the sun is still strong; then your plants will be sturdy and well grown by the time cold weather and short days arrive. I start seeds of leafy crops in September and then transplant the seedlings six inches apart into the deep bed in the greenhouse. Seedlings can often remain in the flat, on hold, for another two months, during which time I gradually transfer them to the deep bed as space becomes available.

A greenhouse allows you to grow plants year round.

I sow Chinese cabbage and Bibb lettuce seeds again in December for late winter harvest. As for cucumbers—one of my favorite greenhouse crops—I start seeds in August for early winter cukes and again in March for vines that will produce until summer.

In general, leafy plants are more productive under glass than those that produce fruit, especially when light levels are low, but cucumbers produce well from fall and spring plantings, and beans will bear in late spring. From my experience, I can recommend Swiss chard, Chinese cabbage, leaf and butterhead lettuce, radishes, cucumbers, cherry tomatoes, beans, and New Zealand spinach as good solar greenhouse food crops, but don't stop there. Other gardeners grow beets, turnips, garlic, watercress, and many other crops in their unheated solar greenhouses.

Design of the Greenhouse

Space doesn't permit a discussion of details of greenhouse construction here. Whole books have been written on that subject alone. For the purposes of this chapter, we'll define a solar greenhouse as a structure that admits enough light to foster plant growth and stores enough heat from the sun to keep temperatures above freezing without using auxiliary heat.

Our solar greenhouse is not a textbook ideal, but it serves us very well. It's small—8 by 12 feet—with a high ceiling. A deep growing bed, enclosed by a concrete block wall, runs the length of the greenhouse's south side. On the north side, we've settled five steel drums painted black and filled with water to soak up and store the warmth of the sun. On top of the steel drums, we created a work surface made of strips of locust lumber. Two locust-strip shelves line the tall north wall. The greenhouse's concrete block foundation is insulated on the outside with panels of 2-inch urethane foam faced by pressure-treated exterior plywood. We used vertically installed patio replacement doors for the glazing and kept the well-insulated west wall unglazed. Our greenhouse has two 21- by 36-inch vent doors, both sandwiches of 1½-inch Styrofoam fastened between panels of exterior plywood. The ceiling is unglazed and heavily insulated with Styrofoam panels.

Preparing the Growing Bed. If at all possible, you'll want to include a large growing bed in your greenhouse. This massive hunk of soil, of course, serves as the growing medium for your plants, but it also provides thermal mass to absorb and retain the sun's warmth. When we built our greenhouse, we poured concrete for that section of the floor where we would walk and where we would store water-filled steel drums, but we left the ground area under our growing bed open so the soil in the bed could drain well. We built the two-foot-high bed wall of concrete blocks; then we gradually filled in the bed, starting with three inches of gravel topped by four-inch blocks of hay peeled from spoiled bales. Then we added four inches of decomposing wood chips, about four inches of compost, and a good eight inches of loose garden soil mixed with several large bags of vermiculite and commercial potting soil. Each year we add another two- to three-inch layer of compost mixed with approximately three quart-sized containers of wood ashes to the bed. We also dig into the bed any clean, fine-textured household scraps like tea bags and coffee grounds. A good population of earthworms lived in the original garden soil we put into the bed, but we also toss in a few new ones from time to time to help digest the tea leaves, aerate the bed, and add rich castings. We bend over backwards to avoid stepping, or even leaning hard, on that rich, loose soil, and five years later, it's still loose enough to scoop up in our fingers. If I must step over the bed area, I lay a plank across the bed to walk on, resting it on the wall rather than on the soil.

Making the Most Use of Space. The top of the wall enclosing the growing bed serves the same purpose as a wide south-facing windowsill in the house of any enthusiastic gardener—it's packed with additional plants in containers of all descriptions. Even in a larger greenhouse than mine, it isn't long before the growing bed is solidly carpeted with plants, and there you stand, juggling a flat of tomato seedlings in one hand and a pot of petunias in the other, wondering where to put them. That's when you appreciate the other dimensions of the greenhouse: the height of the ceiling above the growing bed, the depth and width of shelving on the back wall, even the windowsills. A determined gardener can manage to make creative use of every cranny of growing space. The happy result is not only more food and flowers for the winter table, but also a lusher, softer atmosphere of abundance and growth within the greenhouse.

Take that petunia or New Zealand spinach or sweet potato vine or any one of a dozen other plants whose natural habit is trailing or vining, and perch the plant on a shelf so the leaves and flowers can cascade down into the light. As for tomatoes, if they're a bush variety like Tiny Tim, they'll do well in hanging baskets suspended near the glass over the growing bed. (Be sure to water the baskets more often than the planting bed because they dry out more quickly.) I've also augmented my sunnier growing area by bridging the space between the growing bed and the steel drums with a plank, on which I set basins of leafy plants angled toward the sun. The use of a trellis or some other support of string, poles, netting, or fencing can make good use of vertical space. Our cucumbers climb the fence next to the window, leaving space for a small planting of Bibb lettuce behind them. I've even seen photos of a rock wall in a greenhouse in which herbs were tucked into spaces between the rock—a good productive use of some additional thermal mass.

Greenhouse Conditions

The form of the greenhouse determines much of its efficiency, but there are other factors, too, that you can juggle to influence those critical greenhouse conditions: light, temperature, humidity, and good ventilation.

Light. Let's consider first what you can do to boost the level of light your plants receive, especially in winter when the sun's rays are weak and the days are short. If your greenhouse has fiberglass or acrylic glazing, the light will seem softer, more diffuse than the beams that stream through glass. Don't worry, though. These translucent glazing materials actually scatter rays so that light strikes plants more evenly and without shadows than the more direct light that passes through transparent glass. (Glass has the advantage, though, of being more durable in the face of ultraviolet rays, which tend to deteriorate fiberglass and acrylic panels.) Most vegetable and flower plants need eight hours of light a day to grow properly. If your plants are receiving enough light, they'll be green and stocky with closely spaced leaves. Light-hungry plants will show one or more of the following symptoms:

- elongated stems
- sparse leaves and frail stems
- slow growth
- excessive leaning toward the light

To increase light levels, first check to be sure that there are no trees, wood piles, leaning snow shovels, or any other materials outside that shade the greenhouse. Next, and most effective, paint all interior walls and shelves with white paint. Use exterior-grade paint for durability in the greenhouse's high humidity. If you use heat-storing barrels that face the sun on one side and the plants on the other side, slap some white paint on the side close to the plants but not, of course, on the sunny side.

In my greenhouse, I have to watch my tendency to let bags of moss and potting soil, clay pots, and planting flats pile up and cover parts of the white-painted wall.

This clutter makes a good home for Toad, who likes to take his winter vacation in our little solar space, but it also absorbs a lot of light that would otherwise reflect back on plants.

Shane Smith, whose book *The Bountiful Solar Greenhouse* has been one of my most helpful guides, suggests that for continued summer production when there's plenty of warmth but less light from the sun, which is high in the sky, growers can hang a white sheet over the black barrels to reflect more light onto the plants. Smith also suggests that when light levels are on the low side, you'd be wise to keep the greenhouse as warm as you can and also avoid crowding the plants.

Temperature. On a snowbound February day, stepping into your warm, sunny greenhouse can really lift your spirits. Greenhouse temperatures soar to 90°F (32°C) and above when sunlight is intense. In fact, some folks find that when they are growing the kind of cool-weather crops best suited to the winter solar grow space, overheating can cause more problems than cold temperatures. A carefully sealed, well-insulated greenhouse with plenty of thermal mass isn't likely to freeze in winter, unless someone forgets, as I once did, to latch the door. (It blew open, and that was the end of winter tomato plants for that year!) Short periods of overheating won't hurt most crops, but a long spell (more than four to six hours) of sweltering temperatures, above 90°F (32°C), can make cool-weather vegetables like Chinese cabbage tough and bitter. Even plants that like warmer temperatures, such as tomatoes, will suffer from poor pollination and other problems when the thermometer goes above 100°F (38°C).

To moderate wild swings from cool to hot, add more thermal mass—another water-filled barrel or a brick retaining wall for the growing bed, or water-filled jugs on the shelves. When we installed a tank in our greenhouse to preheat household water, we noticed that the air temperature tended to remain steadier because more heat was being absorbed and slowly released by the 30 gallons of water in the black-painted tank.

At the other end of the scale, plant roots aren't able to absorb water or soil nutrients very well when the soil temperature is below 45°F (7°C). (Soil temperature, by the way, is more critical than air temperature.) Lower night temperatures don't seem to hurt plants as long as they don't go below freezing. A big mass of soil like that in our growing bed is slower to warm up but better at retaining its warmth than the clay pots we have lined up on the shelves. Watering with warm water can help to raise soil temperature. You can use a soil thermometer to check on your growing beds and a maximum-minimum thermometer (filled with alcohol instead of mercury, in case of breakage) to keep tabs on air temperature.

I don't usually try to grow plants in my greenhouse over summer, but those who do often spread cheesecloth on the glazing for light shading, or they paint the glass with dilute white latex paint (70 percent water, 30 percent paint) or with a commercial shading compound to moderate temperatures. Sprinkling or misting will also cool plant and soil surfaces by evaporation. Too much sprinkling, though, can cause mold and disease problems.

Humidity. Plants grow best in moderately humid air: 45 to 60 percent relative humidity. Most do quite well at 70 percent, but at high moisture levels, over 90 percent, the molds and diseases that thrive in all that dampness will slow, and often stop, your plants' productivity. Warm air, you remember, can hold more moisture than cool air. Overwatering can cause humidity problems. I find that in winter, when my plants are growing slowly, most of them thrive on one watering a week. The large growing bed holds moisture, and more important, the plentiful humus in the soil, from annual additions of compost, helps to retain water while gradually doling it out to plants as they need it. The best way to deal with excess humidity is to ventilate the greenhouse.

Good Ventilation. Light, temperature, and humidity are easy to measure with common instruments: light meters, thermometers, and hygrometers. You won't find a handy CO_2 meter at your local hardware store, but you can be sure that a tightly closed greenhouse will be deficient in this compound, which is so vital to your plants. Carbon dioxide, you see, is the plant's only source of the element carbon, which makes up *half* of the plant's total dry weight. It is an essential component in photosynthesis. In fact, in winter, when light levels are low, an abundance of carbon dioxide can, to a degree, actually compensate for the lack of light and thus help the plant to keep growing even when it receives a less than ideal dose of light. In a well-sealed greenhouse, though, plants may use up all the available carbon dioxide by early morning. Scientists who have studied greenhouse conditions say that a thin layer of air deficient in carbon dioxide hovers around the leaves of plants in an unvented greenhouse.

By now you've guessed, I'm sure, that an effective way to get more carbon dioxide to your plants is to ventilate the greenhouse. Ventilation helps in two ways: It encourages air circulation, which breaks up those pockets of stale air, and it introduces a new supply of carbon dioxide from fresh outdoor air. A small fan will help to break up air pockets and improve air circulation.

When weather is too cold for continuous ventilation, a compost pile stashed under a bench will release a steady supply of carbon dioxide. You've noticed, I'm sure, that a well-made pile of compost will shrink as it decomposes. Well, the reason for that decrease in volume is that the material has given off between one-third and one-half of its dry weight in carbon dioxide.

Diseases

Unlike an outdoor garden, a greenhouse is an unnatural arrangement. It is roofed against rain, shielded from ultraviolet light, and protected from wind. Occasionally it becomes too hot or damp. It shouldn't be surprising, then, that in such a setup certain insects and diseases can cause problems. It's a rare greenhouse that doesn't suffer an occasional attack of whiteflies or damping-off disease. The best way to escape such problems is to concentrate on prevention.

Disease is much easier to prevent than to cure. Excess humidity and lack of ultraviolet light, which sanitizes the growing bed, can make the greenhouse a good incubator for disease. (Glass excludes ultraviolet light, but fiberglass does allow a few ultraviolet rays to penetrate.) Occasional temperature extremes in the greenhouse can cause trouble, too. When cool-loving plants like Chinese cabbage and stocks are stressed by high temperatures, and heat lovers like tomatoes take a chill, the stage is set for more unfriendly organisms to proliferate there. In addition, insects can spread disease as they feed on plants. For example, my winter greenhouse crop of Swiss chard sometimes gets sooty mold, a fungus disease that clogs pores and reduces photosynthesis. The mold spores are probably there all the time, but it's the sticky "honeydew" left by whiteflies that supports the mold and allows it to get a foothold.

The following disease-prevention tips should help you to run a reasonably clean greenhouse. They're important enough, in fact, to print up on a little poster to be mounted in the greenhouse, especially if the space is used by several people or frequented by visitors:

- Remove and compost all dead leaves and destroy diseased plants.
- Avoid overcrowding plants.
- Keep the greenhouse well ventilated.
- Never smoke in the greenhouse or handle pepper or tomato plants after handling tobacco. Tobacco mosaic virus is highly contagious.
- Select disease-resistant varieties of vegetables and flowers.
- Don't reuse old potting soil for seedlings. When I did this, I sometimes got away with it, but sometimes lost tiny plants from damping-off, too. Now I save the used planting mix for potting up rooted cuttings of shrubs and other sturdy plants.

Insects

Certain insects like the specialized conditions in the greenhouse. Some of the commonsense measures you use to keep the greenhouse humming, such as maintaining good ventilation, will help to discourage these pests. In my greenhouse, insect problems seem to be worst in midwinter when humidity is high and natural air circulation is sluggish. Now that we've finally cut vents between the greenhouse and the house, we seem to have fewer insects and less disease, probably because air circulation is improved and excess heat isn't retained. Here's a rundown on the insects that have bothered my greenhouse plants.

Aphids. These astonishingly prolific insects are less active than whiteflies and, at one-tenth inch long, not quite so tiny but just as damaging. Their slender mouthparts pierce tender plant stems and leaves to extract the juices. They also excrete honeydew. Aphids often attack plants that have been given too much nitrogen or those that have been undersupplied with organic matter.

Control. If you overdo the manure tea, dig in sawdust to absorb some of that excess nitrogen; however, don't use cedar, walnut, or redwood dust; these contain substances toxic to plants. Soap sprays kill aphids, but they may also kill beneficial insects. An easier and safer treatment is to hose the soft-bodied pests off of your

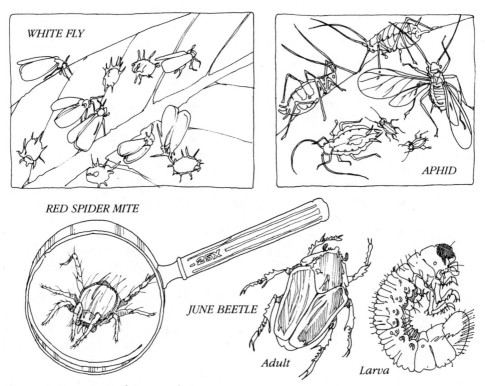

Common insect pests of the greenhouse.

plants. Immature ladybugs gorge on aphids and may be purchased by mail from natural insect control suppliers. Daddy longlegs will also eat aphids and larvae of some other insects.

June Beetles. During the first summer of its use, our greenhouse windows were not yet screened. Several green June beetles, *Cotinis nitida,* flew into the greenhouse and laid eggs, which hatched into destructive grubs. These pests eat roots and seedlings. They continue to plague us because each summer some of the larvae grow into beetles, which lay more eggs and perpetuate the cycle.

Control. Had we installed screens before that first growing season or before filling the bed with soil, the beetles could not have gotten in. To control them, I handpick adult beetles, dig and squash the grubs, and spread diatomaceous earth around new seedlings when I plant them in the bed. Next season, I intend to try an application of parasitic juvenile nematodes to attack the grubs below ground.

Slugs. Sowbugs (described on page 156) nibble away gradually at seedlings and young plants. If you find leaf pieces missing from older plants as well as from seedlings, look for a telltale dried mucus trail on the soil around the plants. Take a midnight safari into the greenhouse with a flashlight. I'll bet you'll find slugs, happy as clams in all that dampness and lush plant growth.

Control. You can trap slugs, or you can spread wood ashes or diatomaceous earth to irritate their soft bodies. For me, the manual approach has worked best. I simply visit the greenhouse several hours after dark, armed with a thick glove, a can, and a flashlight and handpick the slugs. My highest count was 50 in one expedition, and the numbers dropped steadily as I continued to handpick regularly for seven to ten days.

Sowbugs. Also called pillbugs, these are among the more annoying of my greenhouse's uninvited residents. These small, gray, oval crustaceans have segmented bodies and many legs, and roll into a ball when disturbed. They feed on plant debris and hide under boards and in crevices, emerging when you're not looking—and often when you are—to feed on roots and young seedlings.

Control. Prevention is easy to describe but not so simple to practice. Remove fallen leaves, flats, and as many other hiding places as you can from the surface of the growing bed. I've trapped pillbugs under cut pieces of potatoes, apple peelings, grapefruit, and cantaloupe rinds—an effective ploy that also generates a certain amount of vengeful satisfaction. You can handpick them too, of course . . . if you have all day. Crushing even a few of them, on a daily basis, will help to decrease their numbers, though.

I have no illusions that my greenhouse will ever be free of sowbugs; rather, I'll be satisfied if I can keep them from destroying my seedlings. They do perform the useful service of reducing any plant debris on my growing bed to the consistency of coffee grounds so I try to remember that they have their place in the scheme of things. When I set out especially valuable new seedlings, like my greenhouse cucumbers, I often dust around them with diatomaceous earth to fend off those little gray marauders.

Spider Mites. These tiny leaf-sucking pests can live in hot, dry air or in excessively damp, stagnant air. In my experience, they are most likely to bother ornamentals, which are shifted from house to greenhouse. Plants infested by spider mites have mottled foliage, and you can usually find a skimpy, delicate webbing among their leaves.

Control. Your best method of prevention is to maintain good air circulation and adequate but not excessive humidity in your indoor plant nursery. A soil well supplied with calcium will discourage spider mites (good old limestone will do the trick). When you find them on your plants, knock them off with a strong stream of water. If the infestation is severe, you might try swabbing the leaves, including the undersides, with a mixture of half water and half rubbing alcohol. Test the treatment first on a few leaves to be sure it won't harm the plant.

Whiteflies. Whiteflies are notorious greenhouse pests. Although you'll find them outside, too, they're much more damaging in a greenhouse. They suck vital juices from plants and produce a fungus-attracting honeydew. Sometimes they are worse when soil is deficient in phosphorus or magnesium. These tiny, grayish white, flying insects live longer and multiply more at temperatures below 65°F (18°C); in spring, when the greenhouse is warmer, I find fewer whiteflies on my plants.

Control. The easiest way to control whiteflies is to trap them on yellow cards coated with something sticky like Vaseline or Tangletrap. Whiteflies are attracted to the color yellow. I use Stiky Strips—very sticky pieces of yellow plastic purchased from a greenhouse supplier. These are inserted in reusable wire holders and poked into the soil around the plants. When my Chinese cabbage and cucumber vines were badly infested with whiteflies, I sprayed them with Safer's Soap and noticed fairly good results. Other greenhouse gardeners have used solutions of mild soap or Basic H (about two tablespoons to the gallon) for the same purpose. Test the solution on a single plant first, though, because some plants are sensitive to these substances. Because whitefly eggs hatch 10 days after they're laid, the soaps must be applied every 7 to 20 days for several weeks to control succeeding generations.

A tiny predatory wasp, *Encarsia formosa,* controls whiteflies by laying its eggs on the immature whiteflies and also by feeding on whitefly pupae. *Encarsia* needs warm temperatures, though—preferably an average of 75°F (24°C) during the day and no lower than 55°F (13°F) at night. I haven't used *Encarsia* myself because my worst problems with whiteflies are in the winter, when my greenhouse would be too cool for them to be effective.

If you do adopt some *Encarsia* as allies, you might want to keep some blooming plants around the greenhouse to provide some nectar and pollen for the tiny wasps. Calendula, stock, geraniums, and Johnny-jump-ups have all done well for me in the winter greenhouse. At the higher temperatures *Encarsia* prefers, you could keep a wider variety of flowering plants. Remember, too, that these and other predators need *some* insects to feed on if they are to keep living in your greenhouse, so don't try to totally eliminate any pest population. Even if you don't introduce purchased predators, you'll often have some volunteers like syrphid flies and ladybugs working for you unbidden.

Other Uses of the Greenhouse

In addition to starting seeds, your solar greenhouse can also be used to raise food, dry herbs, cure onions, and propagate cuttings and fresh flowers. Think of the space as a little resort to which you can repair whenever you wish, for a breath of balmy, leaf-and-earth-scented air, for a few relaxing moments of pruning, for transplanting, or for a renewing spell of just plain sniffing and looking.

The Cycles of the Greenhouse

If you need more proof that everything depends on everything else, look in the greenhouse. When you vent the space to dissipate excess heat, you also set up a lively pattern of circulating air that will discourage whiteflies and help control fungus diseases. When you overfertilize plants with high-nitrogen feedings, the resulting sappy growth is often both attractive to insects and susceptible to disease. The carbon dioxide you exhale as you tend your plants is to them life breath that helps them flourish. The strong, vibrant plants that result are more satisfying to tend, so you linger longer in the greenhouse—snipping, transplanting, rearranging . . . and breathing, and so it goes.

24 Growing Vegetables in the Greenhouse

You'll notice that my greenhouse vegetable list is not arranged in alphabetical order. The vegetables are, rather, ranked in order of my personal preference—a quirky system, perhaps, but one that tells you what I've concluded from experience. My choices are based partly on results and partly on food value. I have plenty of good tomatoes in jars and beets in the root cellar, but the fresh, nutritious, unsprayed cucumbers and leafy greens are hard to come by in winter, and we eat up all we can grow.

Cucumbers. We love the taste of cucumbers, and they make us feel pampered when the greenhouse vines are loaded, while outside trees are bare. The new European-type cucumbers that produce all-female blossoms are my favorites. The fruits are long, tender, and sweet—better quality than those I grow in the garden. Two good varieties are Dynasty and Superator. Pandex, which I haven't tried, will set fruit at lower temperatures than others.

Greenhouse cucumbers like the good life—warm soil, weekly fertilizing, and frequent watering. (About every other day when in fruit. I use warm water.) They prefer temperatures above 60°F (16°C), but if they get plenty of light and warmth during the day, they will be more able to tolerate the cool night temperatures in the average solar greenhouse.

I grow two cucumber crops a year, planting in August or September for a November to December harvest and again in February or March for harvesting during May and June. To support the vines, I suspend a length of wire fencing from two nails driven into the framework. I don't always get my vines properly pruned, but according to Colleen Armstrong, staff horticulturist at the New Alchemy Institute, you should remove every sucker from the six lowest leaf joints to make the vine stronger. The next eight shoots should be cut back to a single leaf, and the remaining ones permitted to develop two to three leaves. Over-production weakens the vines and reduces fruit size. These special greenhouse cucumbers produce fruit without pollination; in fact, any male blossoms that develop should be removed to keep the cukes tender and free of developed seeds. (Even before they open, female blossoms have a small swelling at the base—a miniature cucumber. Male blossoms lack the "baby cuke" end that characterizes female flowers.)

Swiss Chard. This is the most reliable and long-lasting leafy green I grow in my greenhouse. It is deep rooted, so it will survive (if in a deep-soiled bed) even if you forget to water it some week. The leaves are low in calories and high in vitamins, and the flavor is considerably milder than that of outdoor chard.

158

Vegetables that can be grown in the greenhouse.

I plant seeds in August or September and start picking the succulent seersuckered leaves in December, continuing all winter till spring greens take over. Sometimes my chard gets a plague of aphids or whiteflies in midwinter when cloudy days keep the greenhouse closed and cut down on the ventilation that helps to discourage these pests. However, once I start opening vents regularly again, the bug pox clears up, and the leaves are clean and green again. Chard is a winner!

Chinese Cabbage and Oriental Greens. Loose-leaf varieties of Chinese cabbage will grow better in the greenhouse than types that form heads. I've had good luck with Prize Choy and Spoon cabbage.

I sow seeds in a flat in August or September and again in December or January. I plant some seedlings in the growing bed and keep some in the flat for later transplanting as space becomes available. I've also gotten nice winter crops of Chinese cabbage from plantings made in five-inch-deep rectangular plastic wash basins and refrigerator crispers.

Another Oriental green, Mizuna (also called Kyona), is a first-rate greenhouse crop from which I can always count on repeated cuttings of tender, mild-flavored fringed leaves. I've had no trouble with Mizuna going to seed, but the heading types of Chinese cabbage will go to seed when days turn long or temperatures stay high. I was surprised to discover, though, that when grown in the greenhouse, the stalks and flowers of the bolted Chinese cabbage are not bitter and tough as I had expected them to be.

Lettuce. With its delicate, buttery soft heads, Bibb lettuce is perfect for the greenhouse. I also like Kwiek, Magnet, and Mescher, a wonderful heirloom butterhead I obtained from a fellow member of the Seed Savers Exchange. Leaf lettuce is fine, too. I've grown head lettuce, as transplants rescued from the outdoor garden when black frost threatened, but they were basically being kept on hold and didn't grow much in the greenhouse. I wouldn't count on head lettuce as a greenhouse crop; it takes too long to develop, and besides, its vitamin content is much lower than that of the leaf and butterhead types.

The tricks to growing lettuce are to keep some seedlings in reserve, to replant regularly, and to plant the fall crop early enough so that it can finish its growth in the winter greenhouse. Lettuce likes rich soil, plenty of moisture, and good drainage.

Radishes. Radishes are a good greenhouse crop because their flavor is so much milder than when grown outside. They form roots best in spring when days are lengthening. You can sneak in a row at the edge of the growing bed without disturbing other plantings, or grow them in tubs or pots. Plentiful moisture keeps them crisp and mild.

New Zealand Spinach. This crop surprises everyone who tries it with its luxuriant growth in the greenhouse. You'll be picking the leaves for supper two months after you plant the seeds.

To encourage germination, file a little notch in the tough seed coats, and soak the seeds overnight before planting. New Zealand spinach is a good space-saving greenhouse plant because it can either grow in a tub set on a shelf, with its vines spilling down into the sun, or it can climb a trellis or string to leave bed space free for other plants. Regular spinach (*Spinacia oleracea*) is not as good a greenhouse crop as you might think. Both chard and Mizuna have been much more productive for me.

Onions. Onions won't bulb up during the short days of winter, but onion sets tucked in around the edges of the growing bed will provide a modest supply of savory green spears for winter cutting. Chives makes a good greenhouse plant, too. For continuous production, let the plants freeze outdoors for two weeks before you bring them into the greenhouse. They need a rest after producing all summer.

Tomatoes. Tomatoes don't always produce enough to earn their space in a cool solar greenhouse, but the satisfaction of raising even a few ripe tomatoes in the off-season can be worth the trouble. Small-fruited kinds like Presto, Pixie, and Sweet 100 will be more productive and also faster to ripen under greenhouse conditions. Vendor and Park's Greenhouse Hybrid are good tomato varieties to grow if you insist on full-sized fruits.

To produce fruit, tomato plants need more light and heat than leafy plants—60°F to 80°F (16°C to 27°C) is best. To keep at least a few ripening tomatoes in the greenhouse, start with potted plants in the fall from mid- or late-summer seedlings, or rooted cuttings from established plants. These should start to bear by early January. I've also brought bearing plants in pots into the greenhouse in September or October to continue fruiting for a few more weeks. Then, for a spring crop, start seeds in December for May fruits. I often use the cold-tolerant Siberia and Santa varieties for these spring plantings.

Tomatoes are self-pollinating, but in the absence of wind it's a good idea to jiggle the blossoming plants several times a week to help disperse the pollen. You won't need bushel baskets for your harvest, but the few fruits you do get will mean a lot.

Peppers. When I have extra pepper seedlings in spring, I often pot up several of them, each in a two-gallon pot, keep them on the sunny patio during the summer, and then return them to the greenhouse in the fall to keep bearing for at least an extra few weeks. You can also sow pepper seeds in May and grow them in pots until August; then transplant them to the greenhouse growing bed. Since they are perennials in their native climate, peppers are glad to continue leafing and podding as long as they are free of insects. Whitefly and aphid infestations often lead to disease. Greenhouse peppers tend to be small, but are nonetheless welcome salad material when the outdoor garden is frozen shut.

Beets and Turnips. Both beets and turnips will grow well in the greenhouse. They may be planted in fall, winter or spring, directly in the growing bed or in eight-inch-deep pots or tubs. Eat them while the roots are small, and use the tops, too.

Beans. Greenhouse beans are fun to grow in spring. Plant them in the beginning of March for early June beans. They're self-pollinating, and they like warm soil. You can even plant climbing beans to shade the summer greenhouse, or use them as a green-manure crop to keep your growing bed in good condition. The latter is a good example of crop rotation because the soil-improving beans can be grown after greenhouse greens have been harvested, and they can be turned under before it's time to stock the beds with greens again in the fall.

This list of greenhouse vegetables is not by any means all-inclusive. There are other vegetables and even some fruits and a good many flowers that you might want to try in your greenhouse. George "Doc" Abraham, author of *Organic Gardening Under Glass,* mentions strawberries, figs, bush summer squash, and okra as good greenhouse crops. Shane Smith, author of *The Bountiful Solar Greenhouse,* reports successful harvests of, among other things, globe artichokes and cantaloupes in summer and peas and fava beans in winter.

25 *Starting Wildflowers from Seed*

Every time I walk past the patch of forget-me-nots growing by the side of our pond, I feel a renewed sense of satisfaction. No other wild forget-me-nots grow in our area, and because I love them, I decided to start some from seed. The original plants have spread by root extension, and they've also probably reseeded. I hope they'll continue to spread. It is a good feeling to be able to add something to the landscape that wasn't there before. That is part of the special joy of growing wildflowers from seed. If the wildflower you start becomes well established, it will continue to give pleasure for years to come. In many cases, it will also spread, and even, possibly, reproduce to form new colonies.

Raising wildflowers from seed is an adventure that can last a lifetime. Start with the easy ones, like lobelia, columbines, butterfly weed, and others mentioned in the encyclopedia section in the back of this book. I've also included a few highly specialized wildflowers, such as bunchberries, trailing arbutus, and yellow lady's slipper, to guide those gardeners who may be ready for a fascinating—if demanding—gardening challenge.

Site and Soil Preference

Because they have evolved into their present form unaided by people, most wildflowers have more definite site and soil preferences than the zinnias and petunias we grow in our flower borders. Some wildflowers need shade, others thrive in damp places, and many prefer acid soil. These requirements for certain conditions can be turned to your advantage, though. Wildflowers can be the perfect solution for hard-to-plant areas: the shady side of the house, the damp corner, the bare stretches under high trees, the steep slope. The secret of success with wildflowers is to match the needs of the plants to the conditions that exist in your yard, fields, or woods. To help you decide on suitable plants, study your place at different times of the day. When you start to look around your property for likely places to plant wildflowers, you'll notice that there are different degrees of shade—from the high, light dappling of birches to the heavier shadow cast by maples and oaks to the deep, year-round shade under evergreens. No one wildflower will thrive in all of those spots, but for each form of shade there is a choice of wildflowers. Check after rain to be sure your proposed site is not waterlogged. Plants that thrive in full sun when grown in the north may do better in part shade in the south. In the far north, plants will bloom up to a month later than indicated here.

When you start wildflowers from seed, you can settle them into their permanent niches while they're still small so they won't suffer setbacks from transplanting. Plants raised from a single packet or collection of seeds may vary considerably in size, germination time, even in color and hardiness—unlike garden flowers and particularly unlike hybrids, which are especially uniform in size and growth habit. These variations are a characteristic of many wild plants that helps ensure that at least some members of the new generation will be able to thrive where they find themselves. When you raise wildflowers from seed, you'll often have enough seedlings so that you can try them in several locations to determine the best site.

Collecting Seeds

Once you start looking, you'll find a surprising amount of wildflower seeds available. In addition to the companies that specialize in wildflower seeds, several garden seed companies offer seeds of some of the easy-to-grow varieties, and a couple of seed exchanges offer varieties for the true enthusiast.

Some wildflowers that can be grown from seed.

You can also collect your own wildflower seeds. To do so, you'll need (in addition to the permission of the landowner) to know when the seeds are likely to ripen in your area. In general, seeds are ready to collect about a month after flowering. You'll find it helpful to study the plant and its habit, though. Some, like jack-in-the-pulpit, retain their seeds until fall. Others, like bloodroot, shed seeds very early or, in the case of wild geraniums, disperse them suddenly once they're ripe. Seeds with arils (fleshy appendages) like those of trillium and bloodroot are often carried off by ants and animals. The plant as a whole often changes in appearance as the season progresses. Foliage may yellow and disappear, stalks droop, and seeds trail on the ground. You want the seeds to be mature, because if they haven't developed sufficiently on the plant, they may not germinate. When you collect pods and bare seeds, they should be as dry as possible. Berries should be fully formed and ripe. On the other hand, you don't want to get there too late, when seeds have fallen and blown away. If this does occur, try again the following year. Learning how to collect seeds is a process of patient observation and gradually growing understanding. Watch the plant. Return often to observe the seed formation process. It's a good idea to mark the plant from which you want to save seeds with a small stake or red yarn.

When the seeds are ready, collect a small amount—never all the seeds from any one colony—with respect and appreciation. It goes without saying, of course, that you wouldn't take seeds from a single isolated stand of plants, but only from a flourishing group in a place where you can see more of the plants growing nearby. To help catch some seeds when you're not around, you can tie small cheesecloth bags or pieces of stocking over the ripening seeds, or, with a large plant, clip an envelope onto the plant with a clothespin.

Keep a record of the location of marked plants to which you hope to return to collect seeds, and also record the date, site, and soil characteristics for each numbered sample. This will help you to choose an appropriate site for the resulting plants, especially if you're collecting and planting different species from different locations. You will want to observe, too, what sort of site your chosen flower seems to favor: rocky slopes or boggy meadows; filtered sun, afternoon shade, or a site that is sunny in spring and shaded in summer; oak woods, pine woods, or roadside conditions.

Cleaning and Storing Seeds

Serious wildflower seed collectors screen and sift seeds to remove all debris. You should at least pick out green leaves and stems, which could heat and mold and thus spoil the seeds in storage. Next, air-dry the cleaned seeds for a week or two before storage. There are some exceptions to this rule. Seeds containing beards or tufts should be planted as soon as they are dry and ready to be shed by the plant because they often fail to germinate at all if allowed to air-dry. Some seeds, like hepatica, have such low viability that they should be sown immediately after collecting, and of course seeds of wild aquatic plants should be either sown immediately or kept moist, because dampness is their natural condition.

Dry seeds should be kept cool and dry until you are ready to plant them. You can either seal labeled envelopes in plastic bags or enclose small plastic bags of seeds in jars and keep them in the refrigerator (see chapter 31 on seed storage).

Moist seeds such as those found in fleshy fruits should usually be kept moist until planting. If allowed to dry out, they enter an extended period of dormancy or, in some cases, lose their viability entirely. Cover them with dampened sphagnum moss, put the mossy clump in a plastic bag, and keep it in the refrigerator.

Dormancy
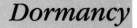

Unlike cultivated garden flowers, many wild plants require a period of dormancy before they will germinate. The depth and extent of dormancy varies greatly in different species of wildflowers. Generally, one to four months exposure to temperatures below 40°F (4°C), usually in the presence of moisture, will be enough to break dormancy in most wildflowers. Even to botanists, many aspects of wildflower dormancy are still a mystery. Inconvenient as it is to the gardener (who suddenly, after a wild plant has been growing on its own for all these centuries,

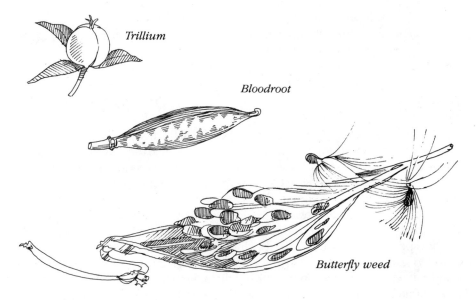

Trillium

Bloodroot

Butterfly weed

A few wildflower seeds.

decides to adopt and cultivate it), the process is really a vital protective mechanism for the plant. In cold climates, if seeds were to germinate in the fall, the seedlings would be killed by frost before they had a chance to develop strong roots or to reproduce themselves.

Seeds retain dormancy in numerous ways, among them by developing tough seed coats, by producing chemicals that inhibit germination, or by requiring a period of afterripening to complete their maturity. Some seeds that possess hard seed coats, like those of certain wild legumes, will germinate promptly if the seed coats are scarified and the seeds soaked in hot water. To scarify a seed, cut a notch in its outer covering with a file, knife, or wire cutters, or rub the seed between two sheets of sandpaper. Avoid cutting deeply into the seed; just nick its coat.

For most other wildflower seeds, you'll have better results if you duplicate natural conditions as closely as possible. Unless otherwise indicated in the specific wildflower directions in the encyclopedia section at the back of the book, wait until after November 15 (a month earlier in the far north) to sow the seeds outdoors so that they will have a good long period of exposure to natural cold and moisture before the warmth of spring encourages germination. The cold and repeated soaking by rain helps to inactivate and dissolve inhibiting chemicals and soften hard seed coats. This exposure to cold, moist conditions as a necessary preliminary to germination is called stratification.

Stratification. You can stratify seeds in several ways:

- As mentioned above, surround them with damp moss or vermiculite and keep them in the refrigerator during winter.
- Plant them in flats kept in a lightly shaded cold frame.
- Sow the seeds directly in a well-prepared bed of finely screened and raked humus-rich soil. Seeds of plants that suffer from transplanting may be sown in individual pots.

Sowing Seeds

Scattering seeds on unprepared ground is usually a waste of effort and good seeds, especially if you do as an acquaintance of mine once did and simply broadcast wildflower seeds collected in a distant state. It's one thing to idly toss out pinches of seeds of aster or black-eyed susan by the wayside while hiking country roads, as I've more than once delighted in doing. But when you want to enrich your woods with a patch of bloodroot or foamflower, then it's well worth taking the time to give the seeds a sporting chance to germinate.

For especially rare or tricky seeds, wildflower expert Bebe Miles, whose books I've found extremely helpful, suggests dividing the seeds into two portions and sowing half outdoors in beds in the fall and the other half in flats planted in February or March (after wintering in the refrigerator). Cover the seeds with soil to a depth of twice their thickness. Keep the flats on a cold porch until spring. Miles also plants seeds of biennial wildflowers in May so they won't be killed by frost just as they're

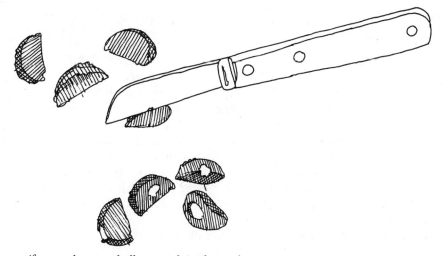

To scarify a seed, cut a shallow notch in the seed coat.

starting to bloom, as they might if planted earlier. In addition, biennials that are planted later go into winter as strong plants and then have time to bloom and set seed the following spring.

If you're working with seeds that you've collected or that for any other reason you'd have difficulty replacing, take a hint from the Seed Savers Exchange: Always reserve at least a pinch of your seeds in case the first sowing doesn't take. Don't discard planted flats that show no sign of life. It's not unusual for wildflower seeds to germinate a year after planting, and two to three years is not rare. Seeds will only germinate, though, if the growing medium is kept moist. For purchased wildflower seeds, which have been in storage, allow an extra year for germination because storage might have sent them into deep dormancy.

For shade-loving wildflowers, be sure to shade the bed or cold frame with evergreen boughs or laths, and when the seeds do sprout, ventilate the cold frame as necessary to prevent overheating. Good air circulation is important, too, to prevent fungus diseases.

At Bowman's Hill Wildflower Preserve near Washington Crossing, Pennsylvania, staff horticulturists raise many wildflowers from seed each year. With the help of volunteers, they plant seeds outdoors between November and late February. They cover the seeds with hardware cloth to keep out pilfering wildlife and with burlap to retain moisture. These are left in place until seeds sprout or for four winter seasons. When sowing seeds in a cold frame, the folks at Bowman's Hill like to bank the sides of the frame with soil to retain more heat. It's also a good idea, they say, to sink the frame four to six inches in the ground for more cold protection. The sides of the cold frame can be insulated, too, by sandwiching one-inch Styrofoam between board walls. Painting the inside white boosts light levels.

Caring for Seedlings

When seedlings have their first true leaves, transplant them to individual pots or to larger flats. Then let them grow some before you set them out in either a nursery bed or in their permanent locations. Many wildflowers grow very slowly. That's their natural pace. Harry Phillips, former curator of native plants at North Carolina's Botanical Garden, and the author of the splendid book *Growing and Propagating Wildflowers,* suggests fertilizing wildflower seedlings to promote rapid growth. Once the plant is growing in place, though, it should seldom, if ever, need fertilizer—if it is in an appropriate place. Overfertilization often encourages weedy competition and causes long, weak stems in many meadow plants. Wildflowers have evolved without our help or interference, and they can make it on their own if the soil type meets their needs.

For details on growing specific species of wildflowers, see the encyclopedia section at the back of this book.

Wildflower Conservation

A good reason to grow wildflowers from seed is that, unfortunately, some nurseries that sell wildflowers have obtained their plants, either directly or indirectly, from the wild. Such depredations reduce even further the sometimes precarious populations of our native plants. Careful wildflower businesses like Native Gardens in Greenback, Tennessee, Yerba Buena Nursery in Woodside, California, and Boehlke's Woodland Gardens in Germantown, Wisconsin, are leading the way by selling native plants that they have propagated themselves. Until the rest of the industry follows that honorable example, gardeners who care about wildflower conservation will want to know the sources of plants they are buying and, better yet, learn how to grow their own.

26 Starting Trees and Shrubs from Seed

Great oaks do indeed grow from small acorns, and participating in that marvelous process can be a great gardening adventure. You might try starting some trees from seed just to see what happens, but there are other practical reasons, too, for planting tree seeds. Perhaps you need many trees and shrubs for a windbreak or hedge, or you might want more trees of a species, like hickory, that develops a taproot and consequently transplants most successfully when young. Possibly you've collected tree seeds during your travels or from an ancestral homestead, and you've decided to grow some living souvenirs. You could raise seedling trees for grafting rootstocks, too, or you might plan to sell trees you've grown from seed. In any case, you'll want to know about the important ways that tree seed starting differs from garden plant seed starting.

Seed Production

Trees and shrubs live for many years, so they have plenty of time to produce the seeds that will grow their replacements. Trees may start to bear seeds as early as the American maple, 5 years, or as late as the sessile oak, 40 years. Although many woody plants, especially the smaller ones, produce regular seed crops, a good many trees produce abundant seed crops only intermittently—some as seldom as every 7 years. Tree researchers theorize that in these cases, the great effort put into seed bearing depletes the tree's supply of stored carbohydrates, which must gradually be replenished before the tree is able to fruit generously again. Sometimes, seeds produced in off years, when carbohydrate reserves are low, are low in viability, and the seedlings may be less vigorous. So if you do have a choice, plant the nuts, fruits, or pods from a season of good production.

Dormancy

Tree seeds are more likely than garden seeds to exhibit a deep and sometimes persistent dormancy. In a fair number of woody plants, germination is naturally delayed until the second spring. (Isolated trees, which out of necessity have self-pollinated, will sometimes bear a higher percentage of aborted or malformed seeds.) This failure to germinate when conditions are otherwise right is one of the

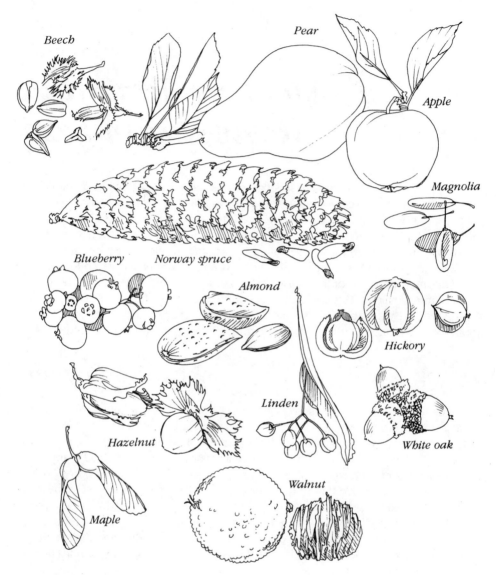

Seeds of some trees.

most baffling dilemmas in the seed-starting process. Seed scientists still have few explanations for many of the hows and whys of seed dormancy. If we empathize with the tree for a moment, though, we can see that dormancy is a survival mechanism that enables the seeds to wait patiently for a favorable time to sprout. Germination of dormant seeds is sometimes spread over several years, increasing the possibility that the new young tree will meet with good growing conditions in at least one year.

Black locust seedpods.

What makes a seed dormant? Its seed coat might be hard or impermeable to the oxygen that's so vital in the germination process, or it might have an immature embryo that needs a period of afterripening. Still another possibility is that chemical germination inhibitors are produced by the seed or, often, by the flesh of a berry surrounding the seed. A great many species of trees produce seeds with combined dormancy inhibitors—impervious seed coats and underdeveloped embryos, for example. Some seeds, like those of the apple and service tree, become more dormant if allowed to dry after planting.

Breaking Dormancy. Under natural conditions, when a tree seed drops to the ground, the forces of nature gradually help to break its dormancy. Bacteria and acids in the soil penetrate the hard seed coat. Weeks of warm weather at the end of summer encourage the embryo to complete its development. Then, scientists theorize, as the chill of fall spreads over the land, it initiates the accumulation of substances in the seed that stimulate cell growth. When these substances reach a certain concentration, they nullify the effect of the inhibiting chemicals within the seed. Some seeds are able to germinate in spring after a winter spent in the cool ground. Others need a second period of warming and chilling before they're ready to germinate, and a few take three or more years to sprout.

Sweet gum seedpods.

The person who plants tree seeds can duplicate some of these natural processes to help break the seed's dormancy. For many seeds, fall planting will subject the seeds to enough of a chill to trigger germination when warm days return in spring. Soil that is rich in organic matter will encourage the decay of the interfering seed coat. Hard seeds like those of woody plants in the legume family are often stimulated to sprout by pouring boiling water over them and allowing them to soak in the water for a day while they cool. Don't let the soaked seeds dry before planting.

Scarification. Stubborn seeds often germinate better after scarification—etching of the seed coats by mechanical or chemical means. Try mechanical first. It's much easier. Simply file or cut a shallow notch in the seed coats or scratch them with sandpaper, then soak the seeds in water as directed above. Chemical scarification involves soaking the seeds in twice their volume of concentrated sulfuric acid (commercial grade 95 percent) for 15 to 60 minutes in order to dissolve some of the seed coat. Be sure seeds are dry before putting them in the acid, and *never add any water.* Water reacts chemically with acid and sometimes produces a violent reaction. Stir the seeds slowly and gently while they're in the acid bath; avoid introducing air, which could cause heating. When the time is up, rinse the seeds in cold water to which you've added a small amount of washing soda (sodium carbonate) and then in running water. Sulfuric acid can disintegrate fabric and skin, too, so be extremely careful if you decide to use this dangerous seed treatment. Always wear rubber gloves and eye protection, and put the acid *only* in acid-resistant containers.

Willow Tea. Water in which willow cuttings have soaked is an effective growth stimulator. You might want to try soaking seeds of woody plants in willow tea before planting. It will not take the place of acid scarification, though.

Early Planting. Some researchers have found that certain tree seeds, especially those encased in pods or berries, will germinate more promptly if collected and planted before the fruit becomes mature enough to produce the inhibiting chemicals that plunge the seeds into deep dormancy. There are no clear guidelines yet on the timing of such early collection. You want to harvest the seeds before they are affected by the ripe fruit's chemical stop signs, yet the seeds must be sufficiently mature to contain a well-developed embryo. The best plan, when experimenting with such seeds, is to gather some when they're still fairly green and others at progressive stages of ripeness and plant them all.

Cold Stratification. You'll remember that we mentioned stratifying wildflower seeds as a way to encourage germination. You can use the same procedure for woody plants. If you obtain your tree seeds late or for some other reason don't have a chance to plant them in the fall, you can stratify them until spring planting time to provide the chill necessary to bring the seeds out of dormancy. Simply mix the seeds with three times their bulk volume of moist sphagnum moss or vermiculite, put the mixture in a plastic bag, and keep it cool—about 40°F (4°C)—for at least one month. Two or three months is even better. The moss should be damp but not soggy, because germinating seeds need a good supply of air. To get the right consistency, slowly add water to the medium until it will just release a drop of water when you squeeze it.

Stratification is, in some ways, a kind of slow motion germination that progresses through three stages: First, the seeds swell as they absorb water; then,

A hickory fruit.

increased enzyme activity revs up the life processes that lead to sprouting; and finally, the embryos grow at an increasingly rapid rate. If you've stratified seeds over winter, either in your refrigerator or in a cool room, it's important to avoid letting them dry before planting, or they're likely to enter a second dormancy, which can then only be broken by repeating the cycle.

Warm Stratification. In most cases, cold stratification will be enough, but for some trees and shrubs, an additional period of warm stratification helps to encourage germination. This seems to be especially true of certain species that have immature embryos, which need to develop further, and hard seed coats, which must be softened before the seeds can start to germinate. In addition, with some tree and shrub species, including viburnum, eastern hop hornbeams, American hornbeams, and white oaks, the seeds first form roots the year they fall from the tree, but stems and leaves don't emerge until the following spring. To speed up this process and get complete seedlings the same year, nurserymen sometimes use warm stratification until roots form, followed by cold stratification to finish the germination process.

For warm stratification, simply plant seeds while weather is still warm, especially in mid- to late summer when they'll soon be subjected to a natural winter chill.

For more control of the temperature, you can plant seeds in flats of damp peat, sand, or another seed-starting medium and keep them where the temperature range is between 55°F and 75°F (13°C and 24°C). Most species that need warm stratification should probably be kept warm for at least two months; then treat them with cold stratification for the recommended period of time without allowing them to dry out. Follow the recommendations for individual woody plants given in the encyclopedia section at the back of the book.

Planting the Seeds

You can plant tree and shrub seeds quite casually, simply scuffing them into the earth as the squirrels do, and you'll often get results. Once, when we were hiking and picking wild blueberries in the rich wilderness of New Jersey's Pine Barrens, we met a forest ranger who also loved blueberries. We still delight in recalling the sight of this burly man driving the back roads of the barrens in his open truck with a big lug of blueberries at his side, alternately eating handfuls and tossing out berries by the roadside in the hope that some would sprout into more of those wonderful highbush blueberries. We hope he's still at it; the bushes he was "planting" that day would be bearing fruit by now.

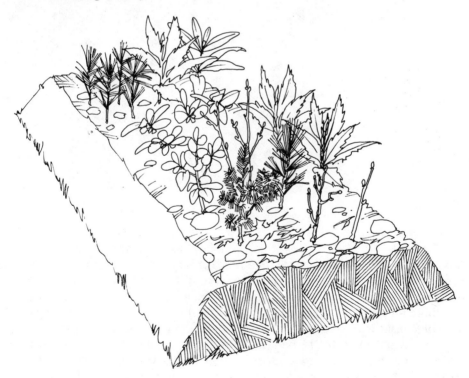

A seedbed for trees.

A magnolia seedpod.

Seedbeds. On our place, we have redbud, black walnut, locust, oak, maple, Russian olive, elderberry, peach, apricot, and other woody plant volunteers popping up regularly. Sometimes we leave them in place and sometimes we transplant them. For more dependable results, though, you'll want to create a special seedbed and possibly a nursery row at the edge of your garden. As long as you're building a tree seed plot, make it a raised bed so the soil will be loose enough to enclose good root growth and to drain well. The bed can be any convenient length, but limit its width to about three feet so you can reach in from both sides without stepping on the soil. If you possibly can, dig some leaf mold or compost into the bed. If you start the season early, you can mix dry leaves thoroughly with the soil and let the bed mellow for several months.

Leaf mold in the soil is especially valuable to trees, which often depend on specific fungi in the soil to pull from the soil the substances they need in order to grow. This seems to be especially true for conifers (pines, spruces, hemlocks, and other cone bearers), which often grow best in soil where other members of their species have been growing. The term "mycorrhiza" is used to describe this mutually helpful association between a fungus and the roots of a plant. Each partner receives nourishment from the other. A single plant may host many different beneficial fungi, many of which associate only with that particular species. For example, if you wanted to grow a stand of hemlock, some investigators think that it's worthwhile to incorporate into your seedbed some soil or duff from a site where hemlocks were growing. Others maintain that since fungus spores fall on the ground continuously, the important thing is to provide plenty of organic matter of any kind to encourage the helpful fungi. If you have access to pine needles for pines and oak leaf duff for oaks, that's great, but don't worry if you don't. Just use whatever other organic matter you have handy.

Sowing the Seeds. As with garden seeds, plant tree and shrub seeds at a depth of about three times the size of a seed. You might need to protect planted beds of nuts or acorns from squirrels by spreading hardware cloth over them. When planting nuts, some determined gardeners bury an open can, with its top cut open but not removed, over each nut. The open top permits the sprout to grow but the encasing metal foils digging squirrels.

When starting tree seeds—holly, dogwood, or whatever—most gardeners plant the whole berry, but there are good reasons for separating the seeds from their fleshy covering. In addition to the germination inhibitors the pulp sometimes contains, it may also cause mold in a wet season. Also, mice or other animals might unearth it for food. Generally, seeds contained in a moist fruit should not be allowed to dry out before planting. Drying can send them into deep dormancy. Purchased seeds of woody plants will necessarily be dry, and will often require an extra season in the ground before they will germinate.

Care of Seeds and Seedlings

Like any other garden bed, the tree seedbed will need to be kept moist and free of weeds for the year or two or even three that might be required for the seeds to sprout. Light shade, either near trees or under a lath roof, a section of snow fencing, or an old slatted crate, will provide the kind of conditions under which the seeds would naturally grow after falling from their parent plant. Most experienced tree seed planters recommend mulching the planted seedbed over winter with sawdust, pine needles, straw, or ground corncobs.

When the seedlings sprout, they'll need to be watered if rain is sparse, and weeded and thinned to give them room to grow. Winter mulch is a good idea, too, for the first few seasons. Many tree seedlings appreciate light shade over the first summer. Wind can damage them by tearing roots and shredding leaves. If there is no natural windbreak nearby, pile evergreen boughs on the windward side or improvise a screen from burlap or some other material. If they're growing in good soil, young

Linden fruit.

tree seedlings won't need much fertilization. In fact, too much high-nitrogen fertilizer can discourage the helpful fungi that are working so hard at the root level to nourish the tree. Watch for frosts, especially during the first few springs, when new growth is tender. In nature, seedling trees are protected by the tree canopy arching over them where they have sprung. Keep some evergreen branches handy for quick frost protection.

Most seedling trees and shrubs can be potted or moved to a permanent location in their second or third year. Those that develop tap roots, like hickory or other nuts, should be moved as early as possible so they can sink these large single roots in their permanent location.

For details on growing specific species of trees and shrubs, see the encyclopedia section at the back of this book.

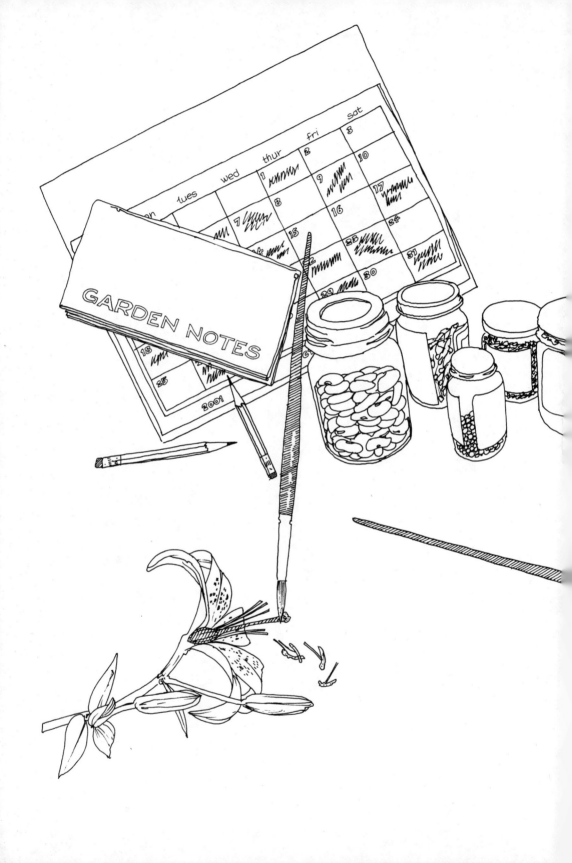

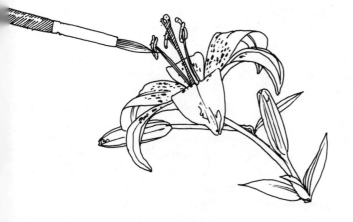

Section Four

Saving Seeds and Making Further Plans

What we call poor is someone who has no animals and no garden.

Patrick Kagoda of Uganda, in an interview with Kathy Duncan of the
New York Daily Record

27 Why Save Seeds?

The gardener who saves seeds from this year's crop to plant in next year's garden has, in addition to the assurance that he or she is prepared for the future, the prospect of experiment, discovery, even surprise. Selection of seed-bearing plants along with deliberate crossing by nurserymen and dedicated amateurs has made possible most of the improvements that have been developed in our favorite garden vegetable and flower plants.

The Benefits

Several benefits can be derived from seed saving that make the whole process worthwhile.

Improvement. Improvement of the yield, earliness, disease resistance, or another quality or constellation of qualities in a certain strain of vegetables can often be accomplished simply by selection—saving seeds from the best plants. Over a period of years, consistent selection of seeds from outstanding plants can give you an improved strain that has better qualities than the original seeds.

Economy. Reducing cost is important to most of us. And, although commercial seeds are modestly priced in relation to their potential, home-saved seeds cost even less—usually nothing, unless you purchase stakes or storage supplies.

Adjustment of Plants to Local Conditions. After a few years of selecting the most frost-resistant, early-germinating, or drought-proof seeds, the plants your seeds produce will be more suited to the environment you grow them in. Even without deliberate selection, the simple act of saving seeds from thriving plants will often condition the plant strain to the peculiarities of the place where it is regularly replanted.

Untreated Seeds. Although seed companies increasingly offer untreated seeds in response to concern among gardeners for the health of their soil, you can be certain yours haven't been treated when you save your own.

Heirloom Strains. Heirloom strains of garden vegetables that are not commercially available can be perpetuated *only* by means of home-saved seeds. "The greatest service which can be rendered any country," wrote Thomas Jefferson, "is to add a useful plant to its culture." It is equally true that the dying out of an irreplaceable variety is a loss not only to the gardener who depended on it, but to other gardeners and eaters, both present and future. Several good vegetables now offered by seed companies—Royalty snap bean and Clemson Spineless okra, for example—

182

were developed from heirloom strains that had been repeatedly grown and saved by isolated families.

Keeping these old strains alive, then, by replanting and saving the seeds is a responsible action that should be greatly appreciated by future generations. It isn't necessary to plant most kinds of seeds every single year in order to keep the strain alive. Renewal planting every second or third year will keep the seeds of most vegetables, flowers, and herbs alive and well, provided they are stored well.

Open-pollinated Varieties. Preference for open-pollinated varieties of vegetables is increasing, with many gardeners choosing to grow corn and tomato varieties that have not been subject to the rigorous inbreeding program that is part of the hybridizing process. Open-pollinated corn, for example, has been shown frequently to contain higher amounts of protein than hybrid corn. In addition, its genetic heritage is more diverse, so that it is less likely to be totally wiped out by disease that would be more serious in strictly controlled, uniform seed crops. Hybrids bear well and many of us will continue to plant them, but there is new respect now for the genetic diversity of open-pollinated seeds. Hybrid plants tend to be highly uniform in both physical characteristics and bearing times. Many gardeners prefer the longer, more gradual maturing periods found in open-pollinated plants.

Experimentation. Saving seeds allows for experimentation, with selection and even with deliberate crossing of certain varieties, that constitutes a challenge for the

Saving seeds from vegetables that your family has grown for years perpetuates old strains that are not available commercially.

experienced gardener. Planting and evaluating the results of each year's careful choices of seeds can be an ongoing adventure. There is even the possibility—remote, but nonetheless real—of discovering a worthwhile new strain.

Although most new vegetable introductions within the last 28 years have been produced by purposeful crossing, simple discovery accounts for the appearance of many garden favorites. The Henderson bush lima, for example, was spotted along the roadside in the 1870s by an elderly man who was taking a walk. Golden Bantam corn, still a favorite, was selected and developed by two Massachusetts gardeners. (Two quarts of the seeds were sold to the Burpee seed company sometime in the 1890s for $25.00.)

Mutation, the sudden altering of plant character caused by a change in the molecular structure of its genes, usually produces undesirable changes in the plant. Beneficial mutations affect about one plant in a million within each generation. The fact that mutations are not reversible makes it possible to breed new generations of like plants from the one that has suddenly changed. Burpee's Fluffy Ruffles sweet pea, introduced in 1928, was a mutation.

The gardener who discovers something new is the one who spends time with plants, observing and recording, with a sensitivity attuned to subtle shades of difference. Granted, the odds against a dramatic plant discovery are high, but the search is a lot of fun.

Self-Reliance. A supply of well-chosen, correctly stored garden seeds will feed you next year, no matter what happens to prices of food, fuel, or postage. Saving seeds will also yield satisfaction on which a price tag can't be put—pride in being able to provide for yourself, and possibly for a few other people as well, and satisfaction in refining and upgrading the crops you grow.

28 How Seeds Are Formed

Your seed-saving efforts will be more successful if you have some idea of how the seeds are formed.

Reproduction in Plants

The process is a cycle, and so we can't say, really, whether it begins with the flower or with the original seed that produced the plant. Since we must start somewhere, let's break into the cycle at the point of flowering.

The Flower. A flower's reason for being is to produce seeds. All kinds of flowers can be found in nature—the large, showy yellow trumpets of the squash family; the small, white stars of peppers and tomatoes; the bright yellow florets of broccoli; the inconspicuous petalless blossoms of spinach. The variations in color and form have evolved as ways of encouraging pollination. Insect-pollinated flowers need to attract attention, whether by producing nectar, smelling sweet, or displaying an inviting color. Wind-pollinated flowers, since there is no need for them to invite insect visitors, are often drab, tiny, or otherwise obscure.

Structure. The pollen-bearing, fertilizing part of the plant is the *stamen.* It is composed of a long, thin stalk called the *filament* and the pollen-containing sacs on the ends of the filaments, called *anthers.*

The receptive, seed-nurturing parts of the flower are called the *carpels.* These include the pollen-receiving region, or *stigma;* the *style,* a long, thin tube that leads from the stigma to the ovary; and the *ovary,* an enclosure containing an *ovule* (egg) or ovules. In flowers containing a cluster of more than one carpel, the whole assembly is called a *pistil* (after the Latin for pestle). The term pistil is also used for a single carpel. The *calyx,* the cuplike outgrowth of the stem, which may be composed of individual parts called *sepals,* and the *corolla,* or petal cluster, are supportive parts without any reproductive function. A flower can be fertilized in the absence of the corolla; in fact, many people remove the corolla when practicing hand-pollination.

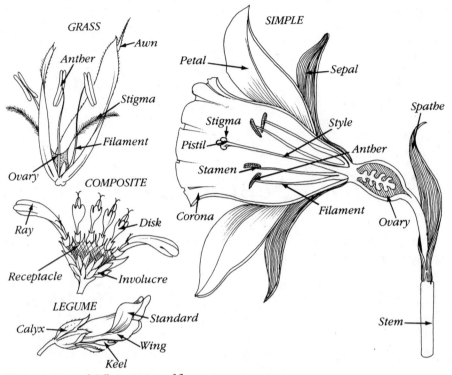

The structures of different types of flowers.

Incomplete Flowers. Most vegetable plants have complete flowers, containing both stamens and pistils as described earlier, but some plants have incomplete flowers, containing either stamens or pistils but not both. A whole group of vegetables, the cucurbits, which include melons, cucumbers, pumpkins, and squashes, bear two kinds of incomplete flowers on each plant. The staminate, or male, flower appears first, followed in a few days by the pistillate, or female, flower. Spinach plants can occur in a bewildering number of gender variations. Some produce both male and female flowers, others only female; still others are either pollen producers or seed-nurturers. Male and female flowers of asparagus are produced on separate plants. Female asparagus plants tend to produce thicker but fewer stalks. Males are thin but more numerous. Berries are, of course, borne only by the female plants. That's why not every frond in your asparagus patch will go to seed.

Fertilization. If the flower is to form seeds, the ripe ovule must be fertilized by a grain of pollen. The speck of pollen, whether it lands on the stigma by the agency of wind, insects, or gravity, has the same fabulous journey to make. It begins by growing an extension of itself, a long, tenuous thread of tissue, which is somehow stimulated by the arrival of the pollen in the right place at the right time. This living thread extends, cell by cell, growing down through the style until it reaches the ovary. There it enters an ovule, penetrating the embryo sac.

The two cells, that of the pollen and that of the ovule, unite to form a single living cell, a zygote, which is able to divide repeatedly and multiply into the complex organism that will be the seed embryo. The ovary grows and toughens into the protective seed coat.

Endosperm. The origin of the endosperm, the supply of stored nourishment for the encapsulated plant, is a bit more roundabout. We have a pretty good idea of what happens, but we don't know why. Mysteries still abound in the plant world. Backtrack for a moment to that grain of pollen. Sometime between the time it is shed from the anther and received by the stigma, the original single grain splits into two cells. One of these cells, as we have just seen, fertilizes the ovule. The other cell unites with two polar nuclei in the ovule's embryo sac and gradually grows into the endosperm. (Some seeds, those of the bean, pea, pumpkin, and watermelon, for example, have no endosperm.)

When the developing seed has reached its characteristic size and complexity, it stops growing and begins the drying, ripening process that precedes dormancy. The

The male flower (left) and female flowers (right two) of a cucurbit.

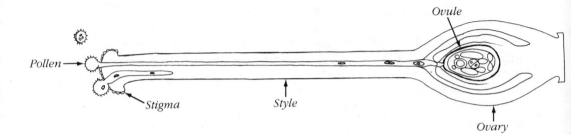

During fertilization, the pollen grain forms an extension, which travels down the style to the ovule.

ripened seed-containing ovary is called a fruit. A bean pod, for example, is a fruit. The seeds it contains are ripened fertilized ovules.

Pollination

In order to fertilize the ovule, pollen must be of the correct kind, and it must arrive at the right time, when the plant has achieved enough vegetative growth to enable it to begin to reproduce. Pollen from unrelated species will not "take." Pea pollen landing on a tomato blossom, for example, won't get anywhere. Even within a species, some plants are not receptive to their own pollen. Many vegetables of the cabbage family cross-pollinate readily but are incompatible with themselves, so you need at least two and preferably four or more flowering plants to produce good seeds. In such cases, a flower will sometimes discharge pollen before its stigma is ready to receive it. Other flowers are sterile to their own pollen.

Self-pollination. Self-pollinating flowers, those in which the flower accepts its own pollen, with or without insect intervention, can be depended on to produce seeds that will grow into plants like the parent since their inheritance is the same. Self-pollinating crops include the following vegetables and commonly grown grains. (Technically speaking, 0.1 to 5 percent of these self-pollinated crops can cross-pollinate.)

barley	endive (escarole)	soybeans
beans, lima	lettuce	tomatoes
beans, snap	oats	wheat
cowpeas	peas	

These seeds, then, are the best to choose for your first seed-saving endeavors, since you can be fairly confident that the plants grown from them will "come true." Most other favorite garden vegetables will cross-pollinate with other varieties of the same group to a greater or lesser extent, depending on variety, weather, insect activity, and other conditions.

Cross-Pollination

Cross-pollination—the acceptance by a flower of pollen from a plant of the same species that has a different genetic makeup—often results in seeds containing genes that differ from those of the parent plant. Plants grown from those seeds, then, may have different characteristics from the parent plant.

Wind Pollination. The pollen of wind-pollinated plants is fine, virtually dust-like, and the number of grains produced is lavish, since the wind can sometimes be more unpredictable than insect activity when distances are great. The following plants are wind-pollinated:

beets	rye	Swiss chard
corn	spinach	

The pollen produced by these plants is so fine and so light that any plantings spaced more closely than one mile apart have a chance of crossing, with the exception of corn, which has relatively heavy pollen and crosses rarely (if at all) at distances greater than 1,000 feet.

Wind Pollination and Plant Design

Wind pollination is a very uncertain occurrence; botanists estimate that 1 grain out of 1,000 might reach a receptive ovule. Recent studies have shown, though, that wind pollination is not as wildly random as it might seem. When scientists tested pine cones, millet flowers, and ovule-bearing parts of 18 other plant species, they found that most of them created air turbulence when struck by wind. This pattern of turbulence increases the chance of contact between airborne pollen grains and the ovules in the cone or flower. Even the pine needles surrounding the cone exert a "snow fence" effect, slowing the wind speed and thus encouraging more pollen to land on the cone.

According to a recent report in *Scientific American,* researchers concluded that many plants are aerodynamically designed to capture pollen from the air. Even more amazing is the fact that the pollen itself is, at least in the examples considered, so shaped that it responds well to the air turbulence created by the shape of the inflorescence of its particular species. Further proof of this theory is offered by studies of fossil ovules, which show that ovules developed increasingly aerodynamically efficient shapes as they evolved over the millenia. One wonders how many other purposeful designs exist in the natural world that so far have escaped our notice.

Insect Pollination. The large insect-pollinated group includes the following vegetables (those marked with an asterisk are biennials):

asparagus	collards*	parsley*
broccoli	cucumbers	parsnips*
brussels sprouts*	eggplant	peppers
cabbage*	gherkins	pumpkins
cabbage, Chinese*	kale*	radishes
carrots*	kohlrabi*	rutabagas*
celeriac*	melons	squash
celery*	onions*	turnips*

Although flowers of these vegetables *may* cross-pollinate within one mile, a quarter-mile separation is generally sufficient to prevent cross-pollination for home garden purposes. Cross-pollination is especially likely if several stands of different varieties of the same vegetable are located on the direct flight line of a bee colony. Bees generally collect pollen from one species at a time.

Keeping Strains Pure. To keep strains of cross-pollinating vegetables pure, you will need to isolate flowering seed crops from any other flowering plants of the same species. You can do this by planting different varieties at different times. For example, if you want to save seeds from two varieties of carrots, plant one early and another late so that they will flower at different times, or isolate by distance as described earlier. (Remember, though, that carrots, radishes, sunflowers, and some other vegetable and flower plants will cross readily with nearby wild relatives.)

If you can't control planting time or distance due to a short growing season or lack of space, you can cage seed-bearing plants. When using wire or cheesecloth

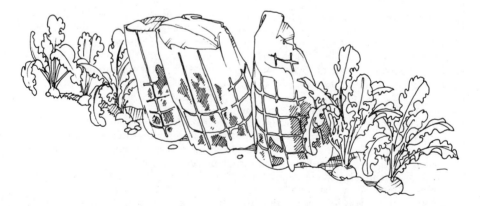

To keep insects away from a plant and prevent cross-pollination, you may have to cover the plant with wire mesh or cheesecloth.

cages to keep out cross-pollinating insects, or muslin to shut out fine wind-blown pollen, it's important to put the cage over the plants well before the blossoms are ready to open, and not to remove it until you see that the plant has produced seeds. For caged insect-pollinated plants, you might need to put some flies in the cage to ensure pollination.

Crossing, however, can occur only within a species. When you know this, you can dismiss the old garden myth that you shouldn't grow cucumbers next to canta-loupes. It's amazing how many people still take this seriously. You'll even hear tales about the cucamelons or watercucumbers that someone grew, but they're just that—tales. If you're not saving seeds, you can grow any of the cucurbits next to each other. Any close-relative crossing that *did* occur would show up only in the next generation planted from seeds of this year's plants. I hope this is good news to the man who remarked to us last year that his wife didn't want him to plant zucchini because he had already planted watermelon, and she didn't want to waste garden space on weird crosses.

Seed-Saver's Caveats

Generally speaking, squash, pumpkins, watermelons, cantaloupes, and cucum-bers will cross-pollinate within their own species. A pumpkin blossom won't accept watermelon, cantaloupe, or cucumber pollen; a cantaloupe can't be pollinated by a cucumber, squash, pumpkin, or watermelon; and a cucumber won't mix with a zucchini, pumpkin, or watermelon. However, some kinds of pumpkin and squash *do* cross, with odd and sometimes picturesque results. The vegetables produced by such crosses are edible, but they may not always be as tender or delicious as the originals.

Here's a quick look at the cucurbit family tree to give you an idea of what to expect in your garden. The cucurbit genus is divided into various species. The species *Cucurbita pepo,* for example, includes pumpkins, all summer squashes (zucchini, scallop, crookneck, and so forth), acorn squash, spaghetti squash, and small gourds. All of these widely varied vegetables can cross, so you can see that there's plenty of potential for mischief within this group alone.

The species *Cucurbita maxima* includes the long-keeping winter squashes like Hubbard, Buttercup, Turban, Banana, and Delicious. These will cross with the species *C. moschata* (Butternut, Cheese, and Melon squashes), but they are less likely to cross with *C. mixta,* which includes the Cushaw and Japanese Pie pumpkin, or with *C. pepo*—the large and diverse group mentioned above. Neither hard-shelled bottle gourds (*Lagenaria siceraria*) nor luffa sponges (*Luffa cylindrica*) will cross with cucurbits or with each other.

Corn, a member of the grass family, presents another special case. A cross between adjacent stands of, say, popcorn and sweet corn, field corn and sweet corn, or yellow corn and white corn, may affect the ears of *this* year's crop depending on wind direction, tasseling time, and other variables. The result of such a cross is not always evident upon eating the corn, because enough sweet kernels usually develop

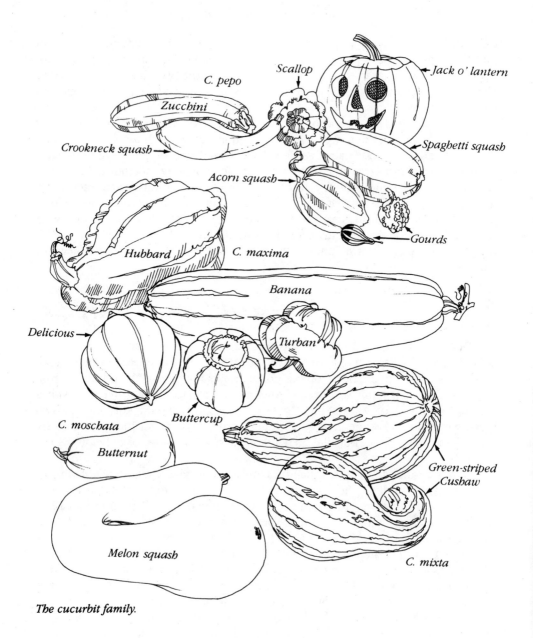

C. pepo

Scallop

Jack o' lantern

Zucchini

Crookneck squash →

Spaghetti squash

Acorn squash →

Hubbard

C. maxima

Gourds

Banana

Delicious →

Turban

Buttercup

C. moschata

Butternut

Green-striped
Cushaw

Melon squash

C. mixta

The cucurbit family.

to cover any odd-pollinated, less flavorful ones. When the kernels are dried, differences—if any—between kernels will be more pronounced.

If you intend to save seeds of more than one variety of a certain vegetable, especially one that is easily cross-pollinated, you'll probably find it more practical to plan for seed development in alternate years to keep the strains as pure as possible. Since most seeds are good for two years after harvest, anyway, this should simplify garden planning, especially for biennials.

29 Choosing Seeds to Save

When you save garden seeds, you can improve as well as perpetuate your vegetable, herb, or flower crops. Simply by practicing careful selection, without any deliberate cross-breeding or hybridizing, you can develop a strain that is suited to your climate and that possesses some of the characteristics you find most desirable. Selection won't introduce dramatic new improvements, but over the years the process can gradually intensify good qualities.

The Bases for Selection

Your aims in selecting plants from which to save seeds will depend on the strain with which you're starting, the peculiarities of your local weather conditions, and your personal tastes. Here are some of the qualities you might want to consider in choosing your parent plants:

- color
- disease resistance
- early bearing
- flavor in vegetables and herbs
- ability to germinate and thrive in cold weather
- insect resistance
- lateness to go to seed (in leafy crops)
- resistance to drought, wind, smog, dampness, or other stressful atmospheric conditions
- size
- special qualities—absence of thorns, spines, strings, and so forth
- storage life in vegetables
- suitability for purpose, for example, paste tomatoes should be meaty, kraut cabbage should be solid, flowers for drying should have pleasing colors
- fruit texture—tenderness, juiciness, seediness
- plant vigor
- yield

Flower characteristics for which you might want to select include:

- color
- disease resistance
- early bloom
- fragrance
- height

Acquired characteristics like fanciful shapes or dwarfing due to drought cannot be transmitted to the next generation. Disease itself is not an inherited trait, but resistance or vulnerability to plant ailments may be inherited. Seed-borne diseases, although they affect the next generation, are not genetically based.

Annuals, Perennials, Biennials

The choice of which seeds you save may depend in part on whether the plant is an annual, perennial, or biennial.

Annuals. The easiest seeds to save, and those you should probably begin with if the process is new to you, are those of the annuals: vegetables such as tomatoes, lettuce, corn, and eggplant; herbs such as dill and borage; and flowers such as portulaca and morning-glories. All of these produce seed the same season they're planted.

Perennials. Perennials are easy to save seeds from, too, since the seeds appear each year once the plant has reached the proper stage of maturity. Vegetable perennials don't always set seed, though. Those that do are listed in the box that follows. The others, Jerusalem artichoke, comfrey, and horseradish, must be propagated by root cuttings or plant division. The Burpee seed company once offered $1,000 for viable horseradish seeds, but had no takers. Comfrey, although it flowers, hardly ever produces viable seeds.

Biennials. Biennials are another story. Vegetables and herbs in this group, which includes most of the root vegetables and the cabbage family, produce an edible crop the year they're planted, but they do not flower and develop seeds until the second year. Biennial flowers, like the English daisy, flower in their second year. In most temperate-zone gardens, biennial plants intended for seed production must be protected over the winter either with a heavy mulch or, if winter cold is severe and persistent, by digging them up, storing them in sand or sawdust, and replanting them in spring. When replanted, they should be buried deeply, right up to the crown of the plant. Biennial vegetables vary greatly in their seeding behavior, and some, like celery and Chinese cabbage, may bolt to seed the first year if chilled early. Don't save seeds from such early bolters, or you'll end up with a strain of vegetables prone to bolting.

The flower stalk that will grow tall and go to seed in the plant's second spring begins to form in the winter. Biennial vegetables that are at marketing size in the fall,

Annual, Perennial, and Biennial Vegetables that Form Seed

Annuals	*Perennials*	*Biennials*
Beans, Lima	Artichokes, Globe	Beets
Beans, Snap	Asparagus	Brussels Sprouts
Broccoli (if planted	Chives	Cabbage
early)	Rhubarb	Cabbage, Chinese
Cucumbers	Sorrel	Caraway
Dill	Watercress	Carrots
Eggplant		Cauliflower
Endive (Escarole)		Celeriac
Fennel		Celery
Gherkins		Collards
Lettuce		Kale
Melons		Kohlrabi
Okra		Mustard
Onions		Parsley
Peas		Parsnips
Peppers		Radishes, Winter
Pumpkins		Rutabagas
Radishes (if planted		Salsify
early)		Turnips
Soybeans		
Spinach		
Squash		
Tomatoes		

neither immature nor old and woody, will be the best seed producers. A month or two of cold temperatures, no higher than 40°F to 50°F (4°C to 10°C), is necessary to promote seed-stalk formation. Hot weather during the short days of winter can interfere with proper development of the flower in beets and cabbage. When you save seeds of biennial vegetables that have kept well over winter, you are selecting for long storage life, which is a valuable quality in these winter keepers.

Hybrids

Should you save seeds from plants that you've grown from hybrid seeds? Probably not—only if you have plenty of time and space to experiment. I certainly wouldn't rely on such seeds for my main planting. The problem is that F1 (first generation) hybrid seeds, the kind usually sold, are the product of the crossing of two genetically different parent plants, both of which have been severely inbred in order to concentrate the desired characteristics. The first generation of a good cross is noticeably, sometimes outstandingly, superior to the parents—a phenomenon known as hybrid vigor. Succeeding generations grown from seeds saved from the F1 plants will tend to revert to the highly inbred, lackluster, recent ancestors. You can see, then, that you have more to lose than to gain from saving seeds of garden hybrids. Still, there's no harm in fooling around with second-generation crosses as long as you don't expect much. Some hybrid seeds are sterile. Do I save seeds from hybrids, you ask? No.

If you intend to save seeds, you'd be wise to either note hybrids at planting or refer back to the seed catalog to check on varieties you've planted to be sure that you save seed only from open-pollinated varieties. Marglobe and Rutgers tomatoes, for example, are open-pollinated; Spring Giant and Big Boy are hybrids. Bell Boy peppers are hybrids; Vinedale and Staddon's Select are not.

Hybrid seeds are commonly available for the following vegetables:

beets	cauliflower	pumpkins
broccoli	corn	sorghum
brussels sprouts	cucumbers	spinach
cabbage	eggplant	squash
cabbage, Chinese	kohlrabi	tomatoes
cantaloupes	onions	watermelons
carrots	peppers	

Seed-grown flowers that have been hybridized include:

ageratum	geraniums	pansies
begonias, fibrous	hibiscus	petunias
chrysanthemums	impatiens	primroses
coleus	marigolds	snapdragons
columbines	nicotiana	zinnias
gazania		

Judging the Parent Plants

Even though you may not be collecting the seeds until toward the end of the season, you'll want to observe the plants from the very beginning so that you have an idea of their overall performance. Although your first impulse might be to save seeds from the largest single specimen of fruit that you can find, the best approach is to consider the plant as a whole. You might not want to intensify the characteristics of a giant tomato from a disease-prone or low-yielding plant. It is the entire plant—its vigor, resistance to disease and insect predation, yield, and other qualities—that gives you the best clue to the nature of the plants that will grow from its seeds. Biennials, which don't produce seeds until the second season, should be selected for eating quality when stored in the fall and then reselected for keeping quality in the spring after winter storage.

When saving seeds from any plant that cross-pollinates, it's wise to collect from more than one plant. Otherwise you may find yourself selecting increasingly inbred stock. Without sufficient genetic variety, the strain will lose vigor. This is especially true of corn. Pumpkins and squash, though, seem to maintain vigor even when inbred, so it's all right to save seeds from a single fruit of these crops.

Self-pollinated vegetables like beans, peas, and tomatoes get along quite well accepting their own pollen. Since they are naturally somewhat inbred, you needn't worry about preserving genetic variety. Go ahead and save seeds from exceptional individuals; they needn't be mixed with others to preserve their vigor.

Saving seeds from superior vegetable plants sometimes involves a bit of sacrifice, particularly if you're selecting for earliness, and your family is counting the days until that first round of sweet corn or salad tomatoes. It's not always easy to let the plants be. Unfortunately, it's not often possible to eat the vegetable and retain the seeds, because in order to produce viable seeds the fruit must mature on the plant. Pumpkins and winter squash may be eaten at the same time their seeds are collected, but most other vegetables must be left on the plant. Naturally, if you're selecting for some other quality or qualities such as yield, flavor, or size, you can indulge your hunger for the first fruits and wait to observe the plant's mid-season performance before setting aside seed-bearers.

Even if you're not absentminded, you'll want to mark your chosen parent plants, especially if other family members do some of the picking. If that super, all-time champion cucumber gets put in the pickle crock, you might have to wait until next year to get another that is equally outstanding. Use a stake, chicken wire, a ribbon or string, a tag—anything that will stay in place and call attention to the plant's special purpose.

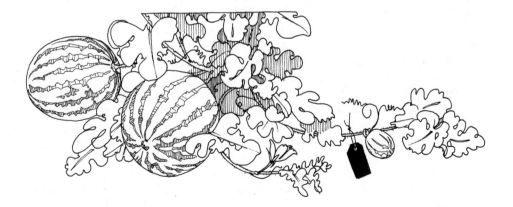

Tag plants from which you want to save seed so that you will remember not to pick and eat their fruit.

Biennials usually form tall seed stalks in their second year.

Care of the Seed-Bearing Plant

Having chosen and marked the plants from which you plan to save seeds, you'll want to take good care of them throughout the growing season. Mulch, water, fertilize, and control insects as necessary.

Stake, mulch, or otherwise protect the fruit from rot-inducing ground moisture and bird or animal predation so that seeds can mature properly. During their second year, biennials develop a long seed stalk, which sometimes needs to be staked so it isn't broken by strong spring winds.

Conditions during seed development have considerable effect on the vigor of the matured seeds. Some influential factors will be beyond your control. For example, the prevailing temperatures 20 to 40 days before maturity have a definite effect on the oil content of soybeans. A sharp rise in temperature affects the protein content and yield of developing wheat seeds. Excess moisture can delay seed development.

You can look at this nurturing process in two ways. You certainly want the plants to be healthy. On the other hand, some gardeners feel, and I agree, that pampering the plant to the point of giving extra care that you wouldn't have time to give to your whole garden is self-defeating. Your aim, after all, is to raise, over a period of several years, a race of cucumbers or whatever that do well under ordinary garden conditions. So give the plant everything it needs, but don't hover over it with second helpings.

30 Collecting and Preparing Seeds

Once you have decided which kinds of seed you want to save, you will need to gather them and prepare them for winter storage.

Deciding When to Collect Seeds

As you remember, seeds depend on stored nourishment to carry them through their winter dormancy so they can live as green plants next spring. Seeds that are harvested too early, even though they may look the same as other seeds, may be deficient in either endosperm or embryonic development or immature in some other way. Such seeds are likely to deteriorate in storage. If they do survive, they may germinate unevenly or produce inferior seedlings. I proved this to my own satisfaction last year by saving seeds from both red (ripe) and green (unripe, though good to eat) cherry peppers. The seeds saved from the green pods had a very low germination rate, although as far as I could tell the few plants that did grow were normal. Germination of seeds saved from the red pods was more rapid and much more complete.

Recent studies have shown that seeds of tomatoes, snap beans, lettuce, spinach, and radishes will germinate satisfactorily even if the seeds are harvested while slightly underripe. Pepper, carrot, celery, and pea seeds, however, germinated poorly when picked before the fruits were fully ripe.

Plants that Shatter. Still, you must be sure to collect the seeds before they rot or shatter. Harvest time is most critical for those plants that release their seeds as soon as they are ripe (shattering). To make things trickier yet, the seed stalks of certain plants that shatter readily also ripen one stalk at a time over a period of weeks—a valuable survival mechanism for the plant, but inconvenient for the gardener. If you're counting on saving seeds from lettuce, onions, okra, or any member of the cabbage family—all of which behave in this way—you can tie small paper bags over the heads of developing seeds to catch them in case you're not able to make daily seed-collecting rounds. Be sure to punch a few holes in the bag to admit air.

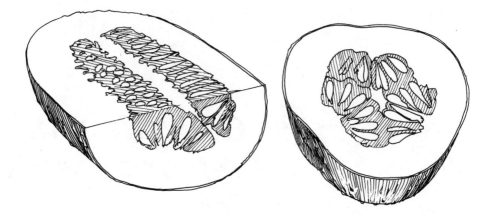

Fleshy fruits should be slightly overripe before picking them for seed saving.

Plants with Fleshy Fruits. The fruit of plants that bear seeds embedded or encased in edible flesh—the cucumbers, eggplant, tomatoes, and peppers—should be slightly overripe before being picked. Let the fruit develop just past the point where you'd want to eat it. Tomatoes should be soft, cucumbers yellow, peppers red and wrinkling. If the fruit actually begins to rot, though, seeds may be damaged by the heat of decomposition. The flesh of the vegetable shouldn't be allowed to dry around the seeds either, or it may form an impermeable covering that will cut down on the gas exchange necessary to the life of the seeds in storage. Because some fungal and bacterial diseases can be transmitted to the next generation by the seeds, be sure to collect seeds from disease-free fruits and plants.

Plants with Edible Seeds. Plants bearing edible seeds, especially corn and wheat, will retain their fully matured seeds for awhile. These grains may be left on the plant for several weeks or until you have a chance to collect them. Snap beans and soybeans also retain their dry seeds for some time, although they do eventually shatter. Seed heads are often cut when the seeds are fully developed, and then piled in a dry, protected place to cure further before the seeds are threshed out.

Collecting Seed from Plants. If possible, seed collecting should be done on a dry, sunny day when the seeds are free of rain or dew. Frost doesn't hurt most seeds, as long as they are dry. The danger in allowing seeds to remain out in freezing weather is that the condensation of moisture that often follows a frost can be damaging to the seeds if another frost follows soon.

Labeling Seeds. It's awfully easy, I find, to get batches of collected seeds mixed up, particularly if you are saving more than one variety of the same vegetable. Putting each batch of seeds into a marked bag, jar, or envelope as you collect it can save a lot of confusion later on.

Cleaning and Sorting the Seeds

You have good, fully ripe seeds collected dry from healthy plants. The way you prepare the seeds for storage can make a difference in their viability.

The simplest seeds to prepare are those that you pluck directly from the seed head of the plant. Such seeds as lettuce, endive, dill, sunflower, and the brassicas, and most flower seeds, need only to be winnowed and then dried. Winnowing—pouring the seeds from one container to another in a stiff breeze or in front of a fan—blows off the lightweight chaff. Screening—passing the seeds through the holes in a piece of mesh—will separate coarser, heavier trash such as small sticks, pebbles, and burrs.

Seeds that are lighter in weight than others of their kind are also often abnormally thin and are likely to contain an imperfect embryo or deficient endosperm. They'll never make it through the winter. Even your hens will probably ignore them. Toss them on the compost pile.

Threshing. Snap beans, soybeans, limas, and peas must be threshed to remove the seeds from the pods. Generations of gardeners have sought easy ways to accomplish this task. Some people simply whack handfuls of the plants, a few at a time, on a hard clean surface. This is an excellent way to spend those awkward aggressive impulses, but it's not too efficient. You'll find beans in the corners for years afterward. I've threshed soybeans by spreading the dry plants on a clean sheet, covering them with another sheet, and treading back and forth over the covered vines. The beans fall out on the bottom sheet. The plants may be lifted off and the chaff winnowed out in the next breeze. You can also swing handfuls of the dried plant stalks so that the seed heads strike a sawhorse set on a sheet or tarp, or swing them

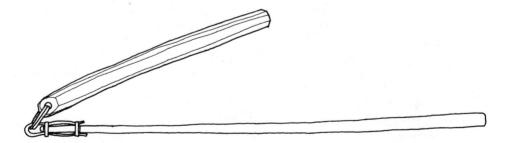

Traditionally, gardeners threshed plants with a flail to remove seeds from their pods.

inside a barrel. This works somewhat better with grains than legumes.

The threshing hand tool of the ages has been the flail, a long wooden handle with a short stick (called the swingle) attached securely to one end with leather. You can use an old mop or broom handle if you have one, and some rope if you can't get pieces of leather. In the traditional peasant household in Poland, Mike recalls, the flail handle would be made from a strong hazelnut shoot, which—unlike many tree saplings—does not taper much from one end to the other. The swingle is always made of hardwood.

To use this tool, you beat the plants with the swingle to release the seeds from the brittle pods. If you remembered to spread a tarp or sheet down first, you'll find it easier to gather the dry seeds. Vigorous as this method is, it injures few seeds. You wouldn't want to use anything heavier or rougher to extract the seeds, though, because violent treatment *can* injure the embryo, with the result that the seedling, if it sprouts at all, may be stunted. Such internal damage is not always evident on the surface of the seeds. Some people find that a plastic bat makes a good threshing tool. Small lots of seeds may, of course, be shelled out by hand.

Still other seeds, those of tomatoes, cucumbers, squash, pumpkins, and melons, must be separated from the flesh that surrounds them. Pumpkin, winter squash, and melon seeds may simply be washed to remove all traces of pulp, then thoroughly air-dried.

Dealing with Seed-Borne Diseases

Tomato seeds will be protected against bacterial canker if you let the seedy pulp ferment for a few days. The fungi that develop during fermentation produce antibiotics that control the offending bacteria. In addition, fermentation dissolves the gelatinous coats surrounding the tomato seeds, which contain germination-inhibiting substances. Just spoon the tomato pulp into a jar, cover it with water, and set it where you'll remember to check on it. In several days you'll notice that the good, heavy seeds have sunk to the bottom, while the pulp and the poor, lightweight seeds have risen to the top. Pour off the matter on top and strain out the good seeds. Some gardeners follow the same procedure with cucumber seeds.

Farmers in England noticed two centuries ago that seed wheat that had been immersed in seawater was free of bunt infection. Observations by Jethro Tull in 1733 led to the fairly routine use of a saltwater soak in areas where this fungus was prevalent.

Other simple ways to control seed-borne diseases include the following:

- Wash seeds in plain water.
- Steep seeds in hot water, 122°F (50°C), for 15 minutes.
- Steam seeds.

These measures are seldom necessary for the home gardener, whose diversified plantings are not likely to build up serious seed diseases, but they are reasonably effective when they are needed. When they are used, however, they must be carefully carried out to avoid overheating the seeds.

To control fungi that may live on stored seeds and lead to seedling losses after planting, you can soak the seeds for 1½ minutes in a solution of Clorox and water (five teaspoons of Clorox in one pint of water). Rinse the seeds after soaking. Do this right before *planting,* not before storing the seeds. To kill weevil eggs that often infest dried beans, freeze the bean seeds for two days *before* storing.

Drying Seeds

Seeds should go into storage as dry as possible. Green seeds or seeds that have accumulated moisture will heat when heaped in a pile. Moisture also speeds up the seed's metabolism, causing it to use up its stored nourishment too quickly.

You won't be able to determine the exact moisture content of the seeds you're storing, of course. The important thing to remember is to give all seeds a thorough postharvest drying period of at least a week, no matter how dry they already look. Seeds of most vegetables may simply be spread on newspapers in a dry, well-ventilated place. Change the papers once or twice if the seeds were damp at first. A seed that is well dried will break when you bend it rather than rebound to its original shape.

If you live in a damp climate and want to be sure that your seeds are drying rather than absorbing moisture from the air, you can subject them to very gentle heat, like that of a light bulb or pilot light. Keep the temperature around 90°F (32°C). Temperatures over 110°F (43°C) will damage the seeds. If seeds dry too rapidly, they are likely to shrink, crack, and develop hard seed coats. The rate of drying is more rapid at first when moisture content is high; it slows down later.

A safer method for drying seeds before you pack them away is to seal them in a jar with silica gel, a special moisture-absorbing product. Put equal weights of the gel, and seeds that have been enclosed in cloth bags or paper envelopes, in a tightly closed jar. In 8 to 12 days small seeds, like those of peppers or lettuce, will dry to a moisture content of 6 to 8 percent. Larger seeds, like those of peas or squash, will take 12 to 16 days to dry. After the drying period, transfer the seeds directly to their storage containers. Don't leave them exposed for any length of time, or they'll reabsorb moisture from the atmosphere.

31 Storing Seeds

The *genetic* vigor of your saved seeds has already been determined by their parentage. Environmental conditions affecting the parent plant during seed formation—temperature, available moisture, weed competition, nutrient supply, and so forth—help to determine the *physiological* vigor of the seeds. Once harvested, the seeds can't be improved upon except in those seeds that afterripen during storage: barley, cucumbers, lettuce, melons, mustard, oats, pumpkins, rice, turnips, watermelons, and wheat. Although improvements can't be made, the vigor of seeds can be drastically reduced by poor storage conditions. In fact, William Crocker and Lela Barton, in their excellent book *The Physiology of Seeds,* go so far as to say that storage conditions are more influential than the age of the seeds in determining viability—the ability of the seeds to germinate.

Keep in mind that the stored seed is alive, with its life processes barely humming. Even in its dormant state, it reacts with its environment. The seed absorbs moisture from the air and carries on the exchange of oxygen and carbon dioxide that is characteristic of life, combining the nutrients in its stored food with the moisture it takes in to make a soluble form of plant food. Oxygen in air taken in from the atmosphere then reacts with this soluble food, with the resulting release of carbon dioxide, water, and heat.

Your aim when storing the seeds, then, should be to keep their metabolism operating at the lowest possible level, to keep the seeds on "hold." To do this, you'll need to control several environmental factors that affect the seeds' life processes.

Temperature

Heat urges the seeds into premature internal activity, which uses up their stored food supply. Seeds stored at home at temperatures between 32°F and 41°F (0°C to 5°C), not freezing but cold enough to retard enzyme activity, usually keep well. Ideally, temperature should fluctuate as little as possible. Under laboratory conditions, seed expert Dr. James Harrington, a professor at the University of California at Davis, has found that between the temperatures of 32°F and 112°F (0°C and 44°C), for every 9°F (5°C) that the storage temperature is lowered, the seeds' period of viability will double. Seeds have even been successfully stored at 0°F (–18°C). If

From Fahrenheit to Celsius and Back

Here's a handy formula for converting Fahrenheit to Celsius (formerly called Centigrade). Subtract 32 from the Fahrenheit temperature, multiply the answer by 5, and divide the product by 9. Example:

$$212°F - 32 = 180; 180 \times 5 = 900; 900 \div 9 = 100°C$$

To change a Celsius reading to Fahrenheit, multiply the Celsius figure by 9, divide the product by 5, then add 32. Example:

$$100°C \times 9 = 900; 900 \div 5 = 180; 180 + 32 = 212°F$$

you do want to try keeping some seeds in your freezer, be sure that they are good and dry. High moisture in frozen seeds will spoil them. Also, be sure the seeds you're storing are ripe. Immature seeds don't freeze well.

Moisture

Moisture revs up seed metabolism. In some cases, it can be even more damaging than heat. Studies at Cornell University, for example, have shown that increasing the moisture content of lettuce seeds by 5 to 10 percent resulted in more rapid loss of viability than an increase in temperature from 68°F to 104°F (20°C to 40°C). Tests on clover seeds yielded similar results.

It is important not only to have the seeds well dried when putting them away, but also to keep them dry. Seeds that have gotten damp and then been redried suffer irreparable damage. It is important not to seal moisture in with seeds stored in a container; if seeds are not thoroughly dry, they will keep better in open storage than when sealed.

Although as a home gardener you can neither measure nor control seed moisture exactly, you might find the following recommended levels useful as a guide. Seeds vary in the permeability of their seed coats, and thus in their ability to take in moisture and oxygen, even under equal temperature and humidity. Legumes, for example, have relatively impermeable seed coats. Peas and beans will last well in storage with a moisture content as high as 13 percent (but no higher), as will corn and most other cereal grains. Soybeans have an upper limit of 12.5 percent moisture, flaxseed 10.5 percent, and peanuts and most other vegetables even less—9 to 10 percent. Seeds to be stored for long periods of time will fare best if moisture is kept around 4 to 6 percent but no lower than 1 to 2 percent, or the embryo may be damaged. In most climates, you'd need to heat-dry your seeds to get the moisture content this low. Dr. Harrington, whose storage temperature studies I have already mentioned, has also determined that each 1 percent reduction in moisture (between 5 and 14 percent) doubles the storage life of the seeds.

If seeds are very dry, they can withstand extremes of both heat and cold that would deteriorate damp seeds. Relatively moist seeds must be kept cool. There is, in fact, a rule of thumb that should help most home gardeners, who can't always control conditions as they might wish: The relative humidity of the atmosphere and the number of degrees (F) of temperature prevailing in the storage area should add up to less than 100. At 60 percent humidity, then, the storage temperature should be no higher than 40°F (4°C), preferably less.

Where humidity tends to be high, seed storage life may be prolonged by packing a dessicant powder with the seeds. Silica gel is the most effective, widely available drying agent. It can be reused repeatedly if it is dried for eight hours at 200°F (93°C) after use. About one part silica gel to ten parts of seeds will be adequate for storage, or put one-fourth to one-half inch of the gel in the bottom of each sealed quart jar full of seed packets. Some seed-savers roll up dried milk in a tissue and seal it in a jar with their seeds. Dried, powdered milk absorbs less moisture than silica gel. Check the jar in midwinter and replace the milk powder if necessary.

Pests and Diseases

While storing your seeds, you'll not only need to control environmental conditions, but you'll need to protect the seeds from pests and diseases.

Bacteria and Fungi. In the process of carrying on normal respiration, fungi and bacteria produce heat, which can damage stored seeds. In addition, some kinds of microorganisms give off substances that can harm the embryos or soften the seed coats, thus making the seeds vulnerable to invasion by other destructive microorganisms. Fortunately, both bacteria and fungi are inhibited by cold, dry conditions. Although individual species of bacteria vary in their temperature preferences, most groups function poorly in moisture levels below 18 percent. Fungi and molds flourish in a moisture content of 13 to 16 percent, but most of them require fairly warm temperatures of 85°F to 95°F (29°C to 35°C). Temperatures below 50°F (10°C) inhibit their growth.

Insect Pests. Insects may impair seed viability by producing heat or by consuming the seeds or portions of them. Most insects need a moisture level higher than 8 percent in order to breed. The majority of insects that are likely to infest stored seeds need temperatures of at least 40°F to 50°F (4°C to 10°C) to maintain activity.

Animal Pests. Rats, mice, raccoons, birds, and other animals can make serious inroads on your seed supply. Prevent this loss by storing seeds in tightly closed glass or metal containers rather than in cardboard boxes or paper envelopes, which are easily broken into.

Containers

Seeds stored in bulk sometimes tend to heat and should be stirred and rotated periodically to keep them in good condition. Small packets of seeds, on the other hand, are more quickly affected by changes in humidity and temperature. Purchased seeds in foil packets shouldn't be tightly resealed in the foil after being open to damp air for a season, if you want to save them for another year.

Stored seeds are safest in securely closed containers such as the following:

- Cans with metal lids.
- Screw-top glass jars or individual envelopes sealed in a large glass jar. Baby-food and peanut butter jars usually have rubber gaskets and seal very well.
- Plastic or metal film containers.
- Vitamin bottles.

Less desirable are wooden boxes or bins or metal cans with plastic lids, either of which may be gnawed by rodents if the contents are inviting. I keep my garden seeds from one season to another in individual envelopes inside a large, tightly lidded lard can in a cold back room.

A Few Reminders

It's easy to forget, during the growing season, that seeds you're holding over for another year may deteriorate if not protected from extremes of heat and humidity during the summer months. Move them to the coolest, driest spot possible, to keep while you're busy hoeing and harvesting outside.

Last but not least, *label your seeds*. Mark each batch of seeds with the variety name and date collected, or you'll have a grab-bag garden next year.

To sum up, store your seeds cool, clean, dry, covered, labeled, and insect-proof. Let the frost sparkle in the grass and the winds blow. You're ready—already—for the next planting time.

Store seeds in well-sealed containers to keep out moisture.

32 Viability of Seeds

Most garden seeds will remain viable (capable of germinating) for several years after harvest, if kept in cool, dry storage. Even the exceptions, onion and parsnip seeds, which ordinarily deteriorate rapidly after ripening, will sometimes last longer than a year under ideal storage conditions.

Table 8, on the facing page, will give you some idea of what to expect of your seeds. Perhaps it will help you to plan your plantings of biennial seed crops so that you can grow a sufficient amount of cabbage seeds, for example, in one season to last for several years. It is, of course, impossible to be absolute about the keeping qualities of seeds because differences in ripening, drying, and storing procedures affect viability. You may find that some seeds will give satisfactory germination for even longer periods than indicated here. That's nice when it happens, but I wouldn't count on it. Even when germination is low, seeds can be used if planted thickly, or, as in the case of lettuce or carrots, which are often overplanted, old seeds may be mixed with new to "dilute" them.

Studies mentioned by William Crocker and Lela Barton in *The Physiology of Seeds* indicate that mutations, caused by the deterioration of embryo cell nuclei, have been found to occur in old seeds. Although it doesn't seem likely that the incidence of mutations would be very high, you might remember this if you discover an off-type plant in a stand of vegetables grown from aged seeds.

Testing Germination

Sometimes you can tell simply by looking at a seed that its quality has deteriorated. One lot of cabbage seeds that I had kept under less-than-ideal conditions for three years looked wrinkled and dented rather than plump and smooth. Sure enough, when I planted them in flats they germinated *very* thinly. Peas, corn, and a good many other seeds are either normally wrinkled and flaky or tiny and hard; they may *look* as good as new when past their prime, but any seed that is ordinarily smooth and round or plump is not likely to germinate well if it has become pocked or wrinkled.

To get a reasonably clear idea of the growth prospects for a particular lot of seeds, you can run a germination test. Count out at least 20 randomly picked seeds. Fifty is better and 100 is the professional way to do it. If you want to be very scientific, mark a paper towel into a grid of squares and place a seed on each square. Since the seeds tend to shift around when they're checked, though, I don't bother to mark the paper, but simply spread the seeds on several layers of premoistened paper towels or paper napkins, roll them up carefully in the paper so they stay separate, tuck the

Table 8

Seed Viability

Approximate age at which seed of good initial viability stored under cool, dry conditions still gives a satisfactory stand of vigorous seedlings with a normal rate of seeding.

Seeds	Years	Seeds	Years
Asparagus	3	Leeks	3
Beans	3	Lettuce	5
Beets	4	Martynia	2
Broccoli	5	Muskmelons	5
Brussels Sprouts	5	Mustard	4
Cabbage	5	Okra	2
Cabbage, Chinese	5	Onions	1
Cardoon	5	Parsley	1
Carrots	3	Parsnips	1
Cauliflower	5	Peas	3
Celeriac	5	Peas, Southern (See	
Celery	5	Cowpeas)	. . .
Chervil	3	Peppers	2
Chicory	5	Pumpkins	4
Ciboule	2	Radishes	5
Collards	5	Roselle	3
Corn	2	Rutabagas	4
Corn-salad	5	Salsify	1
Cowpeas (Southern		Salsify, Black	2
Peas)	3	Sorrel	3
Cress, Garden	5	Spinach	5
Cucumbers	5	Spinach, New	
Dandelions	2	Zealand	5
Eggplant	5	Squash	4
Endive	5	Swiss Chard	4
Fennel	4	Tomatoes	4
Kale	5	Turnips	4
Kale, Sea	1	Watercress	5
Kohlrabi	5	Watermelons	4

Source: Oscar A. Lorenz and Donald N. Maynard, *Knott's Handbook for Vegetable Growers*, 2d ed. (New York: John Wiley and Sons, 1980).

rolled paper (or papers) into a plastic bag, and keep the incubating seeds in a warm place (70°F to 80°F, 21°C to 27°C). Be sure to label each roll. Check the seeds in two or three days, and every day thereafter for a week or so, for evidence of germination. If a root or cotyledon protrudes through the seed coat, the seed has germinated. When some seeds have sprouted, and a one-week wait indicates that no more are about to emerge, you can calculate your rate of germination. Ten seeds out of 20? Fifty percent germination. Five out of 100? After all that counting—5 percent!

Prompt germination (within limits for the particular species, of course, meaning that 18 days would be prompt for parsley but late for corn) indicates vigorous seeds. Allow three weeks at the outside for most seed varieties to germinate.

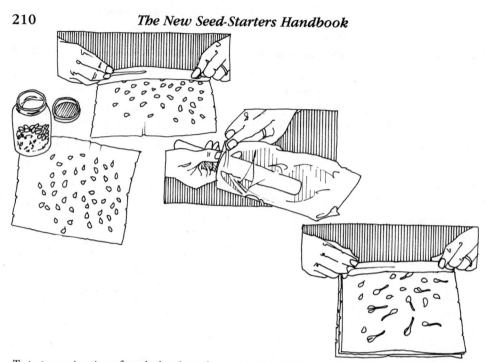

To test germination of seeds that have been stored awhile, spread seeds on moistened paper towels. Carefully roll up the paper, place it in a plastic bag, and leave it in a warm spot. Check the seeds every few days for a week or so to see how many have germinated.

Seed Size

Are large seeds better than small seeds? The answer depends on which scientist or seedsman you ask. There doesn't seem to be conclusive proof that large seeds produce better plants than medium-sized ones. Runts and lightweights, though, are undesirable.

In tests reported by the United States Department of Agriculture in *Seeds, the Yearbook of Agriculture,* small seeds, planted in separate plots but at equal distances, performed as well as plantings of large seeds made at the same population density. In mixed plantings of small and large seeds, though, the plants that grew from large seeds tended to shade out those that grew from small seeds.

After you've been selecting and saving your own garden seeds for a few years, you might like to try what seedsman Rob Johnston calls "quality control." Plant a packet of purchased seeds of the same variety as, preferably side by side with, your home-saved seeds, and evaluate the differences—if any—between the two stands of plants.

No matter how few seeds you have, always save part of your sample for future planting in case the first crop fails. Never plant all your seeds. If lost to frost or drought, they can never be regained.

33 Seed-Saving Tips for Specific Plants

Many of the more popular vegetables are presented here with specific information on how to save their seeds.

Artichokes, Globe. (*Cynara scolymus,* perennial) Commonly considered a fussy exotic vegetable, this delicacy has been successfully grown in several New England states, although it is doubtful whether seeds would ripen there. Adventurous gardeners in warmer climates might like to experiment with saving seeds from these cross-pollinated blossoms (the "choke" *is* the blossom bud). Victor Tiedjens, author of *The Vegetable Encyclopedia,* writes enthusiastically about the possibilities for selection and experimentation with this vegetable. Apparently seed-sown plants exhibit a wide variety of characteristics. The thistlelike artichoke flower will turn to a brushlike head of fluff as seeds form. Either pluck out the seeds as they dry and loosen from the head, or to save more of the seeds before they shatter, bag the whole head until it is thoroughly dry; then cut it from the plant and pull off the seeds.

Asparagus. (*Asparagus officinalis,* perennial) It's easy to save asparagus seeds. In fact, the plants volunteer rather readily from seed. Seeds saved from wild plants, which are usually bird-planted escapees from kitchen gardens, generally produce plants that look and taste just like their cultivated ancestors.

Asparagus flowers cross-pollinate. The seedy red berries are borne only on the female plants, which tend to have thicker but less numerous stalks. The thinner, more plentiful male shoots produce the pollen-contributing blossoms.

To save seeds, cut the ferny plant top when berries are red and when the top starts to bend. Hang it in a well-ventilated place to dry. Soak the berries in water to soften the skin; then wash them and rinse off the pulp. Air-dry the seeds for a week before packaging them for storage.

Although soil and moisture conditions often account for differences in yield, you might want to try selecting seeds from especially productive plants. If possible, choose seeds that are formed early in the season.

Beans, Lima. (*Phaseolus lunatus,* annual) Although limas are self-fertilizing, having both pollen-shedding and pollen-receiving organs in the same flower, they will cross-pollinate when visited by bumblebees. Save seeds in the same way that you would for snap beans, below. After removing seeds from pods, air-dry them indoors for a week or two, but store before mice or insects find them.

Beans, Snap. (*Phaseolus vulgaris,* annual) Beans self-pollinate almost entirely. In fact, pollination usually takes place even before the blossom opens. Occasionally, though, two different closely planted varieties will cross to some extent. If it's impor-

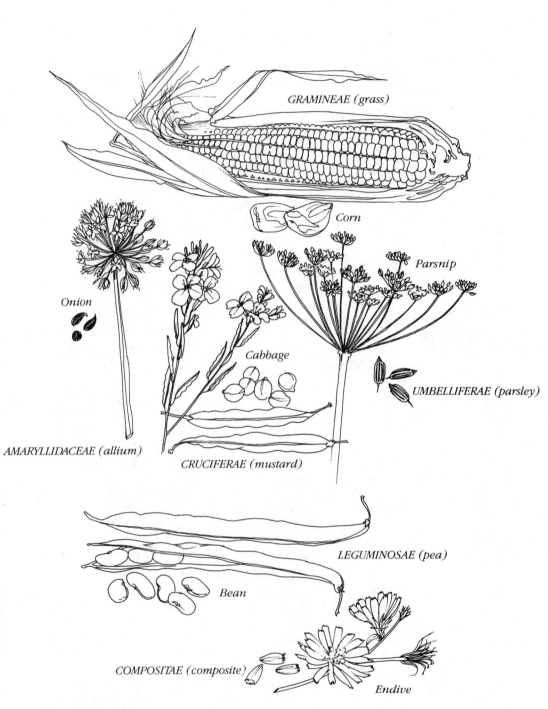

GRAMINEAE (grass)

Corn

Parsnip

Onion

Cabbage

UMBELLIFERAE (parsley)

AMARYLLIDACEAE (allium)

CRUCIFERAE (mustard)

LEGUMINOSAE (pea)

Bean

COMPOSITAE (composite)

Endive

Seed heads of different families of plants.

tant to you to keep the strain unmixed, as with an irreplaceable heirloom strain, plant the different varieties 100 feet apart. Seeds borne by plants that have crossed will sometimes have a different color than the original strain. It's best to separate varieties of beans that have white seeds because crosses might not be evident from seed color. Different varieties of pole beans from which you want to save seeds should also be separated because the vines have a tendency to wander from one pole to another and seeds could be misidentified.

Runner beans, which are a separate species, cross with each other more readily than snap beans do because their blossoms are more accessible to bees. For pure strains of runner beans, you'll need to keep different varieties one-fourth mile apart.

A good supply of zinc in the soil is needed for healthy seed formation. Suspect zinc deficiency if the seeds mature slowly or irregularly.

Leave the pods on the plants until they're dry and brown and the plants are nearly leafless. Seeds are usually mature six weeks or so after the beans were tender and good for eating. To determine whether the crop is dry enough to harvest, try biting a sample bean. You should scarcely be able to make a dent in a bean that has dried properly. Pull the whole plants and stack them loosely in a dry place to cure further. Shell small lots by hand. Thresh out larger lots by beating or flailing, or use a special bean-threshing bag (see Sources for Garden Supplies at the back of this book). The bean diseases bacterial blight and common mosaic are transmitted through seeds, so seeds from diseased plants should not be saved.

Beets. (*Beta vulgaris,* biennial) Beets cross-pollinate freely with Swiss chard and sugar beets. Their fine, airborne pollen can travel for great distances; plantings less than one mile apart that blossom concurrently are likely to cross. Bees sometimes transfer the pollen, too, if the plantings are lined up with their flight path. During the first season of vegetative growth, of course, it's not necessary to isolate different varieties, and you can grow second-season seed beets next to Swiss chard as long as you don't let the chard go to seed.

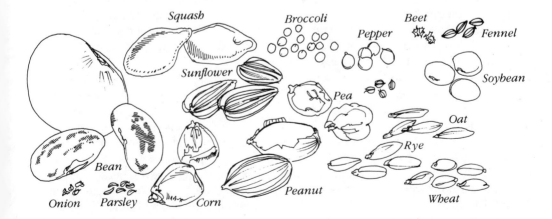

Seeds of various vegetables.

Select roots that aren't too large, and treat them gently to prevent bruises, which could rot. Don't save seed from a beet that goes to seed in its first year. You needn't store a whole row of beets; six or eight can produce several gallons of seeds, if they all live over winter. Before storing the beets, cut off the tops, leaving a 1-inch stub. Chilling (33°F to 45°F, 1°C to 7°C) is necessary for seed-stalk formation, but long periods of below-freezing weather will spoil the beets. Most gardeners who save beet seeds harvest the beets before killing frost, store the roots in sand in a root cellar (or refrigerator, cold basement, or unheated spare room) during the winter, and replant them in spring. Plant at least three; beet flowers sometimes refuse to accept their own pollen and require pollen from another plant to set seed. Where winters are mild, seed beets may be sown in the fall to winter over in the garden, and thinned to 18 inches apart for spring seed-stalk formation. As soon as the seeds at the base of the stalk have ripened, you can pull the plants.

Broccoli. (*Brassica oleracea,* annual) If you get an early start with your broccoli by raising seedlings indoors, you'll be able to harvest seeds the first season. Fall broccoli won't have time to develop ripe seeds unless you live far enough south for it to live over the winter and set seed in spring. Choose an early-bearing plant of good type, and let the tight green flower bud clusters—the part you ordinarily eat—develop into yellow blossoms. Since broccoli flowers aren't usually self-fertilizing, you'll need at least three closely spaced plants for pollination. The flowers are cross-pollinated by insects.

Cut the whole plant when the pods are dry, and place it in a well-ventilated location to dry completely. Pods from just a few plants may be put in a paper bag, crushed, and winnowed to remove chaff.

Brussels Sprouts. (*Brassica oleracea* var. *gemmifera,* biennial) Practice first on a few of the easier vegetables, but don't be afraid to try saving seeds from this hardy member of the cabbage family. The plants often survive cold—though not severe— winters and go on to produce seed stalks, which you should treat like those of cabbage. Brussels sprouts stalks that I have gently bent down to the ground and covered with mulch have survived winter. Storing the plants in a cold root cellar (or refrigerator, cold basement, or unheated spare room) over winter for spring replanting would be a bit chancier but by no means impossible. You could also use a cold frame. Brussels sprouts are cross-pollinated by insects. Collect and store seed from brussels sprouts as you would from cabbage.

Cabbage. (*Brassica oleracea* var. *capitata,* biennial) Cabbage, like its cousins broccoli, brussels sprouts, kale, kohlrabi, and cauliflower, is cross-pollinated by in- sects and will cross readily with these other members of the species *B. oleracea.* For home-garden purposes, keep flowering stands of the seed crops 200 feet apart, or plant a row of tall-growing plants between varieties to throw the bees off course. Commercial seed plants must be much more widely separated.

Plant at least three seed-bearing plants of cabbage and its relatives, because many brassica flowers will not accept their own pollen. Select firm, ready-to-eat heads in the fall. Pull them up, root and all, trim off the largest outer leaves, and store the heads either in the root cellar (or refrigerator, cold basement, or unheated spare room) or in dirt trenches well covered with soil. Keep cellar-stored roots damp and cold.

Replant the heads two to three feet apart in early spring. They'll send out a long seed stalk, which rises directly from the cabbage core. Many growers slash a one-inch-deep cross into the top of each head so that the seed stalk can make its way more easily to light and air. Pick the thin, dry pods when they're brown, or harvest the whole plant when pods are yellow and let it dry further if you're saving a lot of seeds. Cabbage seeds ripen gradually and tend to fall off promptly when they are ripe.

Carrots. (*Daucus carota* var. *sativus,* biennial) Cross-pollination by insects makes it necessary to keep blossoming stands of seed carrots at least 200 feet apart. Even then, you'll get a small amount of crossing. For strict purity, keep plantings 1,000 feet apart. Carrots have a close wild relative—Queen Anne's lace—with which they will cross freely. If you have a lot of this common weed around your garden, you'll need to keep it mowed when your seed carrots are going into bloom, or you could cage your blossoming seed plants and introduce some flies to pollinate the flowers.

Small-cored carrots make good eating, but the large-cored kinds store better. You might want to try selecting a good-keeping strain of small-cored carrots, though. Dig the roots before killing frost, store them in sand or sawdust in a root cellar (or refrigerator, cold basement, or unheated spare room) or leave them in the ground if winters aren't severe and if you don't have a problem with mice tunneling under mulch. Replant them in spring, spacing roots one to two feet apart in the row.

Pick the seed heads when the second set of heads has ripened. If you wait for the third or fourth set of umbels to ripen, the first and second sets will shatter in the meantime. Shattering usually occurs about 60 days after flowering. Small lots of plants may be bagged to catch seeds that would otherwise shatter. Dry the seeds inside for a week or so after harvesting.

Cauliflower. (*Brassica oleracea* var. *botrytis,* biennial) Save this one until you've practiced on a few of the easier biennial vegetables. It is tricky to handle. The problem

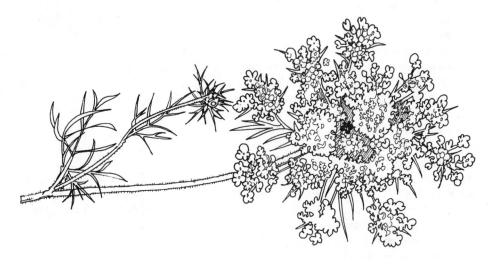

If you are saving seeds from carrots, mow down any queen anne's lace growing nearby or the two plants will cross-pollinate.

is that cauliflower won't live over in the ground where winters are cold, and it doesn't keep in the root cellar either. Yet, like the other biennials, it does need exposure to cool weather to induce seed-stalk formation. Rob Johnston, in his booklet *Growing Garden Seeds,* suggests planting seeds in a cold frame or unheated greenhouse in early fall, no later than October. The young plants that make it through the winter are then set out two feet apart after the last spring frost. The yellow flowers that develop from the cauliflower head are cross-pollinated by insects. (For crossing and isolation specifications, see cabbage.)

For small quantities, pick the seedpods when they turn brown in the fall. If you go into the cauliflower seed business in a bigger way, cut the whole plants when the majority of pods have dried, and cure them under cover. Don't let the seeds freeze before they are thoroughly dry.

Celeriac. (*Apium graveolens* var. *repaceum,* biennial) Slightly easier to store than celery, celeriac otherwise needs the same treatment. Store as you would beets. Varieties will cross up to 200 feet and even to a small extent beyond, as celery will.

Celery. (*Apium graveolens* var. *dulce,* biennial) Although considered a biennial, it can bolt to seed the first year if seedlings are chilled. The umbels of white flowers—a clue to its kinship with carrots, fennel, and parsley—are cross-pollinated by insects. Celery will cross with celeriac.

Select fine, firm, thick-stalked plants in the fall and dig them carefully; bruises and cuts hasten spoilage. Store the plants in soil in the root cellar (or refrigerator, cold basement, or unheated spare room) and keep them moist. Even if outer ribs rot in storage, the plant may be used as seed stock if the roots are sound. Just cut off spoiled parts and plant the celery in rich, moist but well-drained soil. You may need to stake the seed stalk. The seeds shatter very readily. Either make frequent seed-collecting rounds or spread a "fall-out" cloth around the base of the plant to save the seeds that drop.

Collards. (*Brassica oleracea,* Acephala group, biennial) Treat this hardy, leafy cabbage relative as you would cabbage itself, although it will sometimes overwinter in the row if conditions aren't severe. Storage may be a problem; it may do better in a cold frame than a root cellar. The blossoms are insect-pollinated.

Corn. (*Zea mays* var. *rugosa,* annual) Through centuries of selection, American Indians developed six kinds of corn: popcorn, sweet corn, flour corn, flint corn, dent corn, and pod corn. These varieties have been refined by further selection, but the only really new developments added by European immigrants during the last 450 years of cultivation have been hybridizing and the selection of new corn varieties.

Corn is wind-pollinated. The pollen, which is shed by the tassels and received by the silks, is fairly heavy, but you'll still get *some* crossing between varieties planted 200 feet apart. For more complete control of crossing, keep plantings one-quarter mile apart. Save seeds from open-pollinated varieties, not from hybrids. You can isolate corn varieties by time as well as by distance. Early and late varieties, which will tassel at different times, may be grown close together. All corns cross—popcorn with sweet corn, and dent corn with flint corn, for example, so you'll want to check on what kinds of corn neighboring gardeners and farmers are planting. If you're not able to isolate your seed-bearing corn by time or distance, you can hand-pollinate your seed ears.

Cut tip

To pollinate corn, first choose an ear on which the silks have not fully emerged. Cut off the leafy tip and cover the whole ear with a paper bag. Later, when you have collected pollen, you can uncover the silks and sprinkle them with the pollen.

Here's one way to hand-pollinate corn. Choose developing ears on which the silks have formed but have not yet emerged. (Once the silks are out, you can't be sure that they haven't already received some random pollen.) Pull down the leaf that partly surrounds the young ear, and cut off the leafy tip of the ear, leaving the cob intact; then fit a paper bag over the ear. During the next few days, the silks will grow out of the cut end of the ear, and you can gather pollen from tassels on other plants and shake the pollen onto the silks. To collect pollen, choose tassels that have just begun to shed their pollen. Working either in the evening or the early morning, staple paper bags over the tassels, fastening the bags closely around the plant stems. Collect pollen from at least as many tassels as you have ears. Go back at noon to gather the pollen-collecting bags. Most of the pollen is shed in the morning, and if the day is hot and sunny, it may deteriorate if left too long in the bag. Mix together the pollen from all of the bags you've collected, and sprinkle some of the mixed pollen onto the silks of each young ear. Then replace the bag around the ear to keep out extraneous pollen. You can write identifying information on the bag. Pick the ear for seeds when it has dried on the stalk.

Save the fullest, most perfect ears from the earliest-bearing plants. Corn is espe-cially sensitive to inbreeding. If seeds are saved from a limited sampling of ears (4 or 5) for several generations, the resulting plants will tend to be small, late bearing, and low yielding—signs of inbreeding depression. To keep the wide selection of genes that your corn needs in order to continue to produce healthy progeny, save seeds from at least 35 to 50 ears chosen from at least 100 plants. Professional breeders work with lots of at least 100 ears. Mix the seeds thoroughly to ensure a representative plant population, which is especially important if you are trying to maintain a strain of corn without changing it too much. Gardeners who want to change their corn lines by selecting for desired characteristics will be interested in plant breeder Dr. Mark Widrlechner's advice to Seed Savers Exchange members. Easily selected qualities, Widrlechner says, include ear height, earliness or lateness, and plant color. Seed color and disease resistance aren't so easy to select for, and the toughest traits to change by selection are yield, flavor, insect resistance, and nutritional content.

Let the seed corn ripen on the plant. It should be ready to harvest about a month after the corn was right for eating. Well-dried seed corn won't be hurt by frost. Continue drying after harvest by peeling back the husks and hanging the ears in a well-ventilated place. Incompletely cured corn may heat in storage and won't germinate well.

Cucumbers. (*Cucumis sativus*, annual) Your picklers may cross with your slicers, and both will cross with gherkins, but be assured your cukes will have nothing to do with your zucchinis! The separate male and female blossoms transfer pollen through insect activity. As with many other insect-pollinated plants, you can expect a certain minimal amount of crossing with 200-foot separation. One-quarter mile is necessary for absolute seed purity. As with cantaloupes and other cucurbits, you can hand-pollinate cucumbers if you plan to have several varieties in bloom at the same time or if a neighbor grows them. (Follow the directions given under squash, later in this chapter.)

Once in a while, cucumbers set fruit without fertilization—a process called parthenocarpy. Such fruits lack well-developed seeds. To increase your chances of getting good fruit set with viable seeds, dust the female flower with three or four different male flowers rather than just a single one.

Leave the fruit on the vine until it grows fat and turns yellow. Frost won't hurt it, but pick it before it rots. Cut the cucumber in half the long way, scrape out the seed pulp, rinse until the seeds are as clean as possible, then air-dry. Or, to ferment off the pulp and cull worthless seeds, let the pulp stand in a glass for several days. The thick pulp will subside to a thin liquid, and the heavy, good seeds will sink. Strain them off and wash and air-dry them.

Dill. (*Anethum graveolens*, annual) Technically an herb, but practically indis-pensable to many cucumber growers. I couldn't pickle cucumbers without it. Dill is cross-pollinated by insects. Its seeds are one of the easiest kinds of seed to save. Just let the seeds dry on the plant and cut the seed heads off. If you pull them you'll lose a lot of seeds. I keep my harvest of dill heads in an open paper bag in a warm room for several weeks before rubbing off the seeds and storing them.

Eggplant. (*Solanum melongena* var. *esculentum,* annual) When saving eggplant seeds, remember that some of the newer varieties, like Jersey King or Early Beauty, are hybrids. For best results, stick to the open-pollinated kinds, like Black Beauty, Early Black Egg, or Early Long Purple. Although the flowers self-pollinate, they may also be cross-pollinated by insects, up to one-quarter mile or so. Eggplant seeds lose viability more quickly than those of peppers or tomatoes.

The eggplant that has escaped your attention and gone beyond the firm, glossy stage to turn somewhat dull and perhaps a bit wrinkled on the bush is a good candidate for seed harvest. A fruit picked ripe for eating that has spent several days on the kitchen counter will yield viable seeds, too. Scrape out the seedy pulp, rinse the seeds clean, and float off any lightweight seeds. Some gardeners ferment the pulp before extracting seeds, but according to several specialists, this isn't necessary.

Endive. (*Cichorium endivia,* biennial) These may perform like annuals if they are planted early and then bolt to seed in midsummer heat; otherwise they are biennials. Plants to be carried through the winter can be protected by a cold frame or heavy mulching and snow. Don't cover with mulch when the crown of the plant is wet. The cornflower-blue blossom is self-pollinated. Harvest and store seeds as for lettuce.

Fennel. (*Foeniculum vulgare,* annual) Florence fennel, which produces a delicious, semicrisp bulbed stalk, forms umbels of seeds that follow the tiny cross-pollinated flowers. Warm weather hastens seed production. Select seeds from slow-bolting, large-bulbed plants. The seed head formation resembles that of dill and shatters as readily. The seeds are good to eat, but save a handful for planting next year's fennel crop.

Garlic. (*Allium sativum,* perennial) Commonly thought to be incapable of producing seeds, garlic has recently joined the ranks of seed-forming vegetables with the discovery of a few seed-bearing garlic plants—one in California and others in Germany and Japan. Seed-producing garlic is still extremely rare. Should you discover any seed-producers among your garlic plants, notify the horticulture department of your state university. A seed-producing garlic plant would produce a seedpod at the top of the stalk. The seeds should be handled as you would handle onion seeds.

Grains. (annuals) Such small grains as barley, oats, rye, and wheat may be grown in small plots and hand-threshed to provide bread and cereal for the family. The old practice, which still makes sense, was to set aside a portion of the best grains for the next season's seeds right before harvest.

Grains don't shatter as readily as lettuce, onion, or cabbage seeds, but you want to harvest the seed heads before rain beats them down. A cold rain may make the grains sprout while still in the head. Warmth usually induces dormancy. Grains should be well dried before they are picked, with a moisture content of no more than 20 percent. Of course, you won't be able to tell exactly how much moisture remains in the seeds, but if you go by the following guidelines for the different types of grain you should get good seeds.

Barley (*Hordeum sativum*)—It is ready to harvest for seeds when the inside of a grain you've bitten open has a chalky appearance and the grain snaps readily. Too much springiness or doughiness in the grain indicates unripe seeds.

Oats (Avena sativa)—Wait to harvest until a week after grains have assumed a dry appearance.

Rye (Secale cereale)—Pinch or bite a seed to determine whether it is still doughy. Rye pollen is very fine and travels for one mile or more on the wind.

Wheat (Triticum aestivum)—Bend and shake the head. If the seeds are ripe, 75 percent of the grains should fall off after shaking.

Kale. (*Brassica oleracea* var. *acephala,* biennial) Kale was first imported from Scotland, along with rhubarb seeds, by Benjamin Franklin. It is insect-pollinated like the other brassicas mentioned here, with which it will cross-pollinate over distances up to 1,000 feet or so. Like broccoli, the flowers are often self-sterile, so you'll need at least three neighboring plants for adequate pollination. Since kale winters over even in cold climates when protected by mulch or snow, you have a good chance to select your seed stock for hardiness. If the plants have long stems, they'll probably need staking in the spring to support the seed stalk. Kale seedpods, like those of other brassicas, are 1 to 1½ inches long, slim, tubular cases called siliques. Pick the pods when they're brown and dry, but before they shatter, and shell out the seeds.

Lettuce. (*Lactuca sativa,* annual) Begin your seed-saving efforts with this easy annual. Lettuce flowers self-pollinate, so you can usually count on it to breed true, although wild lettuce growing within 200 feet of the garden will sometimes contribute pollen. Save seeds from the plant that is slowest to bolt to seed. Leaf lettuce (*L. sativa* var. *crispa*) is the easiest to handle. Head lettuce (*L. sativa* var. *capitata*) matures somewhat later and must sometimes be cross-cut (by slashing an X one inch deep into the head) to encourage the seed stalk to emerge from the head.

When the yellow flowers have changed into downy white seed heads, pull or shake off loose seeds, winnow them, and dry them indoors for a week. For large quantities, cut the whole plant when enough seed heads have formed, and cure it in an airy place; then thresh out the seeds by shaking and rapping them on a hard surface. A single vigorous plant may bear as many as 30,000 seeds.

Muskmelons. (*Cucumis melo* var. *reticulatus,* annual) What most people think are cantaloupes are actually muskmelons. They won't cross with cucumbers or pumpkins, no matter what you may hear over the back fence. Different varieties of *melons* will cross, though. The results might be odd, but they should be edible and good. To be fairly certain of maintaining a pure strain, keep hills of blooming plants 200 feet apart. Commercial growers maintain a distance of one-quarter mile to ensure purity. Male and female parts are on separate blossoms, which are pollinated by insects, usually bees.

If you want to save seeds from more than one melon (except for watermelon, which is a different species) or if your neighbors have flowering melons closer than one-quarter mile, you will need to hand-pollinate the blossoms. (See the information on squash later in this chapter for the procedure.) Melons are trickier to hand-pollinate than squash or cucumbers because their flowers are small and because only 30 to 40 percent of their female flowers will "take," even under natural conditions. The first female flowers the plant produces are the most likely to set fruit.

You needn't sacrifice good eating to save melon seeds; the seeds are ripe when the melon is ready to eat. Just spoon out the seeds, rinse off the pulp as thoroughly as possible, and air-dry for a week. Rob Johnston of Johnny's Selected Seeds recom-

Cutting an X 1 inch deep into a lettuce head will allow the seed stalk to emerge easily.

mends letting the seed-pulp mixture ferment at room temperature for several days. After fermentation the pulp rinses off more readily, and most of the good seeds sink to the bottom. (Incidentally, all netted melons that slip from the stem when ripe are muskmelons, not cantaloupes. True cantaloupes lack netting on their surfaces and remain on the vines after ripening. Muskmelons are more commonly grown in the United States.)

Okra. (*Abelmoschus esculentus,* annual) Usually self-pollinated, okra is only too anxious to go to seed in a warm summer. Select pods from plants that bear early and heavily. Let the pods ripen completely on the plant unless frost threatens. Shell small lots by hand, and dry the seeds for a week before packaging for storage.

Bees do cross-pollinate okra to a certain extent—potentially enough to mix up a rare heirloom strain. To keep such varieties pure, put a small cotton bag, some Reemay, or discarded nylon stocking over the blossom before it opens and leave it on for two days. Be sure to tag the stem so it won't be confused with other unprotected pods that might have been cross-pollinated.

Onions. (*Allium cepa,* biennial) You don't want to propagate an onion that will bolt to seed early the first year. Insects pollinate the balls (umbels) of small white flowers. Different kinds should be planted at least 100 feet apart. Store good, sound, fall-harvested bulbs in a cold, dry place over winter. Replant in spring, close together, or leave bulbs in the ground where winters are not severe.

The flower-bearing stalk breaks easily, so keep rows wide enough to allow for ease in weeding. If you're growing only a few seed onions, you can support the stalk with a stake. Harvest the umbels as soon as you see the black seeds. They shatter readily. Cure the seed heads for several weeks, then gently rub off the seeds.

In studies reported by William Crocker and Lela Barton in their book *Physiology of Seeds,* the best yields of onion seeds were obtained by giving the plants ample nitrogen at first, then low nitrogen for two months, followed by a final period of high-nitrogen feeding. Onion seeds are particularly vulnerable, in storage, to high temperature and moisture. When well sealed, they have been kept as long as 12 years.

Parsley. (*Petroselinum crispum,* biennial) The delicate flowering umbels reveal parsley's kinship to carrots, fennel, and parsnips. Insects pollinate the blossoms, which form in spring the year after planting. Different varieties of parsley will cross, so let only one kind blossom each year. Parsley winters over fairly well, either under mulch or in a cold frame. Where snow cover is sparse or winds are fierce, you might want to pot up a young plant or two before the taproots grow deep, bring them inside in the fall, and replant in the garden in spring. A single plant will yield plenty of seeds for a household garden, but it's a good idea to keep two in case something happens to one of them. The seeds ripen gradually. Harvest the seed heads individually as they dry, and keep them dry and cool. Parsley seeds lose viability more rapidly than do seeds of many other garden plants.

Parsnips. (*Pastinaca sativa,* biennial) One of the easier root vegetables for home seed saving because it keeps well under mulch right in the garden row even in a bitter cold winter. Insects pollinate the blossoms, and cross-pollination with other parsnip varieties is likely up to 200 feet. How many home gardeners save more than one kind of parsnip seed, though, or have neighbors who save another kind?

Overwintered roots send up leaves and then flower stalks in early spring. The seeds shatter soon after they ripen, so keep an eye on the seed heads and gather the light, dry, browned seeds as they mature. That occurs around the end of July in my Pennsylvania garden.

Peanuts. (*Arachis hypogaea,* annual) Like other legumes, peanut blossoms self-pollinate. Spanish peanuts are easier to work with than the larger varieties, which are more sensitive to an undersupply of calcium in the soil. The developing seeds need plenty of moisture, but harvest should take place during dry weather. Dig the vines and pile them loosely on a rack in an airy but not brightly sunlit place where they can cure gradually. Dampness in a closely stacked pile may bring on blight and mold. Strong direct light, though, often overdries the nuts to the point of brittleness. Boron deficiency in the soil causes the "hollow heart" of peanut seedling cotyledons.

Peas. (*Pisum sativum,* annual) Although the blossoms usually self-pollinate, insect activity may produce a few crosses. Sugar peas shouldn't cross with regular peas, though. To prevent crossing, keep different varieties five to ten feet apart, and plant a tall barrier crop between them.

Pea vines tangle so that individual plants are often hard to identify. The most practical way to save pea seeds is to set aside part of a row and let all the plants in that section ripen pods. You'll need about 15 feet of row to produce a pound of seeds, or a shorter, wider, 5-foot row. Zinc is necessary for pea formation.

Peas should dry thoroughly in the pod. If in doubt, pull the vines and stack them in a well-ventilated spot to ripen for a week or two. If vines are tightly piled or if they become damp, seed quality will suffer. Thresh or hand crack the dry pods. Seeds grown in manganese-deficient soil may have an area of dead tissue on the cotyledon and a damaged tuft of true leaves.

Peppers. (*Capsicum annuum* var. *annuum,* annual) Pepper plants usually self-pollinate, but since insects sometimes transfer pollen, you should keep hot and sweet or small and large varieties about 50 feet apart. Hot peppers are more likely to be cross-pollinated than sweet ones because their stigma and style are more prominently displayed above the anthers. If you're saving seeds of an heirloom or rare pepper, you'll want to be assured of greater purity. For approximately 98 percent seed purity, horticulturist Dr. Jeffrey McCormack suggests separating sweet pepper varieties by 150 feet with barrier crops in between to interrupt the bees' flight. He recommends separating two hot peppers or a hot and a sweet pepper by 600 feet with a barrier crop between them.

The seeds are ripe when the fruit is red. If you live where summers are too short for peppers to turn red, you can pick them before frost and let them ripen in a warm room indoors. To save seeds, simply cut off the top of the pepper and tap the fruit to dislodge the seeds, or spoon them out. They seldom need washing. Dry them for about two weeks before storing. Soaking the seeds in warm water (122°F, 50°C) for 25 minutes helps to control seed-borne bacterial diseases, but I wouldn't do it unless I'd had a disease problem with my peppers, which I haven't. Be sure to rinse the seeds with cool water and dry them well after such a treatment.

Potatoes. (*Solanum tuberosum,* annual) Seed potatoes are the tubers themselves, either large ones cut so that each piece has an eye or, preferably, small whole potatoes saved from last year's crop. Virus diseases of potatoes, usually transmitted by aphids, can be carried by the tubers into next year's plants. Prevention is easier than cure. Don't save seed potatoes from plants that exhibit such virus disease symptoms as curled or yellowed leaves.

Some gardeners reserve an end row for their seed potato crop, planting there all their many-eyed seed potatoes and pieces. They use seed potatoes with no more than two eyes, which will produce larger tubers, in the rows planted for eating. Other gardeners simply collect all the small potatoes that sift to the bottom as the family consumes the year's harvest and replant them next year as seed potatoes. Egg-sized potatoes are good for replanting.

When harvesting a lot of potatoes strictly for seed purposes, cure them in the sun for a few hours. It doesn't matter if they turn green, as long as you don't intend to eat them. Potatoes keep best at 34°F or 35°F (1°C or 2°C).

Then there are potato seeds, which are what Luther Burbank planted in his quest for new and better potato varieties. The seed head, which follows the usually self-pollinated flower and resembles a tiny green tomato (but is poisonous), may be dried and the seeds planted early the following year. Spring-planted seeds, which may be started indoors in March and transplanted to the open garden in June, will form egg-sized tubers by fall. Dig them on a dry day, sun-cure them as suggested above, and replant them the following spring for eating potatoes. This is a project to try just for the fun of it; many seedlings, perhaps most, will be worthless. Although the probability of getting a good one is very low, the possibility does exist.

Pumpkins. (*Cucurbita pepo* var. *pepo,* annual) The longer-stemmed male flowers produce heavy pollen, which is carried to the female flowers by insects (usually bees) over distances of up to one-quarter mile, although a separation of 200 feet or so is satisfactory for most home-garden purposes. Pumpkins of the species *C. pepo,* which includes small Halloween pumpkins, will cross with the other members of the species, zucchini, acorn squash, patty pan, and yellow crookneck squash, and in some studies they have crossed with other species in the gourd family: *C. mixta, C. maxima,* and *C. moschata.* They will not cross with melons or cucumbers.

Collect seeds from fully ripened fruits that have developed a good hard rind. Halve the pumpkins, fork out the seeds, wash off the pulp, and dry the seeds for a week or so indoors. You'll notice a few flat seeds. Since these lack embryos, they'll never grow. You can winnow them off or, if you're handling small batches of seeds, pick them out at planting time next year. British horticulturist Lawrence Hills declares that pumpkin seeds improve with age—up to a point, of course.

Radishes. (*Raphanus sativus,* summer radishes—annual; winter radishes—biennial) The summer radish is one root crop from which seeds may easily be saved. Early-planted radishes produce crisp green seedpods by midsummer. When the pods turn yellow, pull the plants and stack or hang them in a dry place to cure. Small lots of the rather tough pods may be crushed in a paper bag or split by hand to release the seeds. The insect-pollinated flowers will cross over a distance of one-quarter mile or so.

Winter radishes should be treated like beets. They should be stored and replanted the following spring. (Incidentally, the green seedpods of radishes are good to eat.)

Rhubarb. (*Rheum rhabarbarum,* perennial) Seeds saved from your backyard rhubarb patch will produce rhubarb plants, but they may be practical only for quick production of a large number of plants. This can be a valid reason for growing them, but you should know that they might turn out to be early seeders. Division of roots is an easy and dependable way to increase your stock gradually, and one that is less likely to perpetuate early-bolting strains. In addition, seed stalks should be clipped off to conserve the plants' strength. If you do save rhubarb seeds, collect them from a plant that bolts to seed late rather than early, or you may find yourself raising a strain that sends up a flower stalk with the first spring shoots. The seeds are dry and flaky, resembling that of wild dock. In a good season with plenty of rain, seed-grown rhubarb started early will be large enough to pick by summer. The plants should be rigorously culled, however, to eliminate early bolters.

Rutabagas. (*Brassica napus,* Napobrassica group, biennial) Insects cross-pollinate the flowers of this large-rooted biennial, which should be replanted the second spring 200 feet from other flowering varieties for home gardening purposes. This shouldn't be a problem, for there aren't many different strains of rutabaga. Save firm, well-shaped roots from the fall harvest, gathering them before severe frost. Trim the leafy tops to a 1-inch stub, and store the roots in a cold root cellar (or refrigerator, cold basement, or unheated spare room). Replant in spring, reselecting plants for their keeping quality. Where winters are mild, mulch the roots and thin them in spring to stand 12 inches apart for seed production. Gather the pods when they've turned brown and dry. Watch them—they shatter readily, like other brassica seeds.

Salsify. (*Tragopogon porrifolius,* biennial) This vegetable is easily stored over winter right in the garden row under mulch, even by the northern gardener. You'll probably want to dig some roots to store in the root cellar (or refrigerator, cold basement, or unheated spare room), too, for insurance. Replant them 12 inches apart in spring.

Salsify self-pollinates, but the problem of crossing is strictly academic, because only one variety of the vegetable is commonly available. The seed crop may be treated like that of the parsnip. The attractive purple spring flowers are followed by a large fluffy seed head that looks like an overblown dandelion (similar to goatsbeard, which is a wild relative of salsify). The heads ripen successively and should be picked individually as they ripen. Dry them on a rack or shelf for a week or two. Experts are not yet certain whether or not salsify will cross with goatsbeard, but you should probably assume that it will, at least within 200 feet or so.

Sorrel. (*Rumex acetosa,* perennial) Save seeds from the plants that go to seed late. Since flavor is better before seed stalk formation, you don't want to develop an early-seeding strain. Sorrel will cross-pollinate. Cut off the stalk when the flaky seeds are dry, and rub them off.

Soybeans. (*Glycine max,* annual) The self-pollinated flowers appear when days begin to grow shorter. Northern gardeners should choose an early-maturing variety like Altona (100 days). Although some selection is possible, in practice the soybean

Male (left) and female (right) spinach flowers.

seeds I save are what is left after I've picked as many green soybeans as possible during the 10 to 14 days in September when they're at their peak. When the beans start to turn yellow and toughen, I let the rest hang on the plant until they're dry and cut the whole plant when some of the dry pods begin to break open. Mice like the beans, so they should be threshed promptly and dried on a screen indoors for two to three weeks. I thresh all I need for seeds and soup by treading on the vines sandwiched between clean sheets on a cement floor.

Spinach. (*Spinacia oleracea,* annual) The inconspicuous wind-pollinated flowers occur in varied combinations on different plants. A row of spinach will usually include some plants that bear both male and female flowers; others that are only female; small, early-bolting plants that are entirely male and larger, leafier, vegetative male plants, which produce the dustlike pollen that travels in the wind for as far as a mile.

Spinach may be selected for leafiness and late bolting, though, without identifying each graduation of gender. Just look for the undersized, early-seeding plants and grub them out. Most plantings of spinach tend to go to seed, spurred by warm weather and longer days, before the gardener has a chance to get his or her share, but at least selection should help to eliminate any genetically linked early bolting. Pick the seeds in the summer after they've ripened on the plants and store in sealed containers in a cool dry place.

Squash. (annual) Since the species *Cucurbita pepo* includes a variety of squash types—both hard-shelled winter and tender summer squash, as well as acorn squash, small gourds, Halloween pumpkins, and spaghetti squash—you can see that the possibilities for odd and interesting crosses are considerable. Insects often transfer the pollen between varieties, so different kinds of squash from which you want pure seeds should be grown no less than one-quarter mile apart. Bees usually carry squash pollen for only 180 to 200 feet, but they can travel as far as one mile.

Squash won't cross with cucumbers or melons, but pollination of a zucchini by a gourd or an acorn squash can produce a hard-shelled summer squash that isn't too acceptable on the table. All such crosses are edible, though. Some are palatable. Those that aren't make good feed for stock; the animals relish the seeds. If you're saving seeds of a kind of squash no longer offered by seed catalogs, however, such random crosses will spoil the seed strain that generations of gardeners have taken pains to keep pure.

Although a few botanists who have studied squash pollination think that occasional crossing may occur between acorn squash, *Cucurbita maxima,* and butternut squash, *C. moschata,* others disagree. In practice, even the careful members of the Seed Savers Exchange (see chapter 37 for more information about this group) routinely plant one variety of each of the four squash species in the same season. Thus you can safely save seeds during the same summer from a Hubbard squash (*C. maxima*), a butternut (*C. moschata*), a green striped Cushaw (*C. mixta*), and a zucchini (*C. pepo*), but not from an acorn and a spaghetti squash, both of which are *C. pepo.*

If you do want to save seeds from more than one member of a squash species, or if pollen from squash grown by a neighbor is likely to be carried to your squash, you can pollinate the blossoms by hand. Here's the procedure, as described in the *1985*

A male squash flower (left) and two female flowers (right).

Seed Savers Exchange Harvest Edition:

First, observe the flowers and learn to identify the female flower, which has a tiny squash-shaped swelling at its base. The vine's earliest blossoms will be males, which have straight stems and lack the rudimentary fruit.

Inspect the plants in the late afternoon or early evening and find an unopened male and female flower on the same vine. (Flowers that have already opened will droop slightly. Don't use these because they've probably already been pollinated.) Tape both flowers closed. The next morning, after the dew dries, open the female flower. It is all right to tear off the tips of the petals down to the lower edge of the tape, but be careful not to disturb the flower's attachment to the stalk. Pick the male flower, remove its petals, and holding it by the stem, rub the male flower's pollen-dusted anthers on the stigma of the female flower. Some experts recommend applying pollen from several male blossoms to each female flower. Next, tape the female flower closed to keep out bees that may be carrying pollen from other flowers. You can bag the flowers instead of taping them. Mark the hand-pollinated blossom with a loosely tied bright-colored string or plastic ribbon so you'll be able to find the correctly pollinated seed-bearing fruit in the fall! Record your crosses in your garden log, too. To be sure of getting a sufficient amount of good seeds, pollinate several blossoms for each squash variety, always using male and female flowers from the same vine. Squash doesn't seem to suffer inbreeding depression as corn does.

Butternut squash (*Cucurbita moschata*) generally keeps to itself, but occasional crosses with *C. pepo* or *C. maxima* (winter squash) may occur.

You can harvest seeds of hard-shelled winter squash when preparing them for a meal, but zucchini and other summer squash should be allowed to ripen and harden for about eight weeks after they've reached the stage that is good for eating.

To pollinate squash, tape unopened flowers to prevent them from opening and pollinating. The next day, separate the petals of the female flower and rub the pollen-coated anthers of a male flower against the stigma. Then retape the female blossom.

Studies have shown that squash and other cucurbit seeds gain vigor if allowed to afterripen in the fruit for 20 days after the fruit is fully mature. Split the squash, rake out the seeds with your fingers or a fork, and rinse off the pulp. Eliminate lifeless seeds by floating them off in water; the good ones sink. Spread the seeds on screens or newspaper pads to dry for two weeks or so.

Sunflowers. (*Helianthus annuus,* annual) Insects may cross your prize giant sunflower plants with wild sunflowers or with other cultivated sunflowers at a distance of about 1,000 feet. Watch the ripening heads and tie netting over the best ones when the seeds have formed but are still immature—otherwise the birds may beat you to the harvest! Cut the head from the stalk when seeds are dry, but before they begin to fall off, and hang it in a well-ventilated but relatively mouse-proof place to dry for a few weeks. Then rub off the seeds and screen and winnow them to eliminate chaff. We sometimes select seeds for next year's crop by setting aside the largest, plumpest seeds as we snack our way through them. For larger quantities, sift out small seeds through hardware cloth. Package promptly after drying; it doesn't take rodents long to discover and decimate a cache of sunflower seeds.

Swiss Chard. (*Beta vulgaris* var. *cicla,* biennial) The fine, dustlike, wind-blown pollen will carry as far as a mile, crossing varieties of chard with each other and with beets and sugar beets. Chard survives the winter in all but the most northern gardens, and even there a blanket of mulch will often pull it through. The deep, strong taproot stores energy for a second season of growth, when the plant will flower and produce summer seeds. Pluck off the brown seeds as they dry, or cut the plant when seeds begin to dry and allow it to finish drying in the shade.

Tomatoes. (*Lycopersicon lycopersicum,* annual) Seeds saved from open-pollinated (nonhybrid) varieties can generally be counted on to breed true, but although the flowers are usually self-pollinated, they may be crossed by insects about 2 percent of the time. To prevent crossing, keep different varieties of seed stock 10 feet apart. Flowers of some older tomato varieties have longer styles than those of more recently developed cultivars. Because the protruding styles are more accessible to bees, these varieties may have a considerably higher rate of natural cross-pollination. Botanist Dr. Jeffrey McCormack recommends isolation distances of 20 to 25 feet for older tomato varieties with long-styled blossoms. McCormack also suggests planting either tall barrier plants or a pollen-producing crop between varieties. (In their native Peru, where they are perennial, tomatoes are cross-pollinated by native insects.)

Pick the best fruits from your outstanding plants when they are fully ripe, perhaps a bit overripe. To maintain a larger gene pool, save seeds from several fruits. Place the seeds and pulp together in a jar and let the mixture ferment at room temperature. It's best not to put the jar in an especially warm place. Fermentation develops in two days at 80°F (27°C); at 70°F (21°C) it takes three days. Leaving the seeds in the pulp for a longer period, but no longer than five days, permits more complete control of bacterial canker, a seed-borne disease. Stir the brew each day, and during the second, third, and fourth days, pour off the liquified pulp and the floating seeds, retaining the seeds that have sunk to the bottom. Rinse the pulp from the seeds and spread them on paper to dry for three to seven days.

If you're trading seeds with other gardeners in a seed exchange, you should probably use the fermentation process, but I've had excellent results for years with a simpler method. I cut open the tomato, rake out the seeds with a fork, and spread them on newspaper, folded to several thicknesses, on which I write the variety name with a felt-tip marker. After the seeds have dried on the newspaper for two weeks or so, I remove the labeled top layer of paper with the seeds plastered to it, roll it up, and keep it in a tightly closed can over winter. In spring, it's easy to simply scrape the seeds from the paper as I need them or to tear off bits of the seed-studded paper to send to gardening friends.

I've always liked the story (a true one, too) that Peter Tobey related in an editorial in the magazine *House and Porch Gardens,* about the man who grew a tomato plant from the seed in his BLT sandwich. (Incidentally, there are about 2,500 different kinds of open-pollinated tomatoes, and 500 hybrids.)

Turnips. (*Brassica rapa,* Rapifera group, annual, sometimes biennial) Early-planted turnips will form seeds the same season. Fall-planted turnips may live over a fairly mild winter and produce seeds the following spring. If your winters are cold, store your seed turnips in the root cellar (or refrigerator, cold basement, or unheated spare room) until spring and then replant them. The yellow flowers are cross-pollinated by bees, and they'll cross with rutabagas, rape, and other blooming turnip varieties, and occasionally with mustard, radishes, and Chinese cabbage, within 200 feet and often farther. Harvest the dry pods individually and bag them in small amounts for threshing. Larger seed harvests may be handled by cutting the plants when the pods are fairly dry and stacking them lightly in a dry place to cure further.

Watercress. (*Nasturtium officinale,* perennial) Well-established plants will form seed heads by summer's end. Designate a clump from which you will refrain from harvesting leaves. These plants will form seeds sooner. Ducks can eat a small planting of watercress down to the nub in late fall when other greenery fails, so if you have some, you may need to put chicken wire around your seed-stock clump so that it can ripen seeds unmolested. Watercress is cross-pollinated by insects. Collect the slim pods when they dry and break them up to release the seeds.

Watermelons. (*Citrullus lanatus,* annual) The separate male and female blossoms are cross-pollinated by bees, so plantings of different varieties may cross within 200 feet. Since the seeds are mature when the fruit is ripe, you can keep the seeds of especially good melons as you eat the flesh. Melons that remain in the field until they're a bit overripe can also be used as seed sources, if they are still sound. Just rinse the seeds and spread them out to dry for a week.

34 Further Challenges

Growing plants from seed offers challenges for the gardener even beyond those of successful germination and seedling growth. If you truly enjoy the process of seed starting, you might want to consider making your own crosses of the plants in your garden. If you have the room and the time, you could start a small business raising and selling seedlings to other gardeners.

Backyard Plant Breeding

Suppose you want to try making some controlled crosses of seed-bearing plants. How do you begin?

First, you select parent plants that have qualities you'd like to perpetuate and try to combine (see chapter 29 for selection criteria); then you must isolate your plants to prevent accidental wind or insect pollination from unknown sources. A well-anchored cloth cover will keep out wind-borne pollen. A screen-wire cage or cheesecloth drape effectively deters insects.

Hand-Pollination. Having thus taken control of the flower to eliminate random pollination, it is now up to you to carry out the pollen transfer from one parent plant to another. If it's possible to pinch off the blossom petals, do so; you'll have a better view of what you are doing. Transfer the pollen from the parent plant you've chosen as the fertilizing agent to the stigma of the plant you want to bear the seeds. To do this, you can use a twist of paper tissue, a small paintbrush, a fingertip, or a rubber pencil eraser. (See chapter 33 for hand-pollinating techniques for corn, cucumbers, squash, and melons.)

Composite flowers and those of peas, beans, and some other self-pollinating plants are difficult to pollinate by hand, so these would not be good candidates for the amateur plant breeder. If you want to cross-breed tomatoes, which have self-pollinating flowers from which the anthers must be removed in order to prevent self-pollination, consult one of the excellent books on plant breeding recommended in the bibliography.

Saving Pollen. If one of your chosen parent plants blooms before the other one, you can try to save the pollen, according to Ken and Pat Kraft, authors of *Garden to Order,* by tapping it into a small dry vial and then drying it for about 24 hours under a light bulb or in any other spot where a fairly constant temperature of 90°F (32°C) can be maintained. Store the pollen in the freezer. (Pollen grains from corn, and possibly also from other vegetables, often burst when frozen. You could try keeping corn pollen in a humid part of the refrigerator.) Species vary, and pollen is often short-lived at best. Some kinds are potent for only a few hours, but others have lasted for several weeks with this treatment. Saving pollen is a highly experimental venture

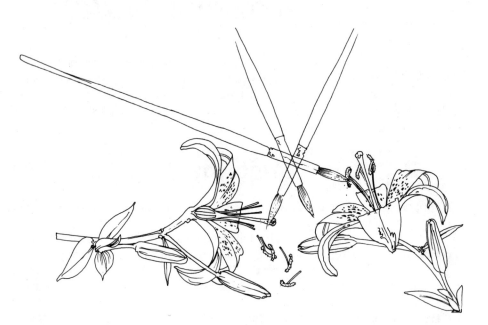

Use a small paint brush to transfer pollen onto the stigma of the plant that you want to pollinate.

even for professional horticulturists, so don't be disappointed if it doesn't always work for you. Record what you did so that you can learn from your successes and failures.

This is a quick look at a complex subject, but perhaps these few hints will help you to make a beginning if you choose to try some deliberate crossing. Don't hesitate to pull apart a few flowers so that you can learn more about their structure. If you become at all serious about this, you will of course want to do a lot more reading about plant genetics, flowering mechanisms, isolation and breeding techniques, and so forth. R. W. Allard's book *Principles of Plant Breeding* and James Welsh's *Fundamentals of Plant Genetics and Breeding* are two good books on the subject. If you label your experiments and record your procedures and results, you'll be able to learn and profit from what you've done.

Selling Your Seedlings

More than one experienced gardener has found it possible to pay for the season's gardening costs—fertilizer, rock powders, mulch, seeds, pots, a new hoe—

by selling well-grown seedlings. A really thriving seedling operation might in a few years even pay for a greenhouse. (The trick is that you'd need the greenhouse to make it possible!)

Deciding What to Sell. Plan your stock around the basic vegetables that many people want to buy: tomatoes, lettuce, cabbage, and peppers, which are staples in most household gardens. Also grow special varieties of those vegetables that often can't be found outside a seed catalog: yellow tomatoes, Jersey Queen cabbage, extra-hot peppers, Buttercrunch and romaine lettuce, burpless cucumbers. Then, to keep things interesting, offer a few plants that are different but easy to grow: vegetables such as leeks, potted chives, cherry peppers, and collards.

Listen to your customers, watch what they buy, take special orders, and you may find that you're able to develop a specialty and a loyal following of busy gardeners who depend on you.

Managing Your Growing Space. You'd probably need at least a three-tiered, double-bank, 48-inch fluorescent light setup or a small greenhouse, good sun porch, or sun pit to raise enough plants for yourself and your customers. A few cold frames, in which you can harden off early-grown plants, can double your seedling-producing capacity. I've been able to grow repeat orders of red cabbage and broccoli by moving them outdoors under glass quite early to make room for succeeding flats of tomatoes and eggplant.

Advertising. Once you become known for quality plants of special varieties, good old word of mouth may bring you as much business as you can handle. If not, or if you're just beginning your venture, a little publicity can help a lot. Advertising can be as simple as a sign on your mailbox or front door, a sentence in the classified section of your local paper, or a notice posted on the supermarket bulletin board. Craft fairs may offer you a chance to set up a table of your wares. A newspaper feature story on your operation would provide free publicity if you have an interesting angle or specialty that would catch the reader's eye.

Cutting Expenses. You can pare expenses by using nonhybrid seeds, which generally cost less; by buying seeds in bulk, once your volume of business justifies such a step; or by using home-saved seeds if you are *sure* they will run true to type. Mix your own potting soil in volume. Use recycled plant containers as much as possible, as long as they are fairly uniform, neat, and provide good growing conditions for the plants. Don't stint on fertilizers, but try substituting manure tea for the heavy feeders in place of the more expensive fish emulsion fertilizer. Remember that the extra electricity used to run your fluorescent lights should be figured into the costs of doing business.

You'll want to keep good records of expenses and income, obviously, so that you can tell whether your seedling business costs or pays. Naturally, expenses in setting up will keep the ledger red for the first few months. It's unlikely you'll get rich at this, but often a little extra effort and space, spent at something you enjoy doing anyway, will help to pay for some household necessities.

35 The Garden Diary

I keep a running record of our gardening activities and results, partly, I guess, because I'm a compulsive note-taker, but mostly for curiosity, comparison, planning, and improvement. The record is a useful tool that helps me, each year, to avoid at least some of the mistakes of years gone by. Garden notes need not be elaborate or even terribly well organized, as long as you can find them.

My own system smacks of the patchwork theme that has become the trademark of our homestead. The heart of the system is a calendar that has space for notations on each day of the month. This is where I record planting dates, both indoors and outdoors, dates of first harvest, peak harvest times for different vegetables, yields, weather and insect problems (excuses, excuses), notes to remind myself to save seeds of certain vegetables, the phone number of the man who shells peas by the bushel (we gardeners always were an optimistic lot), and kinds and amounts of manure and rock minerals and when and where they were spread.

The calendars differ from year to year, but they follow a basic pattern. Notations during the dead of winter center around ordering seeds and planting seed flats. Spring notes proclaim, "Ate first spring onion tops today!" "peepers back," "planted out head lettuce," "hard freeze at night" (April 22). By June the page is scrawled up and running into the margins: "make smaller, more frequent leaf lettuce plantings next year," "mammoth melting sugar peas better than dwarf kinds," "first zucchini today" (June 19). Late summer and fall are crowded with pickling, canning, and freezing tallies, as well as references to prime foraging dates, mushrooms and nuts to watch for, up to and well beyond the day when "black frost . . . 25 degrees last night" proclaims the certain turning of the seasons.

Interspersed with my garden notes, of course, are records of other homestead happenings with goats, bees, hens, sheep, hay, and wildlife, as well as odd tidbits that just should not go unrecorded: "toad returned to the barn today," "found bottle gentians near swamp," "bluebirds nesting," "geese going south," "nine fall ducklings hatched by Crazy Lucy." If we have an extra load of manure and decide to divide it between the corn patch and the grapes, I make note of that fact because I know I'll forget by the next barn-cleaning time which piece of ground got the bonus.

Comparing harvest times from year to year helps me to determine how far back I can push certain planting dates in spring and which varieties work best with this kind of gamble. Yield notes influence the amounts and kinds of vegetables I'll grow next year. Notations of food quantities put into storage, averaged over five to ten years, tell me in a general way how much is enough. Since food quality and nutritional value are highest during the first year in storage, I prefer to can and freeze for one year only.

Two other record sheets figure in my yearly planning guide. One is the garden diagram, that much-revised, out-of-scale sketch of the garden rows and what they grew, including intercropping tricks that worked, succession plantings, and last-

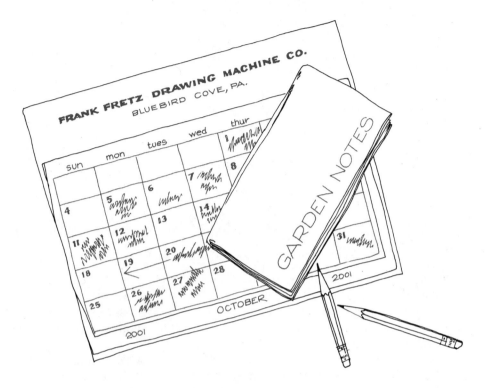

Notes written on a calendar are a valuable record of what you've learned during the growing season.

minute replacements of row or plant wipeouts. (*Who* ate the early beets? I'll never know. Whoever it was never made it back to gobble the chamomile and basil I put in their place, though.)

The other list I keep from year to year is the sheet of scrap paper on which I record kinds and amounts of seeds under the name of the company from which I ordered them. This saves time when ordering next year. Des Vertus Marteau Turnip, for example, is one of my favorites, but only two seed companies carry it. Or I may want to try a different strain of butternut squash from a new seed house or a taller snapdragon variety.

Experiments in plant breeding, seed saving, insect control, and the like will be of lasting value only if recorded: What did you do, when, to what, how long did it take, and what were the results?

Taking notes on the garden gets to be a habit. Keeping the calendar handy, with a pencil nearby, ensures that it's not a chore, and referring to the notes of other years helps me to determine where to start and how to proceed this year. "More kale, more leeks, less broccoli," reads my command to myself. If I ignore it this year, I'll have to record it in red next year. And so the years cycle into one another, each one different, yet grounded in the year before. To leave them unrecorded is to miss much valuable information and many important memories.

36 Seed Catalogs

Recently I read an article by a man who maintained he'd received an excellent education by observing and asking questions in his local hardware store. I couldn't agree more; Mike and I have also learned a lot this way. I feel the same about seed catalogs. In between the vegetable descriptions is helpful information on planting, insect control, harvesting, and even cooking.

As you look for varieties of vegetables that will be most suitable for your garden, you'll learn, too, to evaluate the offerings of the different seed firms. Disease resistance, for example, is important, especially with cucumbers and tomatoes, if you have had previous disease problems in your patch. You must know what your problem was so that you can choose a variety with specific resistance. Even then, resistance doesn't guarantee immunity, and a pea that is resistant, say, to fusarium wilt may still come down with downy mildew, or a scab-resistant cucumber may succumb to anthracnose or mosaic. Generally, a resistant strain is less vulnerable to a disease than one that is termed "tolerant." Breeding for disease-resistant vegetable varieties will increase considerably in the near future.

When shopping for early-producing vegetables, check out the catalog claims for flavor. Some early vegetables have fine flavor, but in others taste is sacrificed to a certain extent for quick harvest. Here again, early varieties are being improved, and you'll find more and more that they taste, as well as look, good.

Should you buy pelleted seeds? I did. Once. The bentonite clay often used to coat fine seeds is sometimes slow to admit water. The pelleted carrot seeds I planted as an experiment cost more and produced less than a comparable number of regular seeds.

If your garden area is small, look for bush varieties of some of the cucurbit space-grabbers like acorn squash and even pumpkins. Flavor is not always quite up to that of the vining crops, but should still compete with anything you could buy.

If you've been bitten by the gardening bug half as badly as I have, I'm sure you'll need no urging to try one or two or three new plants each year. Not every experiment will earn a permanent place on your garden plan, but you're sure to find at least a few vegetables and flowers that you'll wonder how you ever did without. Even if it's just a new variety of tomato, one of the high crimson (extra red) ones perhaps, or the Sweet 100 that bears so generously that it must be staked, treat yourself to something new. Even at today's prices, a packet of seeds is one of the very best buys you can make.

Vegetables and flowers designated as All-America selections have been chosen for their high quality and ability to grow well in different parts of the country. The All-America selection trials have been set up as a nonprofit institution, managed by cooperating seed companies. Member firms have the privilege of selling seeds for new selections introduced by other firms, as long as they offer seeds obtained solely from the original grower—none of their own—for the first three years. This eliminates a lot of the secretive hocus-pocus that formerly attended the introduction of new varieties. The effect is to protect the company offering a good new type of seed, to make it worth their while to share it, and to put good new strains in the hands of home gardeners sooner.

Catalogs differ, too. Some are general, others specialize in northern-grown or open-pollinated seeds, in herbs or cantaloupes, in cutting flowers, or in extra-large vegetables. I don't believe I've ever yet sent for a seed catalog that didn't have *some* different offering I found tempting. Adding to your catalog library might make your January fireside planning a bit more complicated, but I'll bet your June garden will be a lot tastier.

You'll often hear that you should buy seeds from a firm based in a climate similar to yours, or at least as cold in winter, in order to get plants that are acclimated to your regional conditions. That's good advice, as far as it goes, but it ignores the fact that these days most large seed companies buy seeds on contract from large-scale growers. Some seedsmen, however, do raise most of their own seeds, and indicate this in their catalogs. Seeds that I've purchased from Maine, Vermont, New Jersey, California, Canada, England, Maryland, Iowa, North Dakota, and other states have performed well here in my Pennsylvania garden, leading me to conclude that, in this latitude at least, the geographical location of the seed source is not critical. It may be more important for cold-climate gardeners or for those in the Deep South.

One more thing. Keep those catalogs around after you've sent out your order. With the exception of Stokes, whose seed packets are an education in themselves, planting information is usually more completely spelled out in the catalog than on the seed envelope, so you might want to refer to the catalog again at planting time.

In the back of this book, in alphabetical order, is a list of companies selling mail-order seeds. Unless otherwise noted, catalogs are free.

37 Seed Exchanges

Think of all the mountain hollows, city plots, small-town homesteads, and isolated ranches where people may be saving seeds of an heirloom bean or a special meaty tomato or a slow-bolting lettuce. Simply keeping these varieties alive and occasionally improving on them is a good thing. Perhaps the seeds are passed around the family or the neighborhood, but often the strain remains in a relatively restricted pocket, going round and round each year, but not spreading. If the seeds are not replanted regularly, in fact, or if they are lost, the strain will die out and can never be recaptured.

Suppose, though, that some way was found to bring some of these people together, to let them know about each other and the seeds they save, to make it easy, for example, for a gardener in Iowa to try seeds raised in a Georgia garden, and to facilitate the sharing of seeds that might otherwise peter out with people who care about such things. Just imagine all the possibilities that would be generated by such a network!

Now for the really good news: The network exists. Gardeners who save their own nonhybrid seeds now have the opportunity to trade seeds with other gardeners through membership in a seed exchange. These exchanges, which started quite recently as one-man, grass-roots ventures, are growing in size and scope. Their potential for good is tremendous.

The Seed Savers Exchange

How does a seed exchange work? Let's take a look at the Seed Savers Exchange (SSE), originated by Kent Whealy. This exchange was designed to be a "communications network for serious gardeners" devoted to spreading as many good, nonhybrid, and especially old, rare, or highly localized vegetable varieties to as many gardens as possible, before these seeds are lost. Whealy started the SSE as a simple newsletter. He was prompted by the death of his wife's grandfather, who had given him tomato, bean, and morning-glory seeds that had been kept alive by four generations of his family. Whealy realized that it was now up to him to keep those seeds alive, and he began to wonder how many other good old vegetable varieties had already died out with the elderly gardeners who had been keeping them.

The grass-roots exchange that numbered 29 members in 1975 has now grown to a seed preservation effort that is international in scope, with hundreds of members and tax exempt status. Its publications retain the homey flavor of the early back-fence exchanges, but they also crackle with the excitement of networking on the leading edge of a vital effort to safeguard our rich genetic heritage, just before it's too late. The SSE is one of the most hopeful forces for good in home gardening today, and I recommend it to all concerned gardeners as an example of effective positive action.

The SSE runs an annual camp-out—a gathering of members and others with an interest in seed saving and the preservation of genetic diversity. It also has published two books: *The Garden Seed Inventory,* a listing of almost 6,000 nonhybrid garden seeds and their sources ($12.50 postpaid) and *Seed Savers Exchange: The First Ten Years,* a collection of useful articles from previous issues of SSE handbooks ($15 postpaid). Order these books by writing to or calling Kent Whealy at R.R.#3, Box 239, Decorah, IA 52101 (319) 382-3949.

The most heartening news from the SSE is its recent purchase, with help from various foundations, of the 57-acre Heritage Farm near Decorah, Iowa, where the Exchange's collection of nearly 5,000 heirloom and endangered vegetable varieties will be grown regularly to renew the seeds. Large orchards of antique apples and collections of berries and perennial vegetables will be established, and demonstration gardens will show visitors the tremendous variety that still exists in vegetables, thanks to the efforts of home seed-savers.

For $12 a year, the SSE will send you their Winter Yearbook, listing members' names, addresses, and the vegetable seeds offered and wanted by each member, and a Fall Harvest Edition with reports from seed conferences and articles on seed saving and on expert seed-savers. These publications can be ordered by writing to Kent Whealy at the SSE address given above. A member may change his or her wish list each year to correspond with current needs.

Rules are few but necessary to keep all dealings fair. When requesting seeds, members must enclose first-class postage. Nonmembers may order seeds from a member for $1. (Even if you've bought the yearbook, you're not considered a member unless you offer seeds for trade.) When ordering seeds from a member, postage on packages must be paid back to the sender by the recipient. To make possible the widest dissemination of seeds, only a few seeds, enough for a hill or a few feet of row, are sent to each person requesting them. Vegetable seeds are of primary importance; no listings of flowers only will be accepted, and members are asked to limit flower listings to a minimum. "Please don't use the exchange as just a source of free seeds," Kent adds, "because it could easily be ruined by more taking than giving. This should be an exchange among seed-savers. Let's save extra seeds from our best . . . and then share them."

In the exchange's recent catalogs, listings range from high-protein corn, mung beans, Crenshaw melon, Vietnam basil, Sweet Spanish onions, and Swedish pole beans to hop vines, broom corn, sea onions, and sugar cane seeds, with many strains of tomatoes, sweet corn, squash, lettuce, and other vegetables. Membership forms for the SSE are available from Kent Whealy at the SSE address given above. Enclose a loose first-class stamp with your request.

The Abundant Life Seed Foundation

Another exchange, the Abundant Life Seed Foundation, is devoted to preserving and sharing open-pollinated, untreated seeds, especially those of the Pacific North-west. The nonprofit Abundant Life Seed Foundation also offers workshops on raising seeds; it sells seeds of vegetables, herbs, ornamentals, wildflowers, trees, and shrubs; and the exchange distributes surplus and dated seeds to refugees, hunger relief programs, and charitable groups. Members receive both the annual seed catalog and book list and periodic newsletters. Membership costs $5 to $15 a year, according to ability to pay. The Foundation may accept homegrown seeds instead of cash payments for memberships. For more information, contact the Abundant Life Seed Foundation, P.O. Box 772, Port Townsend, WA 98368 (206) 823-5376.

The Grain Exchange

The Grain Exchange was recently established by the Abundant Life Seed Foundation to encourage small-scale growing of cereal grains—oats, wheat, barley, emmer, einkorn, and others, as well as other useful grains—flax and amaranth, for example. Members trade hard-to-find grains like Mandan Indian flour corn and hullless oats. For a $7.50 annual membership fee, members receive two issues of the semiannual newsletter and the privilege of listing seeds offered or sought. For more information, send a self-addressed, stamped envelope to or call the Grain Exchange, The Land Institute, 2440 East Water Well Road, Salina, KS 67401 (913) 823-5376.

Two Flower Exchanges

Two grass-roots flower exchanges publish annual bulletins to help gardeners trade seeds for ornamentals and herbs. The Olde Thyme Flower and Herb Seed Exchange, started by Barbara Bond in 1984, does not charge for listing seed requests. To receive the bulletin, which includes some recipes and lore in addition to the flower and herb seed listings, send $3 to the Olde Thyme Flower and Herb Seed Exchange, RFD 1, Box 124-A, Nebraska City, NE 68410. The Flower, Wildflower and Herb (FWH) seed exchange costs $2 a year. Offerings in recent bulletins include hollyhocks, ornamental popcorn, Texas bluebonnets, cleome, and calendula. Write Ann Taft, P.O. Box 651, Pauma Valley, CA 92061. When making inquiries without enclosing a membership fee, be sure to include a self-addressed, stamped envelope.

Native Seeds/Search

Native Seeds/Search, another tax exempt seed conservation organization, was started by ethnobotanist Gary Nabhan to find, increase, and distribute seeds of native crops of the southwestern United States and northwestern Mexico. Through Search, Nabhan and other ethnobotanists and anthropologists have been able to collect and propagate native Indian seeds and give them back to Indian communities where the seeds had died out. Although not an exchange, Search's valuable rediscoveries of native heirloom seeds make it worth noting. These ancient foods—squashes, lentils, beans, sunflowers, corns—are not only a priceless part of the native American's cultural heritage, but also useful to desert gardeners and devotees of ethnic recipes. Most important, these seeds constitute reservoirs of stress-tolerant genes for plant breeders. A $10 annual membership brings you the handsome seed catalog, a quarterly newsletter, and a 10 percent discount on seeds and books. For more information, contact Native Seeds/Search, 3950 West New York Drive, Tucson, AZ 85745. When making inquiries without enclosing a membership fee, include a self-addressed, stamped envelope.

Other Seed Exchanges

The following is a list of organizations that sponsor exchanges of seeds and plant material. When requesting information from any of these nonprofit organizations, enclose a self-addressed, stamped envelope. In addition, a contribution of a few dollars would help to keep these worthwhile programs operating.

Actinidia Enthusiasts Newsletter, P.O. Box 1064, Tonasket, WA 98855. Subscription $5 per issue. Some members offer seeds of hardy kiwi fruit.

The American Rock Garden Society, Buffy Parker, Secretary, 15 Fairmead Rd., Darien, CT 06820 (203) 655-2750. Seed exchange is open only to members, with membership fee $15 a year.

The Center for Plant Conservation, 125 The Arborway, Jamaica Plain, MA 02130. A network of organizations that collect seeds and cuttings of some of the 3,000 endangered North American native plant species to be preserved in botanical gardens.

Friends of the Trees Seed Service, P.O. Box 1064, Tonasket, WA 98855. Catalog $4. The catalog is a rich source of tree seeds and information about trees of all kinds. It's a one-man effort and worth much more than $4.

Kusa Research Foundation, P.O. Box 761, Ojai, CA 93023. Established to preserve endangered varieties of cereal grass. Send $1 for information.

North American Fruit Explorers, Rt. 1, Box 94, Chapin, IL 62628. Publishes quarterly magazine, *Pomona,* and serves as a clearinghouse for information about heirloom and unusual varieties of fruit.

Northern Nut Grower's Association, Kenneth Bauman, 9870 S. Palmer Rd., New Carlisle, OH 45344 (513) 878-2610. Publishes informative newsletter and sponsors regional get-togethers to share knowledge, seed nuts, and grafting scions.

North Star Seed and Plant Search, Box 1655A, RFD, Burnham, ME 04922. Computerized plant and seed search for reasonable fee. Over 30,000 varieties on file.

The Rare Pit and Plant Council, Debbie Peterson, 251 W. 11th St., New York, NY 10014 (212) 475-2046. Membership $7.50 a year. Offers helpful information to gardeners on growing plants from seeds of fruits purchased in markets.

The Rural Advancement Fund International, P.O. Box 1029, Pittsboro, NC 27312. Offers the Community Seed Bank Kit, developed to aid communities in preserving local varieties of plants. It is available for $4.50 postpaid.

38 Seed Banks

There is a still larger dimension to this business of selecting, saving, and exchanging seeds. Although this concern may seem remote from our neatly bounded backyard gardens, it is a matter that may profoundly affect our children and grandchildren. The prosperity of our future food crops, the corn, soybeans, wheat, millet, barley, rice, beets, and other seed-sown foods on which our agriculture is founded, may well depend on the maintenance of primitive strains of these vegetables.

How can this be? Consider what has happened during the 10,000 years since people began domesticating plants. We have, largely through selection, produced races of food plants that germinate uniformly, yield better than their primitive counterparts, and often taste better. Only very recently, within the last 200 years, has deliberate crossing been used to improve plants. In the process, though, we have made these plants completely dependent on us for their continued survival. Few domesticated food plants could survive in the wild. Moreover, the lack of genetic diversity caused by deliberate inbreeding to produce high-yield hybrids has made some of our primary food crops vulnerable to crop failure, because disease, if it strikes, is likely to affect all the highly inbred plants that lack resistance. The Irish potato famine of the 1830s and the destruction of one-fifth of the corn crop in 1970 by the southern corn blight (which affected only T-cytoplasm corn) are both examples of the disastrous effects a blight or fungus can have on a crop with a narrow genetic base.

After the 1970 corn blight epidemic, the National Academy of Sciences studied the genetic make-up of the major food crops currently being planted. The results were scary. Dr. H. Garrison Wilkes, of the University of Massachusetts, writing in the *Bulletin of the Atomic Scientists,* reported:

For hard winter wheat, about 40 percent of the acreage was planted with just two varieties and their derivatives. In soybeans, the genetic base was limited to just six seed collections. For sorghum, like corn prior to the blight, all then current hybrids used the same cytoplasmic sterility component.

Let's go back, for a moment, to the primitive relatives of our important food crops, those irreplaceable varied strains to which scientists have returned even in recent years for help in strengthening a certain quality that they wanted to breed into a crop. For example, research in developing a frost-resistant tomato, being done by Dr. Richard Robinson of the New York State Agriculture Experiment Station in Geneva, depends heavily on the use of wild tomato varieties from the Andes as sources of the

243

desirable genetic traits. We'll be needing these reservoirs of genetic diversity for qualities we may not at present recognize as important. Where can we find them?

The food crops that have become important to humankind were developed, over the years, in certain ancient seats of civilization. First described by the Soviet geneticist N. I. Vavilov, these are the nine major centers:

- Ethiopia—barley, coffee, flax, okra, onions, sorghum, wheat
- Mediterranean—asparagus, beets, cabbage, carob, chicory, hops, lettuce, oats, parsnips, rhubarb, wheat
- Asia Minor—barley, cabbage, carrots, lentils, oats, peas, rye, wheat
- Central Asia (Afghanistan, Turkestan)—cantaloupe, carrots, chick-peas, cotton, grapes, mustard, onions, peas, spinach, turnips, wheat
- Indo-Burma—amaranth, cucumbers, eggplant, lemons, millet, oranges, black pepper, rice, sugar cane
- Siam, Malaya, Java—bananas, coconut, grapefruit, sugar cane
- China—azuki beans, buckwheat, Chinese cabbage, millet, radishes, rhubarb, soybeans
- Mexico, Guatemala—amaranth, beans, cashews, corn, red pepper, squash, tomatoes
- Peru, Ecuador, Bolivia—beans, red pepper, potatoes, squash (*Cucurbit maxima*), tomatoes

The three minor centers are:

- Southern Chile—potatoes, strawberries
- Brazil, Paraguay—cacao, cashews, peanuts, pineapples
- United States—blueberries, cranberries, Jerusalem artichokes, sunflowers

Until very recently, considerable acreage in each of these areas was devoted to the native varieties of the old traditional crops. Perhaps their yields were something less than spectacular, but there was enough variation in the individual plants grown from seed to ensure that, even under poor conditions, at least *something* would grow.

Today, though, as Dr. Wilkes describes the situation:

Mexican farmers are planting corn seeds from a Midwestern seed firm; Tibetan farmers plant barley from a Scandinavian plant-breeding station; and Turkish farmers plant wheat from the Mexican wheat program. Each of these classic areas of crop-specific genetic diversity is rapidly becoming an area of seed uniformity.

It's a small world. Too small to allow the loss of native strains to continue. In "The World's Crop Plant Germplasm—An Endangered Resource," published in the *Bulletin of the Atomic Scientists,* Dr. Wilkes writes:

The only place genes can be stored is in living systems. And extinction of a native variety can take place in a single year if the seeds are cooked and eaten instead of saved as seed stock. Quite literally, the genetic heritage of a millenium in a particular valley can disappear in a single bowl of porridge. The extinction of these local land forms and primitive races by the introduction of improved varieties is analogous to taking stones from the foundation to repair the roof.

What are we going to do about it? "Positive steps," Dr. Wilkes maintains, "must be taken to bank and preserve this genetic wealth for the future ... when something is burning, you do something about it first and then you talk about it. The current state of genetic erosion is such that we'd better act on it now and talk about it later."

Official efforts at preserving our national and worldwide seed heritage have taken several forms, each of them merely a modest beginning at staving off the disastrous, fast-paced loss of irreplaceable native varieties. These include the following:

- The establishment by individual nations of centers for the protected growth of irreplaceable plant races is a valuable concept on which work has begun, but much more work urgently needs to be done.
- The International Board for Plant Genetic Resources, established in 1974, helps to coordinate plant-conserving efforts in various countries and to initiate new programs.
- The National Seed Storage Laboratory in Fort Collins, Colorado, is building up and maintaining collections of seeds from "unimproved" native varieties. Sad to say, though, the National Laboratory has been underfunded for years, and some experts consider its whole collection endangered.

Samples of native varieties of different crops are preserved in or near their centers of origin by 14 international agricultural research centers. These research centers include, among others:

- The Centers for Tropical Agriculture, in Columbia and Nigeria
- The West African Rice Development Association, in Liberia
- The International Rice Research Institute, in the Philippines
- The International Potato Center, in Peru
- The International Maize and Wheat Improvement Center, known as CIMMYT, a private, internationally funded nonprofit organization, with headquarters in Mexico, that conducts worldwide research on maize, wheat, and triticale

When I observed CIMMYT in operation during a visit to its headquarters in Mexico last summer, I was awed, both by the magnitude of the problem and by the scope of CIMMYT's efforts to rescue and preserve endangered seeds. There at El Batan, north of Mexico City, is the world's largest corn-seed collection: 12,000 different strains of corn from all over the world are being kept alive in long-term storage as a germplasm reserve.

In the recently reorganized CIMMYT seed bank, samples of corn are stored in two rooms—one kept at 32°F (0°C) for a 20-year term, after which time part of the

Saving seeds has an impact that extends far beyond our own backyards; it enables us to pass on the right genetic heritage of our plants to future generations.

collection will be replanted for renewal, and another stored in a new, colder room, kept at 1.4°F (–17°C), meant to maintain seed viability for 100 years, with periodic replanting of small seed samples every 20 years to replenish the active collection. Each year, CIMMYT sends out about 1 million corn seeds from its active collection to researchers around the world. Many of these seeds will produce corn that wouldn't win any prizes at the county fair—irregularly shaped, poorly filled, small, or oddly colored ears—but they are a treasure trove of genes for drought and disease resistance, high protein content, or perhaps for their ability to sprout in dry or cold soil. These genes could never be replaced, once lost.

All this may seem very far from the mulched paths of your own backyard plot, but it seems to me a matter of the greatest practical value to appreciate the vital importance of saving and passing on our rich genetic heritage. "Our whole nation is vulnerable," says Dr. Wilkes. "We risk losing both food crops and germplasm. The brunt of the resulting increase in food prices will fall on people like the young families who are budgeting to make house payments. Their small children are at the most critical point for food quality. Young kids need nutrient density. Our vulnerability in this area is a real, basic problem for the nation to come to grips with."

You buy seeds. You vote. Perhaps you contribute to organizations established to help third world countries improve their agriculture. You have the privilege of writing to your senators and congressmen about matters of importance, and as a gardener, you have a chance to seek out, replant, and save varieties of open-pollinated seeds that are no longer available commercially. At the very least, you have the opportunity to support the efforts of others who, through organizations like the Seed Savers Exchange (see chapter 37), are working to conserve the genetic diversity of the plants we grow in our backyards. In each of these ways, the gardener can make a significant personal contribution to what is, in my opinion, a matter of the greatest concern to all of us. As Professor Edward O. Wilson of Harvard University has said, "The one process ongoing in the 1980s that will take millions of years to correct is the loss of genetic and species diversity . . . this is the folly our descendants are least likely to forgive us."

Encyclopedia of Plants to Grow from Seed

There's not a sprig of grass that shoots uninteresting to me.

Thomas Jefferson

Growing Garden Vegetables and Fruits from Seed

In this section, you'll find hints for handling the seed planting and early seedling care of the full range of popular garden vegetables and a few garden fruits. For plant culture directions spanning the entire season, refer to one of the excellent general vegetable books listed in Recommended Reading at the back of this book. Optimum germination times for most garden vegetables are listed in table 3 on page 33. Time from planting to maturity, indicated after each vegetable heading to follow, varies according to the variety chosen and local conditions. Although most vegetables are day-neutral, some, including okra, onions, spinach, and lettuce, mature in response to the length of the day. You'll find gardening hints for dealing with these photoperiodic vegetables under their individual headings. Except for root crops, cucurbits, leaf lettuce, peas, beans, peanuts, soybeans, New Zealand spinach, celeriac, amaranth, and other crops that are usually direct-seeded, the listed days to maturity are counted from the time the vegetable is transplanted into the garden, not from the time the seed is sown. If you're planting in raised beds, disregard row-spacing instructions and use only the recommended space between plants.

Amaranth
Amaranthus spp.

This nutritious plant is grown in two forms, vegetable amaranth and grain amaranth. *Vegetable Amaranth:* Reaching a height of 2 to 3 feet, this crop is raised primarily for its mild-flavored leaves. Tampala and Chinese spinach (Hinn Choy) are two popular varieties of vegetable amaranth. Amaranth grown for greens makes a good succession crop to follow an early crop of radishes.

Days to Maturity: 40 to 50

When to Plant: Plant seeds outdoors when soil is warm, or start plants early indoors in pots.

How to Plant: Cover the seeds with no more than ¼ inch of soil. Thin to 8 to 10 inches apart in rows 12 to 15 inches apart. Use the thinnings as your first crop.

Growing Conditions: Requires full sun, hot weather.

Remarks: Flavor is best before flowering.

Grain Amaranth: This is a native American plant that was used as a ceremonial food by the Aztecs. It stands 4 to 7 feet tall, with heavy plumes of small protein-rich seeds.

Days to Maturity: 90 to 110

When to Plant: When soil is thoroughly warm.

How to Plant: Sow seeds outdoors, and thin plants to 8 to 10 inches apart, in rows 18 to 36 inches apart. (Some amaranth-growers prefer to leave the plants in a thicker stand, 1 to 2 inches apart.)

Growing Conditions: Amaranth requires well-drained soil. Extra fertilizer will increase yields of vegetable amaranth, but seems to make little difference in the amount of grain produced by seed-bearing types. Once established, grain amaranth plants are highly drought resistant. Vegetable amaranth will be more tender and productive if well watered.

Remarks: Harvest the heads individually as they ripen.

Artichoke, Globe
Cynara scolymus

Although very sensitive to cold weather, this perennial has been successfully grown in the mid-Atlantic and New England states, far from its center of commercial production in California. Chilling during early growth in some cases stimulates bud formation the first year.

Days to Maturity: These bear a year after planting, sometimes the first summer.

When to Plant: Sow seeds indoors at 70°F (21°C) about three months before the last frost, and set the plants out, preferably in a protected place with a southern exposure, when the weather has settled thoroughly—late May in the mid-Atlantic region. Some gardeners rush their artichokes with heavy feedings and early planting under protection about four weeks before the last frost. They do this because the artichoke seems to be stimulated to form buds when days are long, and they want the plant to be sufficiently robust by that time to support sizable "chokes."

How to Plant: Space plants about 2 feet apart, in rows 3 feet apart. Keep soil out of the plant's crown when setting it out.

Growing Conditions: Globe artichokes do well in fertile, humus-rich, well-limed, well-drained soil. Plants need careful protection where winters are cold, and even then some may die from crown rot. If the root lives, it will send up new shoots the following year. Achieving a bud (edible artichoke) the first year makes all the effort worthwhile, though.

Asparagus
Asparagus officinalis

Days to Maturity: Harvest begins three years after planting.

When to Plant: Seeds may be started indoors but are most commonly sown in the garden, often in a special nursery row, either in the fall or in the spring, about the time apple blossom petals begin to fall.

How to Plant: Soak the seeds in water overnight before planting, to soften the tough seed coats. The tiny, feathery seedlings look delicate, but they transplant well. Sow seeds 1 to 2 inches apart in a nursery bed. Transplant the plants out of the nursery row no later than the second year because older plants grow deep roots which must be sacrificed in digging up. Transplant them about 15 inches apart, in rows 3 to 4 feet apart.

We know now that the heroic measures formerly used to plant asparagus on soil mounds in 18-inch-deep trenches are not necessary. Roots planted 2 to 8 inches deep do just as well. You can even set the crowns on the soil surface and mound 3-inch ridges of soil above them, according to some researchers, who say that although the effect of this method on asparagus yields has not been studied, crown survival was excellent.

Growing Conditions: The permanent asparagus bed can be well prepared while the seedlings are growing. Dig in plenty of limestone and well-rotted manure and make certain the ground is well drained and well aerated. Plants that produce male flowers are somewhat more productive than the berry-producing females, so some growers cull out berried asparagus plants. Mulch the bed and wait until the second year to take your first two- to three-week harvest. After five years you can pick for ten weeks.

Bean, Azuki (Adzuki)
Vigna angularis

These small, rich-flavored, deep red beans are one of our favorite vegetable proteins. The variety Express will mature as far north as Maine. Once, when I planted azuki beans purchased from a store, the plants were such a late variety that they failed to mature here in southern central Pennsylvania.

Days to Maturity: 118 for Express, an extra-early variety.

When to Plant: Plant seeds after frost, when daytime soil temperatures average above 60°F (16°C).

How to Plant: Azukis have done well for me in double rows, 4 inches apart, with seeds spaced 2 to 3 inches apart. The seedlings are smaller than those of snap beans at first, but soon develop into strong plants. Don't give up on them; they flower rather late in the summer, but they'll continue to grow and produce flowers and beans until frost.

Growing Conditions: They accept a wide range of garden soils, but prefer soil that isn't highly alkaline.

Bean, Fava
Vicia faba

These frost-tolerant legumes produce the first shell beans of the season for northern gardeners. They need a long, cool spring to mature their pods. When the weather turns warm, blossoms drop, and the plants quit producing. Ipro, a recent introduction, resists heat better than other varieties.

Days to Maturity: 75 to 85
When to Plant: Plant favas when you plant your peas—as soon as your ground is workable.
How to Plant: Sow the seeds 1 to 2 inches deep. Plant them 4 to 6 inches apart, in rows 1½ to 3 feet apart. Thin plants to 6 inches apart. Double rows work well. The bushy 3-foot-tall plants produce 5- to 10-inch-long, tough pods containing large flat beans. Favas are also a first-rate green manure plant.
Growing Conditions: Fava beans survive frost but not heavy freezes. Mulch to keep the soil cool. The plants tolerate poor soil but appreciate good ground, too.
Remarks: Some people are allergic to fava beans, so make your first taste a small one, just to be sure. They are an ancient food, though, having been grown in China for 5,000 years. Amounts of the offending glycosides, vicine and convicine, are lower in mature than immature fava beans and may be further reduced by soaking the beans in dilute vinegar (1 percent acetic acid) for 48 hours, changing the solution after 24 hours.

Bean, Lima
Phaseolus lunatus

Days to Maturity: 60 to 85
When to Plant: Lima beans are real tenderfeet. The seeds rot quickly in cold soil. They are especially vulnerable to damage from low temperatures in the early stages of germination, when they are absorbing moisture. Wait until the soil is thoroughly warm, at least 70°F (21°C), before planting the seeds. This is late May for us in southern central Pennsylvania, when peonies are in full bloom. If you handle seeds carefully, you can presprout them indoors. (See directions for presprouting in chapter 5.)
How to Plant: Presoak seeds in water for an hour or so if you wish, but not too long, or they'll split. Inoculate them with garden pea/bean inoculant. Plant them *eye side down* so that the root and cotyledons that emerge on germination will be headed in the right direction. Space seeds every 5 inches in rows 3 feet apart, with 1 inch of soil drawn over them. Thin to at least 8 inches apart. Don't transplant.
Growing Conditions: Germination is sometimes slow in acid soils. If seedlings decay at the soil surface, excess surface moisture may be to blame, especially in heavy soil. You might fill in over the planted seeds with a small amount of sand in such cases. Pods may drop off if plants are cultivated deeply, so get weeds under control by blossoming time. Limas require more warmth and more calcium in the soil than snap beans.
Remarks: Pole limas are later than bush limas, but have better flavor.

Bean, Snap
Phaseolus vulgaris

Days to Maturity: 40 to 56
When to Plant: Sow seeds when all danger of frost is past. An exception, the Royal

Burgundy purple bean, can be planted as early as three weeks before the last expected frost, because it is less prone to rot in cold soil. The planting of Royal Burgundy we make in late April usually pays off in beans by the end of June.

Make succession plantings of bush beans, which bear heavily but for a short period of time. Neither the bush nor the pole beans transplant well. (I do have a friend who has done it successfully, though, on a damp drizzly day.) I always put in a late planting at the end of July to bear in early fall when cool nights have sent the bean beetles on their way.

How to Plant: Briefly presoak bush bean seeds for an hour or two and plant 2 to 3 inches apart in single rows, 4 to 5 inches apart in double rows, with 15 to 24 inches between rows. Inoculate after soaking.

Growing Conditions: Beans are sensitive to an excess of potash in the soil, a fact I learned the hard way one year when I planted pole beans in a spot where a big, heavy, wet bag of wood ashes had burst and spilled the previous year.

Varieties: Pole beans begin to bear somewhat later but continue longer than bush beans. They should be planted in a circle around each pole, about eight to ten seeds per pole. Thin them to four or five plants. These taller-growing vines should be planted on the north side of the garden to avoid shading other plants.

Scarlet runner beans, which may be eaten when young before they develop strings, accept cooler soil than snap beans and may be planted one to two weeks earlier. Nick and presoak the seeds before planting.

When setting the poles, some gardeners dig a hole, push in the pole, and fill in the hole with compost topped off by an inch of soil. This makes a good starting bed for the bean seeds.

Beet
Beta vulgaris

Days to Maturity: 55 to 80

When to Plant: Plant seeds in the garden about a month before you expect your last frost. Earlier plantings sometimes suffer from freezing. In the far north, beets that are planted too early and chilled may bolt to seed when weather turns warm.

How to Plant: Beet seeds are notorious for spotty germination. Their coats contain a germination inhibitor. Presoaking the seeds for one to two hours before planting will dissolve the chemical as well as soften the seed coats and speed germination. The wet seeds, though, are more difficult to handle. Some gardeners find that crushing beet seeds lightly with a rolling pin also helps to encourage more complete germination.

Beet seeds, for some reason, need to be well tucked in. They often fail to germinate if not in sufficiently close contact with the soil. An accomplished gardener of our acquaintance says "Cover the seeds with ½ inch of soil and then *stamp* on them!" He's used to working sandy soil. Here, on our heavier clay, I tamp the row with the flat blade of the hoe after planting the seeds.

The beet seed that you plant is actually a seedball, an aggregate of two to six individual seeds. Consequently, even when you follow the recommended spacing of 2 inches apart, the seedlings will need to be thinned. Thin plants to 3 inches apart in rows 1 to 2 feet apart. Double rows spaced 3 to 4 inches apart work well, as do solid beds. For an extra early crop, Stokes Seed Company recommends starting beet seeds in the greenhouse March 1 for transplanting to the garden in mid-April. Thinnings may be transplanted, which provides, in effect, a succession crop because the root insult sets them back about two weeks. Some gardeners nip off the long, thready tip of the beet root when transplanting. When I do this, I cut off the outer leaves, too, to balance root and top growth.

Growing Conditions: I often need to cover my beet row with a length of wire fencing bent into a U shape over the plants so rabbits won't eat the tops. Water your beets in dry weather to keep them tender.

Remarks: Beets are very sensitive to toxic substances in the soil and may fail to germinate if planted too close to walnut trees or in soil with herbicide residues. They may fail to grow well in highly acid soil.

Broccoli
Brassica oleracea, Botrytis group

Days to Maturity: 55 to 98

When to Plant: An early start on this cool-weather annual will give you fine-quality green heads before uniformly hot weather reigns. Plant seeds thinly in flats about five to six weeks before planting them outside, which can be up to a month before your last expected frost. Broccoli can withstand frost down to about 25°F (-4°C). It grows best at 60°F to 65°F (16°C to 18°C). Make an early-summer planting— directly in the row if you wish—to produce fresh new plants for fall eating.

How to Plant: Transplant the seedlings at least once into larger flats and set out young plants 18 to 24 inches apart, in rows about 24 inches apart. Broccoli plants spaced 10 to 12 inches apart will also yield well but will have smaller heads.

Growing Conditions: Failure to produce heads can be caused by hot weather, lack of water, and low soil calcium.

Pests: Cutworms, flea beetles, and cabbage moth larvae are common enemies of the broccoli seedling.

Varieties: I like Green Comet or Premium Crop for spring planting. Waltham 29 is still a good fall broccoli.

Brussels Sprout
Brassica oleracea var. *gemmifera*

Days to Maturity: 92 to 120

When to Plant: A cool-season crop, which does best when planted in late spring either in the garden row or in flats in the cold frame and transplanted to its permanent spot when 3 to 5 inches high. May or early June seedlings grow into mature, harvest-sized plants by fall. Don't bother to plant brussels sprouts in early spring for a summer crop. The tiny leafy heads that are so delicious after frost taste pretty murky before being nipped by cold.

How to Plant: Space transplants 2 feet apart. For good solid sprouts, firm the transplants well into the soil. Loose, leafy sprouts often result from drying or insufficient soil-root contact at transplanting time.

Growing Conditions: Brussels sprouts aren't especially particular about soil type, but shortages of potash, phosphorus, or magnesium will hold them back. Gardeners working in cold climates should prune off the top rosette of leaves from their plants in early September to encourage prompt development of sprouts all along the stem.

Cabbage
Brassica oleracea var. *capitata*

Days to Maturity: 60 to 110

When to Plant: Your earliest plantings may be seeded thinly in flats indoors about four to six weeks before planting them outside, which can be a good month before the frost-free date. Make repeated plantings every month or so until July for fresh tender cabbage for the table. The large-headed kraut types requiring a long growing season should be planted in mid- or late spring.

How to Plant: Since cabbage seeds seem to germinate 100 percent, keep a rein on your seed-sowing hand to ensure sturdier seedlings. (For an exercise in self-discipline—and for a fine stand of seedlings—use tweezers to place seeds in the flat at 2-inch intervals. If your seeds are fresh, plenty of them will come up, believe me!) If you do get carried away, transplant those crowded seedlings to a roomy flat before they develop their true leaves, or they'll get hopelessly leggy. Set plants 12 to 18 inches apart, in rows 18 to 36 inches apart. Cabbage appreciates rich soil, well supplied with humus. I give each transplant a handful of compost at planting time and a liquid feeding (manure tea or diluted fish fertilizer) a week or two after planting.

Growing Conditions: Cabbage grows best in cool, damp weather. Although it lives through a fairly severe frost, cabbage that is subjected to a considerable period of very cold weather when the plant is young sometimes becomes fibrous, a stage on the way to developing an early seed stalk. Cabbage started indoors shouldn't be subjected to night temperatures above 60°F (16°C). These are fine points, though. Cabbage is easy to raise. Easier than radishes, I think.

Cabbage, Chinese
Brassica rapa, Pekinensis group

This cool-weather vegetable is extremely prone to bolting when weather turns warm. It does not transplant well, either. Although it is possible to get away with moving the seedlings when they are very young, they may be severely checked, or sent into bolt, especially in spring. I routinely transplant them in the fall. It's better to seed early plantings in pots.

Days to Maturity: 45 to 70

When to Plant: Several new varieties, including Spring A-1 and Nagoda, can be grown in the spring, but for most people the best plan is probably to consider Chinese cabbage solely a fall crop.

How to Plant: Sow seeds in July right in the row, perhaps where the early peas have just been yanked out, and thin to 2-foot spacing. Good, fertile, humus-rich soil will encourage rapid growth of tender leaves. Both heading and loose-leaf varieties are available.

Pests: Flea beetles can be hard on young seedlings.

Carrot
Daucus carota var. *sativus*

Choose seeds of a variety that's suitable for your soil type. The half-long carrots like Danvers are far better choices for heavy soil than long thin kinds like Imperator, which can reach full length only in sandy soil. Carrot flavor is determined mostly by variety, not by soil, so if you're dissatisfied with the flavor of carrots from your garden, try some other varieties.

Days to Maturity: 60 to 76

When to Plant: Plant as soon as the ground can be worked finely, but remember that a severe frost can damage germinated carrots. Second-early plantings made in the beginning of May do better for me than extremely early ones. Carrots usually have better-quality roots when they mature in cool weather. Make several succession plantings up until July, always in deeply worked soil.

How to Plant: This fine-seeded, cool-weather crop is easily overplanted. Dilute seeds by mixing them with sand or dry used coffee grounds. I plant carrots in a "raised row" of loose soil raked onto an 8-inch-wide bed from both sides. Cover the seeds with ¼ inch of soil—never more than ½ inch. Germination is very slow; a month is not unusual for early plantings in cold soil. You might want to dribble a few radish seeds in the furrow to mark the row.

The best time to thin carrots is when the soil is damp. Plants should stand 2 to 4 inches apart, depending on size; the closer spacing is ideal for the smaller Baby Finger carrots, the wide spacing for the Oxheart types.

Growing Conditions: Keep the soil moist and well supplied with organic matter. Crusted soil can suppress tiny germinated sprouts. Carrots don't need rich soil; in fact, excess nitrogen causes hairy roots.

Pests: If the tunnel-boring larvae of the carrot rust fly are a problem in your area, you might find that carrots you plant after June 1 and harvest before mid-September escape both the first and second hatches of larvae. British gardeners plant sage or scorzonera (black salsify) to repel the carrot fly.

Remarks: Young carrot seedlings are easily overwhelmed by weeds, which often seem to get a head start, especially in wide-row plantings.

Cauliflower
Brassica oleracea var. *botrytis*

A little finicky, but not really difficult. The important thing to remember is that any insult—real or imagined—will send cauliflower into a pout. The result is a thumb-sized buttonhead, or none at all. Just keep the seedlings growing steadily.

Days to Maturity: 50 to 85

When to Plant: Plan on setting out spring-started cauliflower seedlings about two or three weeks before your last frost date, not a bit earlier. They grow best at 57°F to 68°F (14°C to 20°C).

Plants for the fall crop may be started in May. Either way, avoid having plants mature in hot, head-stunting summer weather. I plant cauliflower only for a fall crop, setting out seedlings in June or July for September to November heads.

How to Plant: Transplant when the plants are young, 6 or 7 inches tall, firming soil well over seedling roots. Set plants 18 inches apart, in rows 24 inches apart.

Growing Conditions: If possible, water the seedlings daily for several days after moving them. Shade them well, also. For good heads, cauliflower should grow rapidly. Rich soil that is not too acid will help. When the head develops, I break off a bottom leaf and lay it on the curd (the developing head) to keep out the light. In self-blanching cauliflower, leaves grow either wrapperlike or in "stove-pipe" form to shade the heads and keep them white.

Remarks: Southern gardeners whose winters are mild enough to plant cauliflower in the fall for early spring eating should aim for 10-inch plants by frost. Wait until spring to fertilize them.

Celeriac
Apium graveolens var. *repaceum*

Days to Maturity: 110 to 120

When to Plant: Seeds can be planted right in the row in early spring, but since germination is slow and weeds are fast, I prefer to raise celeriac seedlings indoors in flats, setting them out two or three weeks before our last expected frost, about when the apple blossoms begin to fall. Earlier planting-out may cause bolting.

How to Plant: Celeriac transplants well, as long as you remove only one plant at a time from the flat and protect the roots from drying out. Set the seedlings 6 to 8 inches apart.

Growing Conditions: These cool-weather biennials do well in heavy well-limed soil. Keep them moist until they catch on, as evidenced by new leaf growth.

Celery
Apium graveolens var. *dulce*

Days to Maturity: 98 to 130

When to Plant: Celery may be planted indoors in February or March and set out about two weeks after the last expected frost, or sown in April or May for a fall crop. Germination is fairly slow—about three weeks. Celery does best in cool weather, so the fall crop is often the best. Celery germinates best at 60°F to 70°F (16°C to 21°C).

How to Plant: If temperature at planting time is unfavorably high, exposure to light will help to stimulate germination. Alternating temperatures (warmer by day, cooler at night) also helps. At the preferred lower temperature, light doesn't seem to make any difference. Keep the soil evenly moist during germination. When hardening off, avoid chilling the plant below 50°F (10°C), or it may bolt to seed. Celery has a skimpy root system and transplants well as long as the job is done early, when the seedlings are 3 to 5 inches high, before a taproot forms.

Growing Conditions: Because celery's ancestors were marsh dwellers, it needs a steady and plentiful supply of moisture, more than most other garden vegetables, but drainage should be good. Try sinking bottomless cans between plants, which should be planted 8 to 10 inches apart, and filling the cans with water from the hose. Celery is a heavy feeder, too, so put compost in the planting hole and give plants extra feedings of manure tea or fish emulsion about twice a month.

Celtuce
Lactuca sativa var. *asparaginia*

Generally described as a two-purpose vegetable, celtuce produces lettucelike leaves and a crisp, juicy stalk. The leaves soon grow bitter, but the stalk is delicious if harvested when its diameter measures about 1 inch. Later, when the plant flowers, the stalk grows tough.

Days to Maturity: 45 for leaves, 90 for stalks

When to Plant: Plant celtuce in early spring for a summer crop, or in early summer for fall picking.

How to Plant: The seeds are fine and should be sown thinly and lightly covered with fine soil. Eat the greens as you thin them, leaving plants spread 12 to 18 inches apart for stalks.

Growing Conditions: Celtuce is easy to grow. It accepts a wide variety of soil conditions and is seldom troubled by insects or diseases.

Remarks: Be sure to peel the stalk before eating; the rind is bitter.

Collard
Brassica oleracea, Acephala group

Days to Maturity: 80 days

When to Plant: Collards may be started indoors and transplanted, or they may be sown in the garden row in early spring.

How to Plant: As with most members of the cabbage family, germination is profuse, so plant seeds thinly, especially in the flat. Unlike the situation with most cabbage, collards will remain in the garden row all season. They're very hardy, and flavor is really best after frost, so plan your rows accordingly. A few plants are usually enough for the average family. Space them 2 feet apart or closer if you plan to use the thinnings for soup greens.

Growing Conditions: Treat collards like cabbage.

Corn
Zea mays var. *rugosa*

Days to Maturity: 64 to 94

When to Plant: With the exception of the man I once read about who started his small block of early corn in peat pots indoors under lights, gardeners plant corn in the open ground, starting a week or two before the last expected frost for the earliest corn. Corn seeds tend to rot before they sprout if planted in soil colder than 50°F (10°C). Germination is much better above 60°F (16°C). Corn does not transplant well.

The North American Indians, who had no calendar, went by tree signs in planting their corn, making their first plantings when the white oak leaves were as large as a squirrel's foot.

Make successive plantings 10 to 14 days apart. Toward midsummer, sow earlier-maturing corn (65 to 70 days) to sneak in another picking before frost. If you've got 90 days to go until your average first fall frost date, don't plant a 90-day corn. The cool fall nights, which can begin early in September, will slow its growth. You may find that these summer plantings of corn will germinate best, especially in a dry season, at a depth of 2 to 3 inches.

How to Plant: Early plantings should be covered with 1 inch of soil. Sow seeds every 4 inches and thin plants to stand 8 to 12 inches apart, in rows 2½ to 3 feet apart. The first shoots are fairly frost-resistant, but foot-tall corn will be killed by frost. Even if you're planting only a small amount of corn, arrange your rows in blocks at least four rows wide, with a minimum of 16 plants, so that the wind-carried pollen will reach more of the other plants. Four short rows are better than two long ones. Drop two to four seeds in every foot of furrow and thin later to about 10 inches apart for the shorter, early varieties and 12 to 15 inches apart for more robust late corn. Avoid leaving stray seed kernels on the surface of the soil; they'll attract birds.

That old folk saying about planting corn tells it just about right:

> *One for the rook,*
> *One for the crow,*
> *One to rot and*
> *One to grow.*

Growing Conditions: Although corn thrives in many different kinds of soil, it needs plenty of nitrogen and moisture to support its rapid, exuberant growth. Hoe carefully after the corn is 1 foot tall to avoid damage to its many shallow roots. Weed control for corn is probably most critical during the first three weeks after the shoots emerge.

The following tips will help you get the most from your planting of extra-sweet corn: Save the extra-sweet corns for your mid-season plantings, and make early plantings with varieties known to do well in cold soil. (Consult catalogs.) Check the catalog and seed packet carefully to see whether the variety you're planting needs isolation to prevent crossing. If so, plan plantings so that no other corn within 500 feet or so is in tassel at the same time. There should be at least a 14-day difference in maturity dates. For extra protection, plant the earlier corn downwind from the later crop. Barrier plants, like two to five rows of sunflowers, or hedges can reduce isolation distances.

Pests: Corn seeds sown after mid-June should escape corn borer damage. For earlier plantings where corn borers are a problem, try soaking the seeds for two hours in a strong "tea" made from butterfly weed or English ivy.

Varieties: Regular sweet corn, because its shrunken seed kernel contains less stored starch, has lower seedling vigor and disease resistance and a greater need for rich soil than dent, flint, or popcorn. This sensitivity was magnified in the early high-sugar varieties, in which the seed kernels have even less starch. These corns with the Shrunken-2 gene, like Extra Early Sweet and Illini Chief, should be planted less than 1 inch deep in warm, well-drained soil so the seedling leaf can emerge into the sunlight early, before it uses up the scant food supply in its shriveled seeds. Corn with the Shrunken-2 gene must also be isolated from other corn varieties at tasseling time to prevent crossing, which would result in inferior, starchy ears.

Breeders have continued to work on improving the weaknesses in the Shrunken-2 types of corn and on developing new types. One of the most promising of these is the Everlasting Heritage series (identified as E. H. in seed catalogs), which contains genes derived from an American Indian roasting corn. Kandy Korn, despite its coy name, is an excellent E. H. variety. The E. H. corns have solved at least two of the problems associated with growing high-sugar corn: The seeds have more vigor, and it's not necessary to isolate these varieties from other tasseling sweet corns. Kernels pollinated by regular sweet corn will have normal sweetness, but they will not be starchy.

Corn, Broom
Panicum miliaceum

True, you can't eat it, but broom corn is a crop that is lots of fun to grow, and directions for raising it are hard to find.

Days to Maturity: About 120

When to Plant: Plant the small round seeds right after frost.

How to Plant: Sow seeds in the garden, and thin the seedlings to stand about 6 inches apart in rows 3 feet apart. The plant resembles a slender cornstalk.

Making Brooms: Brooms are made from the stiff seed-bearing straws after the seeds have been removed. When the seeds are well formed, bend the top 20 to 24 inches of each stalk so that the straws will hang straight down. If left upright until seeds turn red, the straws sometimes dry in a curved position. Cut off the seed heads at the joint where the stalk was bent, about two weeks after you snapped them over. Broom makers, from whom you can probably obtain seeds for your first planting, often appear at local craft shows. A few seed companies sell broom corn seeds.

Cucumber
Cucumis sativus

Days to Maturity: 48 to 80

When to Plant: Cucumber seeds germinate best around 90°F (32°C). At temperatures below 60°F (16°C) they're likely to rot before they sprout, yet once germinated, they can grow well at an average of 65°F to 75°F (18°C to 24°C). You can make an early chance planting of cukes by presprouting the seeds and planting the germinated seeds in a hill where you've buried a good handful of finished compost under an inch of soil. (Avoid adding unfinished compost as I once did. The carbon dioxide it gives off discourages cucurbit seeds, which need abundant oxygen for sprouting.) Start this early planting about the time the tall late irises bloom—mid-May here in my garden.

I make a second, main-crop planting in late May and a third planting in July to take over when wilt decimates my early vines.

How to Plant: To start the seeds indoors in soil, use individual pots and plant them about three to four weeks before transplanting them to the garden.

Outdoors, plant eight to ten seeds in a hill, with hills spaced 4 or 5 feet apart. For row planting, sow seeds about 8 inches apart and thin later so plants stand at least a foot apart.

Presprouted seeds should be watered in well at planting time. (Always use lukewarm water for cucumber plants.) Some gardeners report improved germination when sowing the seeds thin edge down. Be sure to cover plants if frost threatens. Night temperatures below 40°F (4°C) will retard cucumber plants.

Don't thin the plants too soon. Wait to see how much damage the cucumber beetle will do to the young seedlings. It's better to pinch or cut the seedlings. Pulling them might tear the roots of the ones you want to keep.

Growing Conditions: Whatever your garden plan, try to provide good air circulation for cucumber plants, and keep at least a few hills within reach of the garden hose so you can water them if a week passes without a good rain.

Gynoecious cucumbers, which produce only female flowers, need extra feeding, generous spacing, and regular watering to support their heavier fruit set. Packets of these seeds usually include some seeds that will produce regular plants that bear some male flowers to ensure pollination.

Choose seeds that are resistant, or at least tolerant, to whatever cucumber plagues prevail in your gardening area. I've had excellent results with County Fair 83, a hybrid pickling cuke that is not attractive to cucumber beetles. Cucumbers don't transplant well. It is possible, as I've mentioned, to presprout the seeds, but once set in soil, cucumbers shouldn't be moved. Don't mulch the young plants until the soil is thoroughly warm. Many gardeners fertilize cucumber plants when the vines start to run and again when fruits set.

Pests: Combat disease-carrying striped cucumber beetles with the natural preventive measures suggested above and, if necessary, a dusting of rotenone. Try covering some plants with Reemay early in the season to protect plants from cucumber beetles.

Eggplant
Solanum melongena var. *esculentum*

Days to Maturity: 56 to 76 days

When to Plant: A heat lover, eggplant must be started early indoors, eight to nine weeks before you will plant them out, so they will mature before cool fall weather sets in again. Seeds germinate best between 75°F and 90°F (24°C and 32°C).

How to Plant: I often presprout the seeds in damp paper towels tucked into a plastic bag and kept on the top of our insulated water heater. Keep the seedlings growing steadily. Any wilting or exposure to cold weather can reduce their fruiting later. Ideally, the growing temperature should not fall below 65°F (18°C). Growing the seedlings in individual pots minimizes root shock when setting out. Move them to 4-inch pots when they're about three weeks old so they won't get rootbound. Before transplanting eggplant, block out the flat a week or two beforehand so that new roots will form close to the plant. Since the plants are extremely tender, not only to frost but to cool temperatures in general, they should not be set out until the soil is thoroughly warm and daily temperatures remain in the 65°F to 70°F (18°C to 21°C) range. An old rule of thumb is to wait until the oak leaves are fully developed.

Eggplant seedlings should not be chilled or deprived of water to harden their tissues for a move to the outdoors, as you would do with tomatoes or lettuce. Such harsh treatment can cause woody stems and poor fruiting. Instead, just set them out for increasingly long periods of exposure to the sun. When planting them into the ground, take a good ball of earth along with each transplant, and water it thoroughly as soon as you set it in the hole.

Shade eggplant lightly for a day or so after planting in the garden row. Black plastic mulch helps to warm the soil early in the season, but should probably be removed when summer days turn hot. Cover plants if outdoor temperatures threaten to fall below 55°F (13°C).

Pests and Diseases: Flea beetles like young eggplant seedlings. Watch out for cutworms, too. My worst problem in growing eggplant has been the soil-borne disease verticillium wilt, which stunts the plants and makes the leaves wilt and turn yellow. Finally, I took to planting each seedling in a 3- to 5-gallon pot of compost and commercial potting soil, set on our sunny patio, with excellent results. (Plants in containers need frequent feeding and watering.)

Endive (Escarole)
Cichorium endivia

Escarole is the broad-leaved form; endive, the frilly cut leaf.

Days to Maturity: 85 to 98

When to Plant: Although it is possible to grow a spring crop of these delicious pungent greens by planting seeds indoors in March and setting them out three to four weeks before the last frost, the plants are best, I think, in the fall. Late-spring plantings tend to be tough and bitter and may bolt to seed when days turn hot. For a fall crop, I sow seeds in flats in July and transplant seedlings to gaps in the garden rows.

How to Plant: Cover the seeds with a thin sprinkling of fine soil. Thin the plants to stand about a foot apart, although 10 inches is sufficient if you plan to tie up the leaves to blanch them. Be sure the center of the plant is dry when you tie the leaves together. Blanching takes two to three weeks.

Growing Conditions: Choose a fertile, well-limed, well-drained spot for your endive and escarole.

Fennel
Foeniculum vulgare

Days to Maturity: 60 to 70

When to Plant: The biggest mistake most gardeners make with fennel is to sow a whole row of it at once. Then it all matures at once and grows woody at once when not used. Make small succession plantings instead, beginning in April and early May, skipping the hot summer months, which tend to bring on bolting, and planting a fall crop in July.

How to Plant: Cover the seeds because darkness aids germination. Thin Florence fennel plants to stand 10 to 12 inches apart. The slimmer stalks of sweet fennel may be thinned to a 6-inch spacing. You could start fennel indoors ahead of time, but you should sow seeds in pots and transplant the seedlings before they start to develop taproots. A light frost won't hurt fennel. For more tender stalks, mound the soil a bit around the base of the plant.

Growing Conditions: Fennel likes neutral soil, so add some limestone and compost if your soil is strongly acidic. Water the plants in a dry spell.

Varieties: Fall-maturing Florence fennel, *Foeniculum vulgare* var. *azoricum,* develops nicer crisp, anise-flavored stalks.

Gherkin
Cucumis anguria

Days to Maturity: 60

Botanically the gherkin is slightly different from the cucumber, but it is begun and grown in the same way, usually for pickles. The leaf and vine are smaller than those of the average cuke, and the plant can be more closely spaced—about 3 feet apart for hills, 8 inches apart for plants grown in rows. Gherkin seeds are somewhat slower to germinate than those of cucumbers.

Grains

When to Plant: Oats (*Avena sativa*) are planted in early spring, as early as the ground can be worked. The plants produce grain in 90 to 110 days.

Rye (*Secale cereale*) is often fall-planted. The seeds germinate well at 55°F to 65°F (13°C to 18°C), and some will even sprout at 33°F (1°C). Winter rye, which is planted in fall, matures the following summer, about a week earlier than the local wheat (see below).

Whether you plant your wheat (*Triticum aestivum*) in spring or fall depends on whether you choose winter wheat or spring wheat. Check at your local feed and seed store and buy the kind of wheat that is most commonly planted in your area. To avoid infestation, plant winter wheat after the Hessian fly date in your area—usually mid-September to October 1. (Ask your county agent or the salespeople at your local feed store.)

How to Plant: Although sowing seeds in drills 3 inches apart and covering them with ½ inch of soil will give you the best germination, for a small garden patch of grain you can broadcast seeds and rake them in.

You need about ⅔ pound of wheat seeds to plant 300 square feet, which would yield about ⅓ bushel. Winter wheat, which was planted in fall, ripens in June in the southern states, July farther north, and August in Canada.

Growing Conditions: Oats need a good supply of moisture. Rye tolerates relatively poor soil. Wheat dislikes wet ground and prefers to grow in soil that is not highly acid. In a dry season most grain seedlings can hold their own against weed competition, but in wet weather weeds can overtake a planting if not checked.

Remarks: Recent studies have shown that the rhizobium bacterial inoculant usually used for garden peas and beans can produce measurable increases in yield when applied to wheat seeds. If you have any leftover inoculant, you might want to try using it on grains.

A clever gardener we've visited likes to interplant clover with the winter wheat in his garden. He broadcasts Hubam sweet clover among fall-planted wheat in late February or early March. Then, after hand-harvesting the wheat in August, he lets the clover grow for a few weeks, mows it, and calls in his hens. The hens spend about a week penned in the garden eating the cut clover and return again in October and November to thoroughly scratch up and dig in the clover planting—a neat system of tillage that feeds soil, hens, and people.

Kale
Brassica oleracea, Acephala group

Days to Maturity: 55 to 65

When to Plant: Extra nutritious, fast-growing, very hardy, and nearly pest-free, kale tastes especially good after a frost. To have leafy mature kale plants ready when you need them in the fall and winter, start plants indoors, in a cold frame, or in a special nursery row in May.

How to Plant: Transplant them in June into their appointed well-composted row at 12- to 16-inch intervals. I often interplant fall kale with butterhead lettuce, which matures before the kale is ready to eat and thrives in the light shade.

Growing Conditions: Kale produces abundant tender leaves in well-fertilized soil, but it will survive freezing winters better if not given large doses of high-nitrogen fertilizer.

Pests: Protect the young seedlings from flea beetles.

Varieties: Hanover, a spring kale, develops rapidly during the spring months, but I think it's hard to beat the flavor of well-frosted fall kale.

Kohlrabi
Brassica oleracea var. *gongylodes*

Days to Maturity: 55 to 60

When to Plant: You can start early plants for June eating indoors in flats seven to eight weeks before your last frost and transplant them to the cold frame when they have their second pair of true leaves. Growth is best during cool weather, so for good results get an early-spring start and replant in early summer for fall bulbs.

How to Plant: Set seedlings out in the garden row about three to four weeks before the last frost date, spacing them 6 inches apart. You can also sow seeds directly, but starting with seedlings will give you better control of plant spacing and, in my experience, produce a stronger plant.

Growing Conditions: Plenty of compost in the soil will encourage rapid growth. Kohlrabi doesn't mind a light frost.

Remarks: Standard varieties grow woody when large, so they should be picked when with a diameter of 2 inches or less. The newer hybrids stay tender longer—up until they measure about 4 inches.

Leek
Allium ampeloprasum, Porrum group

Days to Maturity: 70 to 105

When to Plant: I plant the seeds in flats in February, and thin the seedlings to stand no closer than 1 inch apart in the flat. Plant them out in April or May, or direct-seed in the ground in April, preferably in a nursery bed to save space. Leeks need a long growing season (120 to 130 days), but they will accept a variety of growing conditions. Where winters are mild, leeks can be planted in fall for a winter crop.

How to Plant: Plant the seedlings in a trench 6 inches wide and 6 inches deep. When the plants are 6 to 8 inches high, gradually hoe fine soil from the sides of the trench to fill in the space. Do this every three weeks or so, adding an inch of soil to the trench each time. Always keep the soil level at or just below the point where the leaves diverge from the stem. Leeks are biennials, and those that take longer to mature will usually keep well into winter if hilled up with soil or mulched.

Growing Conditions: They do respond well to plenty of humus, and steady moisture, and their extensive roots appreciate deeply dug soil. Leeks left in the ground over winter will form corms—tiny "bulbs"—around the base in late spring. You can plant these corms for an easy second-year crop of this tasty vegetable.

Lettuce, Head
Lactuca sativa var. *capitata*

Days to Maturity: 72 to 96

When to Plant: Start lettuce seedlings six to seven weeks before planting them out, which can be as much as a month before your frost-free date if you cover plants when the weather forecast predicts a severe freeze. Light frost won't hurt them. For fall head lettuce, plant seeds in late July, no later than the first week in August, if you expect killing frost by October.

How to Plant: Head lettuce must be started indoors to beat the summer heat, because it needs 75 to 85 days to head. Head lettuce is less heat-tolerant than leaf or butterhead lettuce or romaine. Plant seeds in flats; then transplant seedlings to larger flats, allowing 2 inches of space between seedlings.

When ready to plant outdoors in spring after hardening off, clip off all outer leaves, leaving a 1- to 2-inch stub of small new leaves. Those outer leaves, when left on, do the plant more harm than good. They tend to be whipped about by the wind and draw off more moisture from the roots than they are worth in feeding power. The new little leaves that grow up gradually will be sturdy, and they have a chance to acclimate themselves to full sun by the time they have grown big. Set head lettuce plants a good foot apart in the row. I usually plant out summer seedlings without clipping.

Growing Conditions: Lettuce has a skimpy root system, so it must have a good supply of water and nutrients. It thrives in well-limed soil enriched with manure. The leaves are, in fact, 95 percent water. If April rains are insufficient, you'll have to take water to the plant yourself, by hose or watering can.

Summer Planting: High temperatures send lettuce seeds into dormancy, which explains occasional poor germination of summer plantings. Older seeds are less responsive to this effect of heat. If you've had trouble getting your summer lettuce plantings to germinate, you might want to save out seeds each year to store for the following summer. Chilling the seeds by keeping them in the refrigerator overnight will also help to break this summer dormancy. Lettuce is one of the few vegetables that germinates better when exposed to light. Varieties vary in their light requirements. Water the furrow well before planting seeds in summer, and just press the seeds into the ground. Leave them uncovered, or sift a scant layer of fine soil or a sprinkling of grass clippings over them.

Remarks: Fast-growing plants produce the best quality lettuce.

Lettuce, Leaf
Lactuca sativa var. *crispa*

Includes butterhead, loose-leaf, and romaine.

Days to Maturity: 58 to 64 for butterhead, 40 to 45 for loose-leaf, 70 to 75 for romaine

When to Plant: Loose-leaf lettuce grows so quickly in the cool days of early spring that few gardeners start it early indoors, although if you have the space to do so, you can beat the season by a week or so or up to a month if you set plants out under protective cloches or tunnels. Plant seeds of leaf lettuce as soon as the ground can be worked. Romaine and butterhead types, which mature a bit later, may be either started indoors or direct-seeded.

How to Plant: Plant seeds no more than ¼ inch deep. Although lettuce seeds you sow indoors usually germinate within a week, early outdoor plantings in cold soil may not come up for two weeks. Thin the plants to 3 inches apart. In another two or three weeks, I pull alternate plants for salad, letting the rest, now spaced 6 inches apart, grow to soft-head or full-leaf size.

Growing Conditions: Lettuce plants will bolt once they're fully developed, no matter what you do, but heat and long days can hasten bolting. Mulch to keep the soil cool and try to shield plants from light at night, such as that from yard and porch lights.

Summer Planting: Since lettuce doesn't keep and can't be canned, plant short rows or blocks every two or three weeks all season, using one of the slower-bolting kinds like Anuenue, Orfeo, or Little Gem for summer plantings. One experienced gardener says, "Plant ten lettuce seeds a week." I've read that you can even scatter seeds on bare ground during a winter thaw for record-early May lettuce. I haven't tried this yet, but it makes sense, because volunteer lettuces usually appear quite early.

Summer heat sometimes sends lettuce seeds into dormancy. To get around this, and raise lettuce to go with your tomatoes, you can expose the germinating seeds to light, refrigerate the seeds for a week or two before planting, or use old seeds, which are less likely to maintain dormancy in hot weather.

Muskmelon
Cucumis melo var. *reticulatus*

Some gardeners raise honeydews, crenshaws, true cantaloupes, and other melons, but muskmelons are the most widely planted.

Days to Maturity: 75 to 90

When to Plant: For sowing directly into the garden, wait until the soil is thoroughly warm (late May here in Pennsylvania). If you start seedlings indoors, sow them in May.

How to Plant: Direct-seed melons in hills 4 to 6 feet apart. Thin to four or five plants to a hill, but not too soon, because some may die of wilt if cucumber beetles strike.

Muskmelons need heat. Plants may suffer a setback at temperatures below 50°F (10°C). The seeds germinate rather unevenly outside, especially if the ground turns the least bit cool, so I've taken to presprouting seeds in a moist paper towel kept in a place where the temperature is 70°F to 80°F (21°C to 27°C). Then I plant the germinated seeds in a peat pot indoors in early May. This gives me many more plants than when I planted the seeds directly in the pot and far more than I got from planting outside.

You can't transplant melon seedlings from flats because they are so succulent. Keep the roots contained by either a degradable pot or a cardboard plant band, or plant the seeds in small plastic pots from which they can be gently turned out into the soil.

Growing Conditions: Site, as well as soil, is important for good melons. A southern slope is excellent. The soil should be rich in humus and not too acid. Fertilize the potted seedlings with diluted fish emulsion twice a week, using a one-half or three-quarter strength solution.

Pests: I plant a ring of radishes around each hill of melons, often sowing the radish seeds seven to ten days before melon planting-out time so the leaves will be up and growing when they're needed to fend off the marauding cucumber beetle.

Okra
Abelmoschus esculentus

Days to Maturity: 50 to 60

When to Plant: These ornamental pod-forming vegetable plants need plenty of heat, but they grow quickly. Wait to plant seeds outdoors until night temperatures stay above 55°F (13°C) and soil temperature is 65°F to 70°F (18°C to 21°C). This means no sooner than the end of May for us here in Pennsylvania; earlier, of course, in the southern states; and well into June for New England and the north central states.

How to Plant: Okra doesn't take kindly to transplanting, but you can get a head start by planting the seeds inside in pots about a month before setting them out. Use deep pots or milk cartons, because okra develops a taproot early.

For direct seeding, presoak the seeds for about 12 hours and plant about three or four seeds to the foot and no more than an inch deep. Later, thin mature dwarf plants, the kind usually planted in the northern states, to stand 18 to 24 inches apart. Tall okra needs 3 to 4 feet of space.

Growing Conditions: Okra is not particular about soil (although some gardeners say it dislikes highly acid soil), but generous fertilizing will encourage quick growth that will increase your yield. Most varieties of okra are sensitive to day-length: Generally, they flower earlier with short days. Some cultivars seem to be day-neutral, and a few (unfortunately not available in the United States) even respond to long days.

Varieties: The best choices for gardeners in the northern states, where summer days are longer, are probably Clemson Spineless, Lee, and Annie Oakley. Evertender is good, too, if you can find seeds for it.

Remarks: Okra seeds have a short period of viability, especially if not kept dry, so you can improve your odds by sowing fresh seeds. Pods form five to seven days after the blossom opens. Pick them at least every other day, because large pods turn woody and signal the plant to stop producing.

Onion
Allium cepa

Easily grown from seed if you get an early start and control weeds while plants are spindly and defenseless. If possible, plant onions to follow either a vegetable that requires clean cultivation or a weed-smothering cover crop.

Days to Maturity: 92 to 115

When to Plant: Sow onion seeds indoors in January or February; outdoors in April or May.

Why is it necessary to get such an early start with onions from seed? Because the onion's ability to form a bulb is influenced by the length of the day. It's the short dark period that makes the onions shape up. Onions suitable for the northern states are called long-day onions. They start to form bulbs when days grow long and nights grow short. If you've gotten a head start with your seedlings, your onion plants will be well developed at bulbing time and therefore vigorous enough to produce a good-sized bulb. When the day length becomes right, a spindly young onion plant will bulb up on cue just like an older one, but the bulb will be puny. Our southern states, being closer to the equator, have somewhat longer summer nights, so growers there choose short-day varieties, which don't require such a short dark period to trigger bulbing.

How to Plant: Plant onion seeds in flats indoors. Use fresh seeds; onion seeds lose much of their viability if not kept cool and dry. For good strong plants, transplant the onion seedlings, leaving 1 inch between plants. If you're short of space, or if you grow so many onions that you end up tripping over flats of onion seedlings scattered everywhere, you can carefully space the seeds ¼ inch apart when planting them, and then leave them in the same flat until time to plant them in the row.

You can also plant onion seeds directly in the ground for summer-bunching or fall-storage onions. Sow the seeds in April or May, no more than ½ inch deep (¼ inch in heavy soils) and thin to 4 inches apart when the top spears have become as thick as spears of chives.

Onion seeds germinate best at 65°F to 80°F (18°C to 27°C), but young plants should be grown in cooler temperatures, near 60°F (16°C) and no higher than 70°F (21°C) by day and 50°F (10°C) at night. Set them out after proper hardening-off a good month or six weeks before your frost-free date. Wide bands of onion plants spaced 4 inches apart make the best use of space.

Growing Conditions: Onions prefer soil that is not strongly acid. When grown in potassium-deficient soil, they will keep poorly, and phosphorus-deficient soil causes thick necks and delayed maturity. For sweeter onions, avoid fertilizing with gypsum, which contains sulfur. Weeds are your worst enemy when plants are young. They sometimes shoulder ahead of seeds planted in mid-spring before the grasslike seedlings can get off the ground.

Bulbing: As summer progresses, days become longer and warmer. Both of these conditions encourage bulbing, which is really the formation of additional storage tissue. If the weather is too cold, onions won't bulb up no matter how long the days are. The size of the onion is also important. As it grows larger, it becomes increasingly sensitive to the bulb-inducing influences of longer day length and warmer temperature. The day length necessary to initiate bulbing varies according to the variety but is generally 12 to 16 hours. A day length considerably longer than the minimum necessary to start bulb formation will exert a very strong impetus toward bulbing.

Within the plant's normal critical day-length range, though, bulbing is more susceptible to the influence of other environmental factors. For example, high soil nitrogen tends to delay bulbing within the critical photoperiod but not in an extra-long day. Warmth alone won't trigger bulbing, but it is necessary for the development of a good-sized bulb. Day length remains constant from year to year but soil and air temperatures change considerably, so even if you duplicate varieties planted and the treatment given your onion plants, crop quality may vary from year to year because of the weather.

Onion Sets: Perhaps you'd like to try growing your own onion sets, those miniature dry bulbs that grow into eating-sized onions when planted in their second spring. Just set aside a bed a few feet square, or a wide row, and in early spring, scatter about an ounce of seeds in a row 2 inches wide and 25 feet long. Don't thin the onions. Crowding keeps them small. Pull the plants late in July before they reach a diameter of ¾ inch. The smaller sets will give you larger bulbs and are less likely to bolt to seed next year. Any sets larger than 1 inch in diameter should be tossed into the pickle crock. Cure the sets in the sun until the tops are thoroughly dry—a week or ten days—and then remove the tops at the neck of the small bulb. Store in a dry, airy, cool but not freezing place.

When planting sets in the spring, push them into the soft earth just far enough to hold them in place, if your soil is heavy. In sandy soil plant them a trifle deeper, but don't cover them. If you have a cat that likes to scratch in the garden, as we do, you might have to do some resetting of bulbs for a week or two until roots grow.

Parsnip
Pastinaca sativa

Days to Maturity: 100 to 120
When to Plant: Plant parsnips in April or May at the latest, but not before the daffodils bloom.
How to Plant: Seeds should be sown outside in the open ground. Be sure to use fresh seeds, and sow even those thickly (at least one every inch) because parsnip seeds are notoriously low in vitality. I've gotten a decent stand from year-old seeds sown in a practically continuous band, but I wouldn't count on it. The seedlings are weak-kneed and easily overwhelmed by a heavy soil cover, so pull no more than ½ inch of fine light soil over the furrow. A light sowing of radish seeds that will emerge early and break the soil crust will make things easier for the young parsnip plants to push through. If a week passes without rain before the seeds have germinated, sprinkle the row with water. Germination is slow; allow at least three weeks.
Growing Conditions: Thin seedlings to stand 3 to 4 inches apart, and keep them well weeded. The plants will form straighter roots if well watered when young. Deeply worked soil will support longer, better-shaped parsnips. They do not transplant well. Once established, the plants need little care.
Remarks: The sweet-flavored, frost-proof roots make up for their wobbly start by feeding your family faithfully through the coldest days of fall and even winter, as long as the ground can be dug. (Flavor is best after frost.)

Peanut
Arachis hypogaea

Although you'd probably have to live in the South to make a living growing peanuts, gardeners can grow these attractive plants as far north as the upper Midwest, New England, and even in Canada. If your garden produces decent melons, you can raise peanuts.

Like other legumes, peanuts harbor nodules of nitrogen-fixing bacteria on their roots. If you want to inoculate your peanuts with these bacteria as you do your peas and beans, you'll need to buy a special peanut inoculant. (See Sources at the back of this book.)
Days to Maturity: 110 to 120
When to Plant: If you're just growing a few peanuts, you can start seeds in pots four to six weeks before your last frost date. Usually it's safe to plant peanuts when the

maple leaves are the size of squirrel ears. Southern gardeners can afford to wait two weeks or so after their last frost to let the soil warm up, since there will still be plenty of time for the nuts to mature before fall frost.

How to Plant: Shelled peanuts germinate more readily than those still encased in the shell. Take care to leave the papery skin intact and to keep the nut whole. Split nuts won't sprout.

Peanuts make good container plants. They do not transplant well, so should be slipped from the pot without disturbing the roots when placing them in the garden.

Plant seeds 1 to 1½ inches deep, every 3 inches, preferably in a 4-inch-high ridge of soil in a wide raised row or raised bed. Thin seedlings to 12 inches apart in the row. Seeds of extra-large peanuts like Park's Whopper should be sprouted indoors because they have a low germination rate if soil is cool and damp. Some northern gardeners presprout all their peanut seeds, and most choose the earlier-maturing Spanish types, although Tennessee Red has also produced well in northern states.

Growing Conditions: Ideal peanut soil is a loose, sandy loam, well supplied with humus. It's better to add manure the fall before planting, rather than in the spring, so the decomposing material won't cause the seeds to rot. You might try prewarming the soil with a mulch of black plastic—just be sure to remove the plastic when the plants start to flower.

Peanuts have their own way of doing things. About six weeks after germination, the now-bushy cloverlike plant begins to produce yellow flowers. The fertilized ovary of each flower enlarges and extends into a peduncle that promptly carries its new plant embryo below the soil surface—a process called pegging. It is there, underground, that the tip of the shoot swells and develops into a pod of peanuts. Now you know why there should be plenty of loose soil and no barriers between the plant and the soil at flowering time. Better yet, mound up each side of the row as you would do with potatoes when the plant is a foot high. Peanuts need extra calcium at flowering time, so some careful growers spread a dusting of gypsum (calcium sulfate) or limestone around the plants then.

Pea
Pisum sativum

Days to Maturity: 55 to 75

When to Plant: You can sow early, mid-season, and late varieties on the same day, or make successive plantings of pea seeds throughout the cool weeks of early spring, but there's no point in sowing most kinds of peas later than two to three weeks before the frost-free date, because the yield of peas maturing in warm weather seldom justifies the space they take. Young plants grow best at 59°F to 68°F (15°C to 20°C). An exception is Wando, a good pea for those who must wait in the spring until their community gardens have been plowed.

For Fall peas, plant seeds in late July or early August. Mature pea plants are more easily killed by frost than the hardier seedlings.

To put peas on your table, you must get them in the ground early, and to do that, you often need to prepare the row in the fall. Some gardeners even plant pea seeds in late fall or during a February thaw. However, although it's safe to plant peas in cold soil, because they can sprout at temperatures as low as 40°F (4°C)—although it may take them a month to do so—it's not wise to work the garden while the earth is still heavily sodden. We get around this, here in our garden, by doing a late-fall plowing, burying all the mulch and leaves and leaving rough mounds. Frost action pulverizes that exposed soil over the winter, and I find that I can usually get out there with a hoe in early March and pull open a furrow of fairly loose soil, going along the top of a ridge left by the plow, *not* in the deeper, colder valley between ridges.

How to Plant: Shake some garden legume inoculant on the moistened seeds before planting. Legume plants growing in zinc-deficient soil have less nitrogen-fixing ability. Plant peas thickly, about one every inch, and cover the seeds with 1 to 1½ inches of soil. Double rows of peas, spaced about 4 to 6 inches apart, make more efficient use of soil space than single rows. Wide rows of peas, up to 3 feet or so, are even more efficient. At one time, I thinned my peas to stand 2 to 4 inches apart, but since I've found that slight crowding doesn't seem to reduce production, I've stopped thinning them. Peas don't transplant well.

Growing Conditions: Pea roots are weak and small, easily dislodged in weeding, so I usually let some weeds grow, close to the plants, to prevent root damage and also to help shade the pea roots, which prefer cool growing weather. They also need plenty of oxygen, so plants grown in compacted or waterlogged soil will not produce as well as those in well-aerated ground. Peas also prefer soil that is not highly acid. They are fairly drought tolerant until flowering, when their moisture needs increase to an inch a week. The first peas appear about three weeks after blossoming.

Staking: Except for leafless kinds like Novella, which is pretty much self-supporting when grown in a triple row, your peas will need some support, even the low-growing ones. Cuttings of brush that have lots of twigs are excellent for all kinds of peas and the best choice for wide rows. For tall-growing vines like Sugar Snap and Mammoth Melting Sugar, grown in single rows, I supplement the brush with binder's twine strung the length of the row between three steel poles. Netting, chicken wire, garden fencing, and string supported by stakes are also successfully used by many gardeners.

Varieties: Last year I planted a row of Tall Sugar Snap peas faced on either side by a closely planted row of short-vined peas. Smooth-seeded peas like Alaska are the hardiest, but I think the sweeter wrinkled-seeded kinds are so much tastier that they are worth waiting for.

Pepper
Capsicum annuum var. *annuum*

Days to Maturity: 55 to 80

When to Plant: Although the traditional timetable for starting pepper plants is 6 to 8 weeks before outdoor planting, I've had excellent results with much earlier seeding in late January or early February, which for me is a good 12 to 14 weeks before the frost-free date.

How to Plant: If you can keep the soil temperature of your planting flat around 80°F to 85°F (27°C to 29°C), the peppers will germinate more rapidly than at 70°F (21°C), at which they usually take two or even three weeks. At cooler temperatures, 55°F to 60°F (13°C to 16°C), the seeds may rot before they germinate. Once when I let a flat of planted pepper seeds dry out, the seedlings emerged just as I was ready to give up on them. Later I learned that peppers, unlike most other vegetables, can germinate with low soil moisture.

Transplant the young seedlings at least once to a roomier flat and harden them off a week or two before your last frost date. If they're in blossom when you set them out, you might find it worthwhile to cover the plants at night until night temperatures are warmer. Cool nights in May and June can cause blossom drop, as can hot weather—especially over 90°F (32°C)—in summer. The ideal temperature for these natives of the moist South American tropics is 70°F to 80°F (21°C to 27°C).

Local garden sages advise setting out pepper transplants after the dogwood blossoms fall. Mine usually go into the ground about a week after our mid-May frost-free date. If you want to get really technical, the soil 4 inches deep should measure 65°F (18°C) at 8 o'clock in the morning.

When setting out pepper plants, I bury them 2 to 4 inches deeper than they were in the flat, but not on their sides in a trench like tomatoes. Recent studies have shown that pepper plants spaced 15 inches or even 12 inches apart produce as much per foot of row as those more widely spaced, and suffer less sunscald.

Growing Conditions: Peppers do well without much added nitrogen, but they need a good supply of magnesium. They are also more tolerant of acid soil than many garden vegetables. Hot peppers are less likely than sweet peppers to object to the low level of aeration in heavy clay soils. Water them well in hot and dry weather.

Pests and Diseases: Protect them from cutworms and avoid soil where related plants—tomatoes, eggplant, or potatoes—have recently grown if you've had disease problems.

Don't let tobacco users handle your pepper seedlings without washing their hands first. The virus that causes tobacco mosaic, which affects peppers, survives cigarette manufacturing processes.

Remarks: According to a Dutch horticulturist who tried removing all but four to six of the first blossoms from pepper plants, this practice results in higher mid- or late-season yields and also produces larger peppers.

Potato, Sweet
Ipomoea batatas

You might need to begin your sweet-potato growing career by purchasing plants—actually rooted slips—from a southern grower, or you might be lucky enough to find them locally at a farmer's market or neighborhood store. Once you have sweets, though, and learn how to keep them over winter (cure for two weeks in a hot, fairly humid place; wrap them individually in newspaper; and store them in a warm, dry, well-ventilated spot protected from rodents), you can raise all the plants you'll need by rooting your own potatoes. I've gotten 50 slips from one large potato.

Days to Maturity: About four months of frost-free days

When to Plant: Start your slips indoors in March in the mid-Atlantic states, February in the southern states, and April in the more northern states.

How to Plant: Begin by choosing a large, sound potato, or two or three. Find a jar with an opening the right size to admit about two-thirds of the sweet potato. Put enough water in the jar to cover the bottom inch or so of the root. Keep it in a warm place (75°F to 90°F, 24°C to 32°C), day and night if possible, to induce sprouting. Shoots will appear in two to four weeks. At lower temperatures they will form much more slowly.

As soon as a leaf cluster forms above the roots on each slip, it may be twisted off the parent potato and planted directly outdoors, if the weather is warm enough. If temperatures outside are still too cool, plant the slip in a deep flat filled with potting soil (I use an old refrigerator crisper) to wait for warm weather. Keep the planted slips in a sunny window, under lights, or in a greenhouse. A farmer's wife who grows sweet potato plants to sell at the local farmer's market has told me that she grows her slips in a hotbed heated with horse manure. Other gardeners start their slips indoors by placing the potatoes on their sides in damp sand or vermiculite and keeping them in a warm place until they sprout.

There's no point in putting sweet potato plants out in the garden until the soil has become thoroughly warm: 70°F to 85°F (21°C to 29°C) is ideal. That should be about 10 to 14 days after the last spring frost, the end of May in my garden. Draw up a ridge of soil by hoeing inward toward the row from both sides, and poke the slips into this ridge, spacing them a foot apart. Use the largest plants first so that inferior plants can be discarded when you get to the end of the row, if you run out of space. Mound soil up around the plants again just before the vines grow out into the path.

Growing Conditions: Sweet potatoes grow well in slightly acid soil because the acidity helps to keep down disease. They don't need much nitrogen; in fact, heavy applications foster excessive vine growth and late maturity. Although they should be well watered in when the slips are planted, the growing vines are amazingly drought resistant.

Potato, White
Solanum tuberosum

Days to Maturity: 56 to 70 days

When to Plant: Since cool weather favors tuber growth, white potatoes should be planted as early as the soil can be worked. A later planting, in mid- or late spring, will produce fall potatoes.

How to Plant: If you've grown potatoes the previous year, chances are you have some small spuds left over. If they've begun to sprout, that's just fine, although it's better if they don't have long white sprouts. If they do, you can break off extra-long, weak sprouts as long as there are several eyes remaining that can grow. Egg-sized potatoes are good for planting whole. Those that are *much* smaller may not have enough stored nourishment to develop a strong plant. Seed potatoes stored at low temperatures (around 40°F, 4°C) will grow into vigorous plants that produce more large tubers than potatoes that have been kept in warmer storage.

Large potatoes with many eyes should be cut, exposing as little of the cut surface as possible, into pieces containing one to three eyes apiece. Too many eyes will produce a leafy plant with small potatoes. (The eye is a dormant bud.) Let the cut pieces dry for a day or two before planting so the surface can heal over. They are less likely to rot in the ground then. Unsprouted seed potatoes may be nudged into growth by being exposed to the light for a few weeks in a cool room.

When planting, place the potatoes or the pieces with the cut side down every 10 to 12 inches in trenches that are 2 to 3 feet apart. Cover them with 3 inches of soil. When the tops are 9 inches high, draw up loose soil around the plants.

You can also grow potatoes in mulch. They will be clean and easy to harvest and should yield at least as well, sometimes better, than those grown in the ground. One of the advantages to using mulch is that scab and potato bug damage are seldom a problem. The one catch is that you must use large amounts of mulch; a good foot of mulch must be spread over developing tubers and replaced as it settles, or the tubers will turn green. The green parts of potatoes and the sprouts contain an alkaloid poison, solanine. Except for the greening problem, I've been delighted with the potatoes we've grown under mulch. The only trouble we have is that it's difficult for us to find enough hay or leaves that early in the season to cover all the potatoes we want to grow.

Growing from Seed: If you're experimenting with planting seeds from a seed ball that one of your potato plants might have produced last year, be prepared for a wide variety of plants, most of which will be worthless. Plant the seeds in a marked row or seedbed in early or mid-spring. The best temperature for potato seed germination is 68°F (20°C). If good tubers do form on any plants, save them and plant them to increase your stock of starter potatoes for cutting up the following spring. My trial plantings of Explorer, the first potato to come true to seeds, were disappointing. The plants, which I started indoors like tomato seedlings, grew slowly, had low vigor, and yielded poorly. Other gardeners have had similar results.

Growing Conditions: Soil for potatoes should be on the acid side, well-supplied with humus and enriched with wood ashes or greensand to supply the potash that makes for a good mealy potato. Only well-rotted manure should be dug into the potato patch, because fresh manure encourages scab.

Varieties: New and newly rediscovered kinds of especially good-flavored potatoes like Yukon Gold, Caribe, Butte, and Yellow Finnish are now available as seed potatoes to home gardeners. As a result of our home garden trials of these "gourmet" potatoes, we've added rich-flavored Yukon Gold and Butte (a baking potato high in vitamin C) to our standard annual planting of all-purpose Kennebec seed potatoes.

Pumpkin
Cucurbita pepo var. *pepo*

Days to Maturity: 100 to 115
When to Plant: Plant seeds in the ground after the last spring frost.
How to Plant: Start pumpkins like squash, with a generous shovelful of compost or well-rotted manure in each hill. When the plant starts to develop vines, anticipate the

squash borer by firming two or three shovelfuls of soil over several vine nodes to encourage auxiliary rooting.

Try the new semibush pumpkin, Funny Face, if space limitations have kept pumpkins out of your garden. Other kinds need lots of space; a hill will ramble over an 8-by-8-foot square of ground by summer's end. Planting at the edge of the corn patch works well. The vines wander among the corn and help to discourage raccoons.

Seeds of naked-seeded pumpkin varieties like Lady Godiva, Streaker, and Triple Treat have seed coats that are just a thin film. Lacking the thick protective coats, which make regular pumpkin seeds harder to get at, the naked seeds tend to rot more readily in cool, damp soil. You'll get more plants from a packet of seeds if you presprout the seeds and then plant them in individual pots, setting them out when warm weather has come to stay. If you do plant the seeds directly in the ground, wait until warm weather has settled in.

Growing Conditions: Pumpkins appreciate soil well supplied with organic matter. Mulching helps to control weeds that would be difficult to hoe out from the spreading vines.

Radish, Summer
Raphanus sativus

Days to Maturity: 20 to 30

When to Plant: Start planting as soon as the soil can be worked, but make small weekly or biweekly plantings rather than large infrequent ones, for radishes mature all at once and become fibrous and bitter when old and on their way to going to seed. I seldom plant summer radishes as late as June or July, but usually resume planting in late August or early September, using the fastest-maturing varieties, for a fall crop.

How to Plant: Drop a seed every inch or so in drills spaced a foot apart or in wide rows or beds, and cover with ¼ to ½ inch of soil. When seedlings appear, thin them to 2 to 3 inches.

Growing Conditions: For best radish quality, promote rapid, even growth. That means grow them in cool weather, plant them in sandy or loosely worked soil, and supply ample moisture. Dryness, even for a day, and warm weather make radishes tough and strong flavored.

Remarks: Some gardeners plan most of their spring radish plantings as row markers for carrots, parsnips, parsley, and other slow-germinating crops.

Radish, Winter
Raphanus sativus

Winter radishes grow more slowly than spring radishes and are slightly more solid, and less delicately crisp than a well-grown early radish. I find, though, that their quality is more consistent. They keep better in the soil without turning bitter or tough, and they don't start going to seed overnight.

Days to Maturity: 50 to 65
When to Plant: Start them in late July or early August so they'll finish in cool weather.
How to Plant: Plant as for summer radishes, but thin them to 4 to 8 inches apart according to variety.
Growing Conditions: Loose soil, well supplied with potash, encourages the growth of well-shaped roots. Radishes that are watered well will be more tender.
Varieties: Some winter radishes, like Miyashige, grow over a foot long. Our favorite is China Rose.

Rhubarb
Rheum rhabarbarum

The most desirable kinds of rhubarb are propagated by root division. Seeds are usually available only for the green-stalked Victoria, and plants grown from seed seem to have an unfortunate tendency to bolt right back to seed early in the season. Strawberry rhubarb, for which you might also be able to find seeds, is red, as its name implies. Since I have not grown it from seed, I can't say whether it goes to seed as eagerly as Victoria, but my guess is that any seed-grown rhubarb would tend to perpetuate seed-prone stock over that which concentrates on vegetative growth. Having said all that, and adding, as I feel I must, that a professional rhubarb grower once wrote to me that he'd rather have a field of Johnson grass than one of seed-grown rhubarb, I still think there's a place for rhubarb grown from seed as an annual, for a quick crop of fruit.
Days to Maturity: 90 to 100
When to Plant: Sow seeds a month before the last frost.
How to Plant: Plant rhubarb seeds 2 inches apart and ½ inch deep. Firm the soil well over them. Thin seedlings to 1 foot apart. Seedlings transplant well.
Growing Conditions: Provide rich soil and fertilize the plants twice a month.
Remarks: I gave my seedlings several feedings of manure tea. We were able to eat crisp, succulent stalks of rhubarb by midsummer. The stalks were excellent that first season. It wasn't until the second year that they started producing seed stalks before the end of April. Grown as an annual, then, and generously fed and watered, rhubarb is worth raising from seed if you recognize its limitations. There aren't many other plants that will give you fruit the first season.

Rutabaga
Brassica napus, Napobrassica group

Days to Maturity: 90 to 100
When to Plant: This is primarily a fall crop, unless you usually have cool summers, since the roots grow tough in hot weather. Sow the seeds in the garden in mid-June.
How to Plant: Plant seeds 2 inches apart in rows 15 to 24 inches apart. When the seedlings are 3 inches high, thin them to stand 6 to 8 inches apart.
Growing Conditions: Rutabaga is one root vegetable that does well in heavy soil.

Although I've never tried transplanting rutabagas, and know no one who has, I'd guess that you might get away with it, though perhaps not quite so easily as with turnips, which have a less extensive root system.

Pests: Flea beetles can be hard on young rutabagas, especially if soil is dry.

Salsify
Tragopogon porrifolius

The foliage is flat and ribbonlike, resembling that of garlic, but larger and bushier.

Days to Maturity: 115 to 120

When to Plant: You can sow seeds of the oyster plant, as this tasty root vegetable is also called, as soon as you can dig the garden in the spring. Ordinarily there's no need to rush, though. Although salsify needs 120 growing days, it's at its best after frost in the fall. If you plant seeds around the time of daffodil bloom or even a week or two later, the roots will have plenty of time to develop by fall.

How to Plant: Plant the seeds no more than ½ inch deep. If you're using seeds that are more than a year or two old, sow them thickly. Thin seedlings to stand 3 to 5 inches apart in the row.

Growing Conditions: Loose, well-dug, fairly light soil produces good roots.

Scorzonera
Scorzonera hispanica

Sometimes called black salsify, scorzonera has white flesh under a black skin.

Days to Maturity: 120

When and How to Plant: Planting and culture are the same as for salsify. Frost improves the flavor of both.

Sorrel
Rumex acetosa

Also called sour grass, this tangy-leaved perennial green is easy to grow from seed. Somewhat sour when eaten alone, the leaves are delicious mixed with other, milder greens.

Days to Maturity: 60

When to Plant: Start plants indoors a month before the last frost, or sow seeds directly in the row in mid-spring. A late-summer sowing will be ready to eat the following spring.

How to Plant: The plants do well spaced 6 to 8 inches apart in rows 15 to 18 inches apart or in a special perennial bed. Once established, sorrel produces harvest-sized leaves quite early in the spring.

Growing Conditions: Sorrel is one vegetable that accepts partial shade. Remove seed stalks to conserve the plant's energy. Fertilize at least once a year.

Soybean
Glycine max

Choose one of the varieties offered especially for home gardens. Field soybeans, I've been told by one seedsman, have a higher oil content. The flavor of vegetable soybeans is sweet and nutlike in the green stage, more ordinary when they're dried.

Days to Maturity: 75 to 110

When to Plant: Plant the seeds when apple blossoms start to fall, no earlier. If spring weather is mild in your area, wait until around the time of your last frost.

How to Plant: Drop a seed every 2 inches in rows 3 feet apart and cover with 1 inch of soil. Thin seedlings to 3 or 4 inches apart.

Growing Conditions: Soybeans like warm weather, plenty of lime, reasonably good drainage, and a good supply of moisture when pods are forming.

Varieties: The variety Butterbeans is our favorite for eating green. For dried soys, black beans are better-flavored than yellow.

Spinach
Spinacia oleracea

Days to Maturity: 43 to 50

When to Plant: Sow seeds in the ground as early as possible because warm weather and long days all too soon will trigger seed-stalk formation. Spinach seeds germinate well in cool soil, and the plants prosper in day temperatures of 60°F (16°C) dropping to 40°F to 45°F (4°C to 7°C) at night.

If your garden is slow to dry for early-spring digging, you can prepare your spinach row in the fall, plant the seeds, and mulch the row lightly. Gradually rake off the protective covering the following spring. In many areas, fall-planted spinach that has grown to a height of an inch or so will winter-over under a straw or hay cover. Plant extra-hardy varieties like Winter Bloomsdale or Cold-Resistant Savoy for this purpose.

How to Plant: Plant the seeds no more than ½ inch deep, and thin the seedlings to stand about 6 inches apart in rich soil, 8 inches in leaner soil.

Growing Conditions: Spinach is a heavy feeder and needs a well-aerated, well-limed soil. Strongly acid soils often make available substances that are toxic to the plant. In highly alkaline soil, the plants may develop a symptom of manganese deficiency: yellow-speckled, curled leaves. Large doses of highly soluble, nitrogen-rich fertilizer can increase the oxalate content of spinach—another good reason for applying slow-release manure.

Diseases: If you have trouble with spinach seedlings dying from fungal attack (the fungi like the seed's mucilaginous coating), soak the seeds in a 3:1 chlorine bleach and water solution for ten minutes before planting.

Remarks: Picking the oldest, largest spinach leaves can help to postpone bolting by reducing the amount of the hormone that stimulates production of the seed stalk.

When making late-summer plantings for fall eating, you may find that the seeds are reluctant to germinate in warm weather. You can usually induce them to sprout by spreading them between damp paper towels and keeping them in a plastic bag in your refrigerator for five to seven days.

Spinach, New Zealand
Tetragonia expansa

This thick-leaved green produces well all summer and into the fall, unlike conventional early spinach. It thrives in hot weather, but oddly enough, germinates readily when it's cool. In fact, it needs exposure to temperatures lower than 55°F (13°C) in order to germinate well.

Days to Maturity: 70

When to Plant: You can plant the slow-germinating seeds about two or three weeks before your last frost date.

How to Plant: Soak the seeds in tepid water overnight to hasten sprouting, and plant them either in hills 3 to 4 feet apart or every 6 inches in rows. Thin row-planted seedlings to stand at least a foot apart. You might be surprised to see how far they'll sprawl by summer's end. Volunteers from seeds ripened by last year's plants often appear in my garden in mid- to late April.

Growing Conditions: Fertile, well-drained soil that is not too acidic is best for New Zealand spinach. It is seldom bothered by disease or insects.

Squash, Summer
Cucurbita pepo var. *melopepo*

Includes zucchini, patty pan, and yellow crookneck.

Days to Maturity: 43 to 75

When to Plant: A warm-weather crop, but one on which I often gamble an early- or mid-May planting with presprouted seeds a week or so before our usual frost-free date. Main-crop plantings should go into the ground about a week after the last frost. Once sprouted, the plants will grow well at a lower temperature than they needed for germination, so presprouting of seeds indoors works well.

How to Plant: Plant the seeds of vining types, whether presprouted or not, in hills 3 to 4 feet apart, or space them every 4 to 8 inches in rows that are 3 to 4 feet apart. I bury a shovelful of compost in each hill, top it with an inch of soil, and then plant the seeds. Hills should be planted with six to eight seeds and thinned to the best three plants. Allow about 15 inches between bush plants in a row. Squash seedlings do not transplant successfully.

Growing Conditions: A steady water supply and soil rich in organic matter encourage continuous fruit production.

Pests: Since the squash borer often attacks my early plantings by midsummer, I make a second planting, especially of zucchini, in mid-June. Radishes planted with the squash may help to deter insects. Rotenone is a more aggressive defense if your insect problem is serious.

Remarks: Pick the fruit when it's young and tender, 4 to 5 inches for yellow crook-neck, 4 to 8 inches for zucchini, and the size of silver dollars for patty pan. Do some comparison-shopping in the seed catalogs to find the earliest-bearing zucchini.

Having too much zucchini is less of a joke around here now that I've found they can be pickled by the same recipe I use to make dilled cucumber pickles, and they are delicious. (See the *Ball Blue Book* put out by the Ball Corporation, Muncie, Indiana, for recipes.)

Squash, Winter
Cucurbita spp.

Includes acorn, buttercup, butternut, and Hubbard, among others.

Days to Maturity: 85 to 110

When to Plant: Wait to plant winter squash until the weather is good and warm—about a week after the last frost, when the late irises are in bloom. With winter squash, you're not after a quick summer crop but a well-matured fall harvest. Soil temperature should be about 60°F (16°C).

How to Plant: Hills should be spaced 5 feet apart, with five or six seeds planted 1 inch deep in each. Thin them later to three seedlings. For row planting, sow seeds every 6 inches and no more than an inch deep. Space rows 8 feet apart. Some gardeners sow squash and pumpkin seeds with the thin edge down so the sprout can emerge more easily. After you've seen how much damage the insects are going to do, pull alternate seedlings, leaving plants 12 to 18 inches apart. Winter squash doesn't transplant well, but may be started indoors in pots if you have an extra-short season.

Growing Conditions: Rich, well-composted soil is good for winter squash. Before the vines spread, apply mulch to control weeds.

Pests: Some gardeners report that potatoes interplanted with squash help to deter the squash bug.

Strawberry
Fragaria spp.

Days to Maturity: Strawberries are perennials. Fruit is produced a year after planting unless seeds are started indoors in midwinter.

When to Plant: Plants started indoors seven to eight weeks before the last frost transplant easily to the garden row or rock garden, at or shortly before the frost-free date. If you start seeds in January you'll have bearing plants the first year.

How to Plant: The important thing to know about raising strawberry plants from seed is that the seeds germinate at a low 60°F to 65°F (16°C to 18°C); at temperatures over 70°F (21°C) they refuse to sprout. They also need some light in order to germinate, so if you cover the seeds at all, use only a few pinches of fine soil or vermiculite dust. Better yet, simply press the seeds into the surface of the soil. Spread clear plastic wrap over the planted flat. Germination takes about four weeks, so be sure to keep the soil in the flat evenly moist during that time. After hardening off, seedlings can be planted out when day temperatures are consistently in the 60s (16°C). If they've been gradually accustomed to cool temperatures, a few light frosts won't hurt them. Space plants 6 to 12 inches apart in rows 3 feet apart. Be sure to set the plants in the ground at the same depth they were growing before transplanting, no lower or higher.

Growing Conditions: Soil well supplied with humus, lightly limed if heavy clay, and close to the hose for irrigation is a good choice for your strawberry transplants. If you put them in a spot where they'll receive lightly dappled shade after bearing, they won't mind.

Remarks: Except for new varieties like Sweetheart, most regular large-fruited garden strawberries are propagated by transplanting runners. The Sweetheart strawberry is an everbearing variety for which seed companies have recently started to offer seeds. Baron Solemacher alpine strawberries, which may be grown from seed, produce few runners.

Sunflower
Helianthus annuus

Days to Maturity: 100 to 120

When to Plant: Plant the seeds around the time of the last frost, because the plants thrive in heat and seedlings are frost-tender. Volunteers, however, which often seem the most vigorous, were obviously planted before that, so if you have plenty of seeds you can toss out a few seeds even earlier. Not for row crops, though; too much work could be wasted.

How to Plant: Large plantings should be made in rows 3 to 4 feet apart. Plant seeds every 6 inches, and thin seedlings to stand 2 to 2½ feet apart. For just a few heads, spot-plant the seeds around your home place—a few by the bird feeder, the mailbox, the back porch, the garage. Remember that they'll grow tall and shade nearby plants, which can sometimes be an advantage, but not always.

Growing Conditions: Rich soil produces strong stalks and large heads.

Remarks: Sunflower roots produce a substance that can suppress the growth of other, nearby plants.

Swiss Chard
Beta vulgaris var. *cicla*

Days to Maturity: 50 to 60

When to Plant: Sow seeds when maples are in bloom, or about two weeks before your last frost.

How to Plant: Sow about eight seeds to the foot in rows 18 to 24 inches apart, and cover seeds firmly with ½ inch of soil. Thin plants to an eventual spacing of 8 to 10 inches apart, but don't be in a hurry to do so. Let them grow till they crowd each other; then eat the thinnings. Thinning gradually in this way, you can get several early meals from the row while the plants are still growing. Young seedlings, measuring 2 to 3 inches or so, may be transplanted, but older ones suffer loss of deep root tissue.

Growing Conditions: This easy-to-grow, loose-leaf beet relative puts down a long, strong root, so it can make do even on fairly poor soils, although better soil, with plenty of organic matter, will give the best results. With some protection, chard will often winter-over.

Tampala
Amaranthus tricolor

Days to Maturity: 42 to 55
When to Plant: This nutritious leafy member of the large amaranth family likes warm weather. Sow seeds after frost, when the soil has warmed up, about the time the late irises are in bloom.
How to Plant: Plant seeds ¼ inch deep. Thin the young plants gradually, eating them as you go, to an eventual spacing of 18 inches.
Growing Conditions: Soil enriched with compost will support rapid growth.

Tomato
Lycopersicon lycopersicum

We were amused, recently, to hear from friends who live in a proper central Pennsylvania town that the custom there, for quite some years, had been to get free tomato seedlings from the sewage plant. (Tomato seeds are well known for their ability to escape from the digestive tract unscathed, as anyone who has ever fed the fruits to pigs can testify.) These free seedlings were, of course, a grab bag of varieties, and no one knew until they bore fruit whether they had planted 25 cherry tomatoes or 25 different varieties. Now that hybrids are more generally grown, such free seedlings are less promising because the second generation from hybrid parentage is unlikely to be as good as the first.

Days to Maturity: 45 to 87 (Fruits ripen 45 to 55 days after blossoming.)
When to Plant: Plant seeds indoors in flats six to eight weeks before your frost-free date.
How to Plant: Cover the seeds; they germinate better in the dark. Keep the soil temperature as close as possible to 80°F (27°C) for prompt germination. Transplant at least once, into larger flats with 2 inches between plants, setting the seedlings deeper than they grew before, especially if they have gotten leggy. Tomato seedlings growing indoors should be kept near 60°F (16°C) to prevent overly rapid growth that is difficult to harden.

Vernalization—chilling to induce early bloom—often works well with tomatoes, but it must be done early, when the first true leaves are opening up and the plant is only about 1½ inches tall. Night temperatures of 50°F to 55°F (10°C to

13°C) for two to three weeks are usually effective. Not all vegetable specialists agree that this method makes sense. If you decide to try it, treat only part of your crop so you'll have some unchilled plants for comparison.

Wait until you're reasonably sure you've had your last frost before setting out your tomato plants. That's mid-May for us, about the time the barn swallows return. Another old rule of thumb is to set out tomato transplants when dogwoods are in full bloom.

Set unstaked plants about 3 feet apart in rows 3 feet apart. Staked tomatoes may be planted 2 feet apart, and you can space cherry tomato and some early varieties as close as 15 to 18 inches. Remember to protect them from cutworms. Since spring winds can still be punishing during the last two weeks in May, I like to bury my tomato plants, parallel to the soil surface, right up to the top tuft of leaves. Dig a long, shallow trench for this, rather than a deep hole. It's too cold way down there in the ground for the roots of the warmth-loving tomato, which is at its best between 70°F and 90°F (21°C and 32°C).

Direct Seeding: As an alternative to prestarted transplants, you might like to try direct-seeding some kinds of tomatoes. The small early cherry tomatoes and the new cold-resistant subarctics are obvious candidates for this treatment, but any except very late varieties are worth a try. Seeds planted at the time the maples bloom will wait out the early cold and germinate when the weather's right, just like all those volunteer tomatoes that grow well without any help from us. Such plants will be a bit later to come into bearing, but they'll often take over at the end of the season when the main planting is on the decline.

Growing Conditions: Good drainage and soil aeration, reasonable fertility, and an ample supply of potash and phosphorus are important. Purple-leaved tomato seedlings are most likely to be suffering from a phosphorus deficiency. Dig in rock phosphate and bone meal before the final transplanting. A seedling that develops purple leaves just before it is planted out often returns to normal when transplanted to the garden row, if the soil is adequate. Avoid giving too much nitrogen during early growth; it will promote vining at the expense of fruiting. The tomato's nitrogen requirements rise at blossoming time, though, and that's when you can give the plants a boost of manure tea or diluted fish emulsion.

Mulching helps to promote even tomato growth throughout the season, but it should not be done until the soil warms up. It's often mid-June, at least, before I have my tomato patch mulched.

Staking: Staking of tomatoes isn't necessary, but it keeps the fruit clean and promotes slightly earlier ripening. Since we've reduced the size of our tomato patch (to more than enough for two people rather than more than enough for four), we now stake our plants. Formerly, though, we let our plants sprawl on a deep mulch, trading some spoiled fruits and a tangled patch for the time we'd have to spend tying the plants and pruning the suckers that appear in the leaf axils. Staked tomatoes produce more fruit per foot of row, though less per plant, than those that are not staked.

If you do stake, it's a good idea to sink the support soon after planting so you remember where you've buried the long stem of the tomato plant. Use soft cloth strips of old sheets or knit fabric to tie plants. Encircle the plant stem once, make a loose knot, and then tie a second circle with the same strip of cloth around the stake. This prevents damage to the plant from rubbing. Many gardeners report excellent results using either wooden frames set over the plants for support or surrounding each plant with a circle of hog wire or concrete reinforcing wire with holes large enough to reach into for fruit picking.

Remarks: Avoid planting tomatoes near walnut trees; they are extremely sensitive to the toxin juglone, which is exuded by walnut tree roots. They are also sensitive to unnatural chemicals in the soil and vulnerable to diseases spread by tobacco users. Indoors, natural gas leaking from cooking stoves or other appliances may retard tomato seedlings.

Turnip
Brassica rapa, Rapifera group

Rapid growth in cool weather produces good turnips. Those maturing in hot weather tend to be more fibrous and strong flavored. Turnip seeds vary widely in size. For some reason, small seeds tend to produce flat-shaped roots.

Days to Maturity: 30 to 60

When to Plant: Sow seeds as soon as you can work the soil in spring and again nine to ten weeks before your fall frost date for a late crop, which for many gardeners is the main turnip planting.

How to Plant: Plant seeds ¼ to ½ inch deep. Particularly for the early spring crop, when you're racing the calendar to beat the hot weather, soil should be loose and humus-rich. Thin early turnips to stand 3 inches apart, fall-storage turnips to 6 inches. Turnips can be transplanted, with reasonable care, according to an experienced market gardener with whom I've corresponded. If you're short of space, you might like to try the old gardening practice of broadcasting turnip seeds between the corn rows.

Growing Conditions: Soil should be loose and humus-rich. A steady supply of moisture is especially important for fast-growing roots.

Varieties: People who think they don't like turnips usually change their minds after a taste of des Vertus Marteau, our favorite, which is sweet and mild. The Gilfeatner Turnip, an heirloom variety, is excellent too, and of course the leaves of any turnip make fine nourishing fare.

Check catalog descriptions. Some turnips are better suited to fall than spring production. The Japanese hybrid Just Right has excellent flavor, but tends to bolt to seed when the weather warms up, so it often fares best as a fall planting.

Watercress
Nasturtium officinale

Days to Maturity: 60 to 150

When to Plant: Seeds may be planted outside a month before the last frost or indoors one to two months before frost-free weather.

How to Plant: The seeds are fine and sure to be overplanted. Transplant the seedlings to flats as soon as they can be handled, spacing them 1 to 2 inches apart. Transplant them to pots when they reach 2 to 3 inches in height and then to stream or pond banks about a month before the last frost. Once established, watercress is easy to propagate from cuttings, which root quite readily, and if you want a *lot* more, you can gather the seedpods when they ripen in the fall.

Growing Conditions: This nutritious perennial likes cool weather, plenty of lime, and a good supply of well-aerated water. You don't need a pond to grow it, though. I've kept an 8-inch clay pot of watercress growing on a porch that gets some east sun, with the pot set in a pan into which I poured water several times a week.

Remarks: Although watercress reaches picking size in five months, thinnings may be eaten at two or three months.

Watermelon
Citrullus lanatus

A space-grabber unless you plant one of the new bush types or grow small melons on a fence, but lots of fun to grow.

Days to Maturity: 74 to 90

When to Plant: Many northern gardeners start watermelon seeds indoors in pots for setting out when weather is warm, a good ten days or so after the last frost. The more warm days the vines experience, the more fruit they'll bear. Southern gardeners have enough time between frosts to plant the seeds directly in the ground. The seeds need consistent warmth, 75°F to 80°F (24°C to 27°C), for five to ten days, to germinate well. Seeds for the "seedless" types of watermelon germinate less readily than the regular kind and are even less tolerant of cool temperatures. Transplanting doesn't work well, so seeds started indoors should be planted in individual pots, not in flats.

How to Plant: Hills should be 6 to 8 feet apart, with one or two plants to a hill. Thin when the plants have five leaves.

One grower recommends preparing a rye cover crop in the watermelon bed the fall before planting. In spring, then, you can till a 3-foot furrow down through the center of the rye, let it settle for a week or two, and plant your watermelon seeds (or set out plants). As the plants vine, continue to till under the rye until the whole cover crop has been turned under. If you mulch, wait until the soil is good and warm, in late June or so, unless you use black plastic.

Growing Conditions: The best site for northern gardeners is a southern slope that is well drained. Watermelons do well in sandy, humus-rich soil that is more acid than that needed by muskmelons, and they don't require much nitrogen.

Witloof Chicory
Cichorium intybus

Days to Maturity: 21

When to Plant: For this winter delicacy—tender, pale salad hearts or chicons—sow seeds right in the garden row about two weeks before your last frost.

How Much Seed Do You Need?

Amaranth—$\frac{1}{25}$ ounce (about 1,000 seeds) sows about 50 feet of row.

Artichoke, Globe—$\frac{1}{4}$ ounce yields about 75 plants.

Asparagus—1 ounce yields about 10,000 plants if started inside at 80°F to 90°F (27°C to 32°C); 1,000 if sown in rows outdoors.

Bean, Azuki—1 pound sows about 500 feet of row.

Bean, Fava—1 pound sows 130 feet of row, except for Windsor and other large-seeded varieties, 1 pound of which sows 90 feet of row.

Bean, Lima—1 pound sows 100 to 150 feet of row, depending on size.

Bean, Snap—1 pound sows 150 feet of row.

Beet—1 ounce sows 100 feet of row; most packets about 25 feet.

Broccoli—1 ounce yields 3,000 to 4,000 plants; most packets about 200.

Brussels Sprout—1 ounce yields about 3,000 plants.

Cabbage—$\frac{1}{25}$ ounce (about 240 seeds) sows about 120 feet of row.

Cabbage, Chinese—$\frac{1}{25}$ ounce (about 330 seeds) sows about 100 feet of row.

Carrot—1 ounce sows 250 to 300 feet of row; most packets sow about 30 feet.

Cauliflower—1 ounce yields 2,000 to 3,000 plants; most packets about 150.

Celeriac—1 ounce yields about 7,500 plants; most packets about 400.

Celery—1 ounce yields about 10,000 plants; most packets 400 to 500.

Celtuce—1 ounce sows 300 feet of row; most packets 30 feet.

Collard—most packets sow about 100 feet of row.

Corn—1 pound sows 800 to 1,000 feet of row.

Corn, Broom—1 pound sows 1,500 to 2,000 feet of row.

Cucumber—1 ounce sows about 50 hills; most packets, except for hybrids, contain 100 to 150 seeds.

Eggplant—1 ounce yields about 2,000 plants; most nonhybrid packets contain about 60 seeds; hybrids 25 to 30.

Endive (Escarole)—1 ounce sows about 200 feet of row; most packets 20 to 30 feet.

Fennel—1 ounce sows about 300 feet of row; most packets 30 to 40 feet.

Gherkin—1 ounce sows about 80 hills; most packets about 20.

Grains—*Buckwheat:* 1 pound sows 300 square feet.

   Oats: 1 pound sows 250 square feet.

   Rye: 1 pound sows 200 square feet.

   Wheat: 1 pound sows 200 square feet.

Kale—1 ounce sows about 200 feet of row; most packets about 40 feet.

Kohlrabi—1 ounce sows about 200 feet of row; most packets about 40 feet.

Leek—1 ounce sows about 200 feet of row; most packets about 25 feet.

Lettuce, Head—1 ounce sows about 400 feet of row; most packets about 40 feet.

Lettuce, Leaf—Same as head lettuce.

How to Plant: Cover seeds with $\frac{1}{4}$ inch of soil. Deeply worked soil will encourage strong roots. Thin seedlings to stand 4 to 6 inches apart in rows a foot or more apart. **Growing Conditions:** Keep the weeds down, water in a drought, and in the fall let a frost or two nip the plant. Then dig up the roots, cut off all but 1 inch of the leafy top, and replant the roots, standing up, in buckets of soil, sand, or hardwood sawdust. (A

Muskmelon—1 ounce of regular-sized seeds yields about 900 plants; more in the case of small-seeded melons like Takii's Honey; most packets will sow about 20 hills.

Okra—1 ounce sows about 50 feet of row; most packets about 15 feet.

Onion—1 ounce sows about 200 feet of row; most packets about 20 feet.

Parsnip—1 ounce sows about 200 feet of row; most packets about 15 to 20 feet.

Peanut—1 pound sows about 200 feet of row; most packets of Spanish (small seeds) about 45 feet of row; large-seeded Virginia about 15 feet of row.

Pea—1 pound sows about 100 feet of row. Most packets plant about 20 feet.

Pepper—1 ounce yields about 1,500 plants; packets of nonhybrid varieties contain about 150 seeds; most hybrids about 50 to 100 seeds.

Potato, Sweet—100 slips will plant about 100 feet of row.

Potato, White—About 7 pounds of seed potatoes are needed to plant 100 feet of row.

Pumpkin—1 ounce sows about 25 hills; most packets about 5.

Radish, Summer—most packets sow about 20 feet of row; 1 ounce sows about 100 feet.

Radish, Winter—1 ounce sows about 150 feet of row.

Rhubarb—1 ounce sows about 100 feet of row; a packet 15 to 20 feet.

Rutabaga—1 ounce sows about 250 feet of row; a packet 40 to 50 feet.

Salsify—1 ounce sows 100 feet of row; most packets about 15 feet.

Scorzonera—1 ounce sows 100 feet of row.

Sorrel—1 ounce sows about 100 feet of row.

Soybean—1 pound sows 100 to 150 feet of row.

Spinach, New Zealand—1 ounce sows about 75 feet of row.

Squash, Summer—1 ounce sows about 100 feet of row; most packets 15 to 20 feet.

Squash, Winter—1 ounce sows 60 to 100 feet of row, or 20 to 30 hills; most packets sow 6 to 8 hills.

Strawberry—50 plants will fill 150 feet of row; an average packet of the Sweetheart variety contains about 50 seeds; a $1/128$-ounce packet contains about 200 seeds.

Sunflower—¼ pound sows 500 to 800 feet of row; most packets about 50 feet.

Swiss Chard—1 ounce sows about 100 feet of row; most packets about 25 feet.

Tampala—Most packets sow 25 to 30 feet of row.

Tomato—1 ounce yields 3,000 to 6,000 plants (seed size varies); most packets of nonhybrid types contain 100 to 150 seeds, 25 to 75 seeds for hybrids.

Turnip—1 ounce sows about 200 feet of row; most packets about 50 feet.

Watercress—Most packets yield about 1,000 plants—seeds are very fine.

Watermelon—1 ounce sows about 30 hills; most packets 5 to 6.

Witloof Chicory—1 ounce sows 250 feet of row; most packets 30 feet.

new variety, Zoom, may be forced in several inches of water.) Keep the buckets in a cold place and bring them, one at a time, to a cool room to sprout (50°F to 65°F, 10°C to 18°C). Water the plants and cover them with a carton or paper bag to exclude light, so the chicory that sprouts from the roots will be creamy white and mild flavored.

Growing Herbs from Seed

Once you start growing some vegetables from seed, the progression to herbs is easy and fun. You could even raise a whole herb garden from seed—if you have plenty of time and patience. To start herbs from seed indoors, you'll use the basic procedures described in Section One of this book. Most herbs need excellent drainage; for some, like thyme, this condition is critical. A few herbs grow better when sown directly in the garden. Several have seeds that germinate erratically and seedlings that grow slowly. When you develop a sure hand with the familiar basils, chives, and sage, you can even branch out to try sowing seeds of vanilla grass, lemon catnip, copper fennel, goldenseal, and other unusual herbs that are difficult to find.

There are two kinds of herbs, though, for which I'd advise you not to plant seeds. One is French tarragon, which doesn't produce viable seeds. Any tarragon seeds you may see offered in catalogs are from the inferior Russian tarragon, an entirely different plant that lacks true tarragon's subtle anise flavor. Mints are not good candidates for seed starting either, because they hybridize readily, and seeds listed in catalogs often fail to come true to type. This isn't a problem if you simply want a stand of mint—any mint—to flavor your iced tea, but if you have your heart set on apple mint, spearmint, or any other specific cultivar, then you'll need to buy plants to be sure of what you're getting. Apart from these two exceptions, there is a whole gardenful of herbs that you *can* grow from seed.

Angelica
Angelica archangelica

A hardy biennial, angelica makes an impressive backdrop presence in the garden. It can grow to 6 feet, with sturdy hollow stems and yellow-green flowers that resemble huge dill heads. All parts of the plant are aromatic; even the root is used for tea.

Days to Germination: Within 4 weeks at 60°F to 70°F (16°C to 21°C), but seeds must be fresh

When to Plant: Angelica seeds are very short-lived. They've been known to keep over winter in an airtight refrigerated container, but the best plan is to sow them in late summer as soon as they mature.

How to Plant: The seeds need light for germination, so just press them into the soil surface. A period of moist chilling also helps to germinate angelica seeds. Fall-sown seeds will receive this treatment naturally outdoors. For spring sowing, plant the seeds on a bed of damp sphagnum moss and keep the planting in the refrigerator for several weeks before exposing it to warmth and light. Angelica appreciates rich soil. Set plants 3 feet apart in a spot sheltered from wind, if possible.

Basil
Ocimum basilicum

A tender 1 to 1½-foot annual, basil likes warm weather.

Days to Germination: About 5 days at 70°C (21°C), indoors

When to Plant: Start plants six to eight weeks before your last frost date.

How to Plant: The seeds are small, so don't bury them deeply; a thin layer of soil will do. Seedlings will grow best in temperatures above 65°F (18°C). Basil seeds may also be sown directly in the garden when the soil is warm—about the time you'd plant beans. Thin basil plants to stand 10 to 12 inches apart.

Growing Conditions: Basil appreciates well-drained, fairly fertile soil. Keep young plants pinched back to encourage bushy growth.

Varieties: You can grow a whole tapestry of basils from seed, from the dramatic, purple-leaved Dark Opal to cinnamon, lettuce-leaf, lemon, and bush basils. For marginal conditions, Fine-Leaf basil is more cold tolerant and needs less light than Sweet Basil. Bush basil is, unfortunately, subject to the plant disease botrytis, so pot up a few extra seedlings for possible replacements.

Borage
Borago officinalis

A quick-growing 2-foot annual with hairy leaves and cornflower-blue star-shaped flowers, borage self-sows generously.

Days to Germination: 5 days at 70°F (21°C)

When to Plant: Sow seeds directly in the garden a week or so before the last frost. There's not much advantage in growing transplants, because borage doesn't like to be moved. For a steady harvest of the decorative (and edible) flowers and the edible leaves, sow small batches of seeds every four weeks.

How to Plant: When planting seeds, be sure to cover them because they need darkness to germinate. Thin seedlings to 1 foot apart.

Growing Conditions: Borage tolerates poor soil and does well in any decent porous soil that is moderately well supplied with water.

Burnet
Poterium sanguisorba

An attractive 1- to 2-foot perennial with sharply toothed leaves, burnet makes a decorative border and a good salad plant.

Days to Germination: 8 to 10 days at 70°F (21°C)

When to Plant: You can plant burnet seeds either in spring or fall. When starting early indoors, sow seeds in individual pots. Burnet may also be directly sown in the garden on or near your frost-free date.

How to Plant: Seedlings may be transplanted, but mature plants don't take well to being moved. Space plants 12 to 15 inches apart.

Growing Conditions: Burnet likes full sun and regular watering, but does tolerate some drought and relatively poor ground. It prefers alkaline or neutral soil.

Remarks: Burnet is a good container plant. Established plants usually self-sow readily.

Caraway
Carum carvi

A hardy biennial, caraway grows 2 feet tall and produces seeds during its second summer. Caraway is one of the oldest culinary herbs; the seeds have been found in an excavated Swiss lake dweller site from 5000 B.C.

Days to Germination: 14 days at 70°F (21°C)

When to Plant: Sow seeds outdoors two to three weeks before the last frost.

How to Plant: Because it develops a taproot, caraway is best planted directly in the ground. If you do transplant caraway, move it while it's still a small seedling. Caraway seeds have a short period of viability and plants grow slowly at first. They self-sow generously, though.

Growing Conditions: Well-drained, ordinary garden soil suits this herb.

Catnip
Nepeta cataria

A 2- to 3-foot perennial with scalloped-edged leaves and white flowers, catnip is easy to grow from seed.

Days to Germination: 7 to 10 days at 60°F to 70°F (16°C to 21°C)

When to Plant: Start seeds indoors six to eight weeks before your last frost or sow seeds in the garden around the time of your last frost.

How to Plant: Space plants 10 to 15 inches apart in full sun in any well-drained soil.

Growing Conditions: Catnip will grow even in poor, dry, sandy soil and in partial shade.

Remarks: Its rather musky aroma is not always agreeable to people, but cats love it when the leaves have been bruised to release the essential oils.

Chamomile
Chamomile nobile

Roman chamomile, *C. nobile,* is a low-growing perennial, and German *Matricaria recutita,* a 1- to 2-foot annual. The tiny daisylike flowers of German chamomile make good tea. Roman chamomile is most often used for rock gardens and as a ground cover.

Days to Germination: 10 to 12 days at 55°F (13°C) for German chamomile; 10 days at 70°F (21°C) for Roman

When to Plant: Outdoors, you can plant German chamomile seeds as soon as the ground can be worked. Indoor plantings may be made two to three months before your frost-free date and seedlings set out two to three weeks before that final frost. Start Roman chamomile seeds two to four weeks later and set out at or before the last frost.

How to Plant: Thin or plant seedlings to 6 inches apart.

Growing Conditions: Average garden soil is fine as long as it is well drained. Chamomile appreciates a decent supply of moisture and doesn't mind a bit of light shade. It self-sows freely, too.

Chervil
Anthriscus cerefolium

Chervil is a 1- to 2-foot annual.

Days to Germination: 14 days at 55°F (13°C)

When to Plant: Chervil likes cool weather, so you can sow seeds outdoors in early spring, or plant them in the fall.

How to Plant: Chervil's delicate seedlings don't transplant well, so it is usually sown directly in the garden. Cover the planted seeds with a thin sprinkling of soil. When seedlings appear, thin them to stand 9 to 12 inches apart. Chervil grows quickly and its finely cut, parsleylike leaves are at their best before flowering and in cool weather.

Growing Conditions: It reseeds readily and appreciates fairly rich, moist soil and a protected location where the sun isn't too hot.

Chives
Allium schoenoprasum

Chives are a hardy perennial member of the onion family, with slim, tubular, quill-like leaves that grow to a length of about 8 to 10 inches.

Days to Germination: 10 to 14 days at 60°F to 70°F (16°C to 21°C)

When to Plant: It's best to start seeds in January for spring planting, but you can also sow seeds in the open ground as soon as the soil is dry enough to work, if you don't mind keeping the thready little seedlings well weeded for the two months or so that it takes them to begin to look as though they'll amount to something.

How to Plant: Plant clumps of chives 8 to 10 inches apart in rich soil. They like full sun but don't mind a bit of light shade.

Growing Conditions: For best quality, cut leaves near the base to encourage tender new growth, rather than snipping tips. Fertilize plants once or twice a season, divide them when they become crowded, and remove spent blooms.

Coriander
Coriandrum sativum

An annual herb, coriander grows 12 to 30 inches tall. It is grown both for its leaves (in which case the name cilantro is used) and for its round white seeds.

Days to Germination: 10 days at 60°F (16°C)
When to Plant: Sow seeds after your last frost.
How to Plant: Sow either directly in the ground or in pots from which seedlings can be easily removed without disturbing the roots. Like other members of the Umbelliferae family, coriander develops a taproot and bolts to seed sooner if trans-planted. This is fine if it's the seeds you're after, but not so desirable if you had planned to use the leaves in a big batch of salsa. The seeds need darkness to germinate, so cover them lightly but firmly. Plant seedlings 8 to 10 inches apart.
Growing Conditions: Coriander does well in average well-drained soil, in either full sun or partial shade.

Dill
Anethum graveolens

Dill is a 2- to 3-foot annual with feathery leaves and aromatic seeds. It is an important ingredient in pickles, potato salad, and cold cucumber soup.
Days to Germination: 21 to 25 days at 60°F to 70°F (16°C to 21°C)
When to Plant: Direct-seeding, as early in spring as the ground can be worked, works best with dill, and you can plant it either in spring or fall.
How to Plant: Sow the seeds one every inch or so. Thin the feathery seedlings to 4 to 6 inches apart for use as greens and 10 to 12 inches apart for seed production. I like to make one or two extra plantings during the season for a continuous supply of the tender thready leaves that are so good snipped on potatoes or fish. Dill seeds need light to germinate, so they shouldn't be covered with more than a light dusting of soil. Perhaps this is why volunteer plants from seeds scattered on the surface of the garden sometimes beat those we've hand-sown.
Growing Conditions: Dill grows well under a wide variety of conditions, as long as it is reasonably free of weeds.

Lemon Balm
Melissa officinalis

A 2-foot hardy perennial, lemon balm has veined leaves with finely scalloped edges.
Days to Germination: 14 to 21 days at 70°F (21°C)
When to Plant: Plant seeds outdoors in early spring or indoors about eight weeks before your last frost.
How to Plant: Plant seeds uncovered because they need light to germinate. Just press them into the soil surface. Set plants 12 inches apart in well-drained soil. Hardened seedlings may be planted out two to three weeks before the last frost.
Growing Conditions: Lemon balm tolerates both dry soil and light shade, but the plant will be bushier when grown in full sun with a good supply of moisture. Lemon balm self-sows freely. Top growth dies back in winter but reappears in spring. Protect young seedlings from heavy frost.

Lovage
Levisticum officinale

Lovage is a hardy perennial celerylike plant that can grow 5 to 7 feet tall. It's a good soup herb, and the seeds may be used in place of celery seeds. Cyrus Hyde of Well-Sweep Herb Farm in Port Murray, New Jersey, says that his family never eats potato salad without lovage.

Days to Germination: 10 to 20 days at 70°F (21°C)

When to Plant: When possible, sow lovage seeds in the ground in late summer or in fall. The alternate freezing and thawing of the ground helps to trigger germination. You can also start plants indoors in spring, planting seeds thickly, eight to ten weeks before the last frost, and setting them out two to three weeks before the frost-free date.

How to Plant: Use fresh seeds; viability decreases sharply with age. Plant lovage seedlings 2 to 3 feet apart and keep them at the back of the bed.

Growing Conditions: Lovage does well in cold climates; in fact, it seems to need a winter freeze in order to break dormancy in spring. Lovage thrives in good soil well supplied with moisture, and it doesn't mind a bit of shade. When well established, it reappears quite early in spring.

Remarks: The flowers attract beneficial wasps that prey on destructive insects.

Marjoram
Origanum majorana

Marjoram is a frost-sensitive 1-foot annual with small, pungently flavored, slightly fuzzy leaves.

Days to Germination: 8 to 14 days at 70°F (21°C)

When to Plant: You can start seeds indoors, in either spring or fall, but fall-planted seedlings should be kept from freezing over winter. Seeds can also be sown directly in the ground after danger of frost is past.

How to Plant: Light improves germination, so don't cover the tiny seeds. Scatter them thinly on finely raked ground and press them into the soil surface. Thin plants to 8 to 12 inches apart. Set seedlings out after the last frost.

Growing Conditions: Marjoram likes well-drained soil, perhaps a bit on the dry side, and plenty of sun.

Remarks: Marjoram, also called sweet marjoram, is often confused with the perennial herb oregano (*O. vulgare*). You can grow oregano from seed, too (treat it like marjoram), but don't bother with seeds of *O. vulgare*, which lack flavor. Look for *O. vulgare* subspecies *hirtum,* sometimes labelled *O. heracleoticum*.

Parsley
Petroselinum crispum

An 8-inch biennial that is usually planted as an annual, parsley is one herb that almost everyone grows. It is easy to grow but slow to start. Just when you're ready to reorder seeds, those perky little scalloped-edged seedlings begin to emerge. Parsley seeds, and those of several other related plants as well, contain furanocoumarins—compounds that block germination, especially in the presence of sunlight.

Days to Germination: 21 days at around 70°F (21°C)

When to Plant: Plant parsley seeds outdoors as early in the spring as you can work the garden soil.

How to Plant: Sow two seeds to the inch, planting them ¼ inch deep. Many gardeners grow a few radish seeds along the row to mark the spot. Weeds often appear first and must be kept under control. Sometimes I beat the weeds by starting parsley seedlings early in the greenhouse and seeding them out in April. Be sure seeds are covered—not deeply but thoroughly—because they germinate more completely in the absence of light. Thin plants to 6 inches apart.

Growing Conditions: Fertilize once or twice during the growing season.

Remarks: To knock a week or more off of parsley's usual three-week sprouting time, soak the parsley seeds in water for 48 hours, changing the water twice, before planting them. This dissolves some of those inhibiting compounds (furanocoumarins). Discard the water, though. Don't do as I used to do (before I knew about the furanocoumarins) and pour it on the planted seeds. I've gotten relatively prompt germination of late-spring sowings after pouring very hot water over the planted seeds. Once planted, though, parsley seeds germinate best when soil temperature is below 85°F (29°C).

It was an old country custom to hang a parsley seed head on a tree, arbor, or fence adjoining the garden so that the windblown seeds would be scattered in the garden. If you try this bit of lore, don't count on it for your whole crop. Plant some seeds in the ground, too.

Rosemary
Rosmarinus officinalis

A shrubby, half-hardy perennial, rosemary can grow to 3 feet high and 3 feet wide as a mature plant. It is one of the more challenging herbs to grow from seed. Rosemary seeds often have low viability and a tendency to mutate, and the seedlings grow slowly.

Days to Germination: Up to 21 days at 70°F (21°C)

When to Plant: Seeds may be started indoors in late winter or early spring and seedlings set out after the last frost.

How to Plant: My friend Bertha Reppert, who owns Rosemary House in Mechanicsburg, Pennsylvania, and has been raising rosemary from seed for years, says: "Be patient. Rosemary seeds can take up to three weeks to germinate. Rosemary really *demands* sharp drainage, both in order to germinate, and after germination. Soggy soil is sure death to rosemary seeds and seedlings. Avoid peat in your starting mix, and add vermiculite or sharp sand, or both, to your seed-starting medium to improve drainage. But don't let the soil dry out, either. The safest thing to do is mist it, and also mist the young seedlings."

Set out young plants about 8 inches apart. They grow slowly. Established plants should be set 12 to 18 inches apart. Although rosemary can be direct-seeded around the time of final frost, it is best to start seeds indoors because germination is erratic.
Growing Conditions: Rosemary plants thrive in full sun and well-drained soil, which can be either slightly acid or slightly alkaline. They make excellent container plants. Although I've known five gardeners who have kept rosemary alive outside over winter, you can't count on it north of the Mason–Dixon line. It can stay outside in areas with mild winters where light frosts but not heavy freezes may occur. I keep mine in clay pots and bring them in the greenhouse over winter.

Sage
Salvia officinalis

A hardy perennial, sage forms a 2½-foot woody bush that should be trimmed often to keep it shapely. Even the tiny seedling leaves have the characteristic gray-green color and pebbly texture of the mature sage plant.
Days to Germination: 21 days at 60°F to 70°F (16°C to 21°C)
When to Plant: Start seeds indoors one to two months before the last frost or outdoors one to two weeks before the frost-free date.
How to Plant: Set seedlings out a week or so before the last frost, setting them 20 inches apart.
Growing Conditions: Tiny seedlings need full sun and a steady supply of water. Well-drained soil is probably their most important requirement; they'll die in waterlogged soil.
Varieties: There are golden, variegated, purple, and tricolor forms, too.
Remarks: Sage is an easy herb to grow from seed. The seedlings grow quickly, and the established plants tolerate drought and poor soil.

Savory
Satureja spp.

Two species of savory are commonly grown. The annual summer savory, *S. hortensis,* grows 14 to 18 inches tall, and the perennial winter savory, *S. montana,* reaches 8 to 15 inches. Both have narrow leaves. Winter savory has stiffer stems, smaller, more pungent leaves, and a more spreading habit than the summer form. Germination is often erratic, and it grows more slowly from seed.
Days to Germination: 10 to 15 days for summer savory and up to 20 days for winter savory at 70°F (21°C)

When to Plant: For either winter or summer savory, you can start transplants early, a month or so before your frost-free date, or sow seeds right in the garden after the last frost.

How to Plant: Light promotes germination of savory, so don't cover the seeds with soil. Place clear plastic wrap or glass over the flat to keep the soil moist. Space winter savory seedlings 10 to 12 inches apart.

If you've grown summer savory before, look for volunteers from last year's plants. Seedlings need 6 to 8 inches of space. Their stems tend to be weak and they sometimes sprawl.

Growing Conditions: Both savories appreciate full sun and good drainage. The plants thrive in lean soil, but summer savory likes somewhat richer soil than winter savory.

Sweet Cicely
Myrrhis odorata

Sweet cicely is a 3- to 5-foot-tall perennial with fernlike leaves and umbels of white flowers. We met this delightful, hardy herb at Well-Sweep Herb Farm in Port Murray, New Jersey, where owner Cyrus Hyde told us that his wife Louise makes a great coffee cake seasoned with the anise-flavored leaves, which are a natural sweetener. Every part of the plant is edible. The seeds are also good for flavoring baked goods.

Days to Germination: 20 to 25 days at 60°F to 70°F (16°C to 21°C)

When to Plant: Fall planting works well, or start seeds indoors in early spring. You can also direct-sow the seeds about three months before the last spring frost.

How to Plant: Like angelica, sweet cicely develops a deep taproot, so seedlings should be transplanted before they grow large. When sowing seeds directly into the garden, plant them in fertile soil and cover with no more than ¼ inch of soil. Thin plants gradually to stand 2 feet apart at maturity. Eat the thinnings.

Growing Conditions: The plant likes cold winters, self-sows generously, and obligingly accepts partial shade.

Thyme
Thymus vulgaris

A perennial with many different forms that vary in height from 2 to 12 inches, thyme has tiny pungent leaves.

Days to Germination: 21 to 28 days at 70°F (21°C)

When to Plant: Start seeds indoors in late winter or early spring or outdoors two to three weeks before the last frost.

How to Plant: Fairman Jayne of Sandy Mush Herb Nursery in Leicester, North Carolina, who often starts hundreds of herb plants each week in his greenhouse, says, "I keep the temperature around 60°F (16°C) day and night and run them quite dry. I don't cover the seeds. I sow them in a little furrow and enough soil washes over them when they're watered." Set out started plants a week or two before the last frost, spacing them 8 to 12 inches apart.

Growing Conditions: Mature plants of common thyme are winter-hardy. All thymes need excellent drainage. Older plants that get woody suffer more winter injury. To keep the plants bushy, cut back three-fourths of their new growth during the growing season. Thyme grows slowly.

Varieties: About 60 different named cultivars of thyme grow in the United States. They hybridize readily and thus are difficult to identify; many that bear some of the 200 names you'll find in catalogs are actually duplicates. Some thymes produce only female flowers.

Growing Garden Flowers
from Seed

Flowers are a necessity of the spirit, not a luxury. If you're living in a place you love, planting flowers will come naturally, as an expression of contentment and—in the case of perennials—of an intent to root oneself in that favored spot. In a temporary home, too, though, planting flowers will help you to feel more settled. A few bright annuals can soften harsh corners and camouflage unattractive areas, and even attract new friends. One of the most appreciated hollyhock patches I've ever grown was a screen of the flowers in front of the bare wall of a rented house. When you start flowers from seed, you'll have dozens of plants for the price of a single six-celled market pack, and you can spread them lavishly around your home and garden. Home-seeded flowers usually come in a greater variety of colors than seedlings offered by nurseries, and you can grow unusual kinds like balloonflower and schizanthus, which aren't widely available as plants.

Most vegetable seeds germinate well in darkness, but many flower seeds need light in order to germinate. This needn't be direct sunlight; in fact, in most cases, it shouldn't be. Indirect light is fine and less likely to over-dry the soil, or you can keep flats of planted seeds under fluorescent lights. Some flowers have extra-fine, dustlike seeds. Sometimes seeds for different colors of a flower will germinate at different rates. If you've sown a packet of mixed-color seeds, allow at least an extra week for possible stragglers to sprout if you want the widest possible color sampling. Otherwise, except for avoiding high-nitrogen fertilizer, which can retard blooming, follow the same principles in starting flower seeds as you would for vegetables.

You can save seeds from garden flowers, too, as long as they're not hybrids; however, both annual and perennial flowers will bloom for a longer time if you remove spent blossoms as they fade. If you do wish to save seeds from your flowers, let the seed capsules or seed heads dry on the plant. In some cases you might need to bag them so they won't shatter. Flower seeds are easy to save because they are usually in dry form, not in fleshy fruits. You'll have to assume that they've been cross-pollinated with others of their species. The colorful, often perfumed, pollen-rich flower is, after all, the plant's invitation to the insect world to assure pollination for a new crop of seeds.

The following directions for starting garden flowers from seed will give you specific hints for the needs of some of the most frequently planted flowers. Germination temperatures refer to the temperature of the soil or growing medium. Unless otherwise specified, cover the seed with soil to a depth of three times its diameter.

300

ANNUALS

Annual flowers bloom the year they're planted, usually all summer until frost. They generally live for just one season, but a few of the flowers we grow as annuals in cold-winter climates, like geraniums, are normally perennial in their native tropics. I've kept alyssum, petunias, geraniums, lobelia, coleus, impatiens, and stocks alive over winter in the greenhouse for a second season of bloom. Some annuals self-seed freely, among them calendula, cosmos, nicotiana, portulaca, snapdragons and spider flower. I've grown and liked all of the following annuals.

Ageratum
Ageratum houstonianum

Also called flossflower, ageratum has lavender-blue or white tufted flowers and grows 6 to 8 inches high. Seeds are available for both hybrid and open-pollinated cultivars.

Days to Germination: About 7 days at 80°F (27°C)

When to Plant: For bedding plants, sow seeds under lights or in a greenhouse in February or March.

How to Plant: Ageratum needs light to germinate, so don't cover the seeds; just press them lightly into the soil. Right after your last frost, set transplants in the garden, spacing them 9 to 10 inches apart.

Growing Conditions: Ageratum prefers rich, well-drained soil in full sun.

Alyssum
Lobularia maritima

The delicate flowers of the 3- to 6-inch-high alyssum plant may be white, rose, lavender, or purple.

Days to Germination: About 7 days at 75°F (24°C)

When to Plant: Sow seeds outdoors two to three weeks before the last frost date.

How to Plant: You can start seeds early indoors, but the easiest planting method for alyssum is to simply scatter the seeds on the ground where you want them to grow. Leave seeds uncovered, because they need light to germinate. Thin plants to stand 4 to 5 inches apart.

Growing Conditions: The plants are fairly frost-hardy and do well in most soils except those that are highly alkaline. Full sun is best. My plants thrive in an eastern exposure with plenty of light, but with direct sun only in the morning.

Aster
Callistephus chinensis

These lovely daisylike flowers come in rose, pink, white, lavender, scarlet, and occasionally yellow. You can buy seeds for 8- to 10-inch-tall bedding asters and 18-

to 24-inch-tall cutting varieties.

Days to Germination: 10 to 14 days at 70°F (21°C); germination is sometimes as low as 55 percent.

When to Plant: For a head start, plant seeds in small pots six weeks before your last frost, or sow seeds directly in the ground after danger of frost has passed.

How to Plant: Cover seeds lightly with fine earth. Give seedlings plenty of light and protect them from severe cold when hardening off. Space plants 12 to 15 inches apart, and pinch them back in June to encourage bushiness. Asters don't like to be transplanted.

Growing Conditions: They'll do better in soil that isn't too highly acid and seem to thrive in both poor and dry soil.

Pests and Diseases: Unfortunately, asters are bothered by several plant diseases, including wilt and yellows. Plant them in a different spot each year to escape wilt, and destroy affected plants. Leafhoppers and other insects spread yellows disease.

Bachelor's Button
Centaurea cyanus

Also called cornflowers, these cottage-garden favorites have slightly shaggy blue, rose, or white blossoms.

Days to Germination: About 7 days at 70°F (21°C)

When to Plant: Most gardeners plant the seeds in the open ground several weeks before the last frost.

How to Plant: Cover lightly with fine soil. Thin plants to 8 inches apart. They bloom best in cool weather.

Growing Conditions: Bachelor's buttons tolerate cold weather, crowding, and poor soils, but they do not transplant well.

Varieties: Both tall (30-inch) and dwarf (12- to 16-inch) varieties are available. A large-flowered perennial bachelor's button is available. Plant this in early spring also, about a week before your last frost, and look for seedlings in a month. The seeds need darkness to germinate, so they should be covered. Germination of perennial *Centaurea* is often erratic.

Balsam
Impatiens balsamina

Also called lady's slipper by the grandmothers of my childhood, this is a traditional favorite that's easy to handle for both seed-starters and seed-savers. The 1- to 2-inch pink, rose, or cerise camellialike flowers peek out from under stems and leaves of bushy 18-inch plants. A 12-inch top-blooming strain is also available.

Days to Germination: 8 to 14 days at 70°F (21°C)

When to Plant: Plant balsam seeds right in the garden after all danger of frost has passed, or for an early start, plant seeds in flats six to eight weeks before last frost.

How to Plant: Space seeds about 2 inches apart, and thin seedlings to 10 inches.

Growing Conditions: Balsam likes good soil, well drained but with a steady supply of moisture. It thrives in part shade as well as in sun.

Bells-of-Ireland
Moluccella laevis

These green charmers send up 2- to 3-foot spires of tiny white flowers surrounded by green cup-shaped bracts. Bells-of-Ireland are cutting-garden favorites. They're good for drying, too.

Days to Germination: 25 to 35 days at 55°F (13°C)

When to Plant: For direct seeding outdoors, press seeds lightly into soil about two weeks before the last frost date, or start indoors six to eight weeks before the last frost and set out seedlings around the time of the last frost.

How to Plant: The hard seed coats delay germination. To speed things up, soak the seeds in room-temperature water overnight. Stokes Seeds recommends prechilling the seeds at 50°F (10°C) for five days before planting. Alternating temperatures between 50°F and 85°F (10°C and 29°C) seems to help promote germination, too. Press seeds into soil but don't cover them—light helps them germinate. Space plants a foot apart.

Growing Conditions: Bells-of-Ireland need good drainage, but don't mind lean soil or light shade. Pick often for continued production. They self-sow readily.

Calendula
Calendula officinalis

Also commonly called pot marigold, this plant blooms best in cool weather, producing daisylike flowers in shades of orange and yellow at the top of 15- to 18-inch stems. Calendula volunteers readily and blooms through the first frosts of fall.

Days to Germination: 10 to 14 days at 70°F (21°C)

When to Plant: Start early spring plants indoors in February or March and a fresh batch for fall in June, or sow seeds right in the ground soon after you plant your spring peas.

How to Plant: Cover the seeds; they germinate better without light. Thin seedlings to about a foot apart.

Growing Conditions: Full sun is best, and lean soil is fine if not waterlogged.

Remarks: Some newer mixes contain a wider color range. Double forms are also available. Calendula petals are edible—good in soups and salads.

Celosia
Celosia spp.

This genus includes *C. cristata,* crested celosia, and *C. plumosa,* plumed celosia. Both the plumed and the crested celosia (also called cockscomb) come in warm shades of yellow, gold, cerise, crimson, and wine, and in a variety of heights, from 6-inch dwarf forms to plants 2 feet tall.

Days to Germination: About 14 days at 70°F (21°C); sooner at 80°F (27°C)

When to Plant: Start seeds indoors about a month before your last frost date. If you plant seeds directly in the garden, wait until all danger of frost is over.

How to Plant: Celosia seedlings suffer from uprooting, so if you start seeds indoors, plant them in individual cells or pots, not in flats. Barely cover the seeds, letting some light shine through to help them germinate. Celosias are sensitive to cold, so water the seedlings with lukewarm water. Space dwarf varieties 6 inches apart, taller ones 12 to 15 inches.

Growing Conditions: Avoid chilling young plants or disturbing their roots while hoeing. Celosia responds to such insults by refusing to flower.

Coleus
Coleus blumei

Coleus is great fun to grow because the plants show their true colors even as young seedlings—and what colors! The leaves display marvelous shades of salmon, burgundy, rose, copper, and green.

Days to Germination: 10 to 12 days at 70°F (21°C)

When to Plant: Coleus may be started indoors 2½ months before planting out, or at any time of the year for houseplants.

How to Plant: Coleus seeds respond to light, so press them gently into the soil without covering them. Keep the soil moist, preferably by bottom watering or misting, until seeds sprout. Always use lukewarm water for coleus seedlings. Set seedlings out after the last frost, spacing them about 1 foot apart.

Growing Conditions: Plants will grow in full sun, but their colors are more brilliant when they're planted in shade or part shade. To encourage more side shoots, pinch off the central shoot when plants are about 8 inches high.

Varieties: Catalogs offer a range of heights from 4 to 24 inches and a choice of leaf types—fringed, mini, spinach-leaf, oak-leaf, sword-leaf, and the new dwarf base-branching Wizards, which are bushier than older types.

Cosmos
Cosmos bipinnatus

Easy to grow from seed and strong self-seeders, these daisy-flowered plants have two common forms: Besides the rangy 3- to 5-foot *C. bipinnatus,* with pink, crimson, or white blossoms and thready, delicate foliage, you can also consider the 1- to 2-foot *C. sulphureus,* with orange or yellow flowers.

Days to Germination: 15 days at 75°F to 80°F (24°C to 27°C)

When to Plant: Sow the seeds outdoors around the time of your last frost. You can start early transplants of cosmos, too, of course, beginning in March or April, but they grow so easily from a scattering of seeds that I always direct-seed mine.

How to Plant: Cover the seeds very lightly to allow some light to penetrate. Space plants 1 foot apart.

Growing Conditions: Cosmos likes full sun and well-drained soil, and blooms even in poor soil. Tall cosmos plants often need staking.

Dahlia
Dahlia spp.

Glorious colors, from silky apricot to shell pink to clear lemon yellow to deep rich burgundy, make seed-grown dahlias one of my annual favorites. Available varieties range in height from 1 to 3 feet. They bloom with abandon in the cool early fall weather just before frost.

Days to Germination: About 7 days at 75°F (24°C)

When to Plant: You can direct-seed dahlias after frost, but for a longer period of bloom, start seeds indoors in flats a month or two before planting them out.

How to Plant: Space dwarf varieties (up to 15 inches tall) about a foot apart; tall varieties 18 to 20 inches apart.

Growing Conditions: Dahlias need full sun and appreciate soil that is rich in organic matter. Provide ample moisture in summer. Tall varieties may need to be staked.

Remarks: You can dig up the tubers right after frost and store them in peat moss, or dip them in paraffin for replanting in spring.

Four-O'Clock
Mirabilis jalapa

An easy old favorite, four-o'clocks are actually perennials if not touched by frost. They have two intriguing characteristics: A single plant will bear flowers of several colors or color combinations—white, yellows, and pinks—plain, bicolor, and tri-color. On sunny days, flowers remain closed until midafternoon. They stay open all day on cloudy days.

Days to Germination: 10 to 12 days at 75°F to 80°F (24°C to 27°C)

When to Plant: Plant seeds in the garden after frost or four to six weeks early indoors.

How to Plant: Space plants 12 to 18 inches apart.

Growing Conditions: They bloom obligingly even in poor soil, hot weather and polluted air, but they do appreciate well-drained soil and respond well to good soil or occasional feeding.

Geranium
Pelargonium ✕*hortorum*

For summer color that will follow you right into the house in the winter, you can't beat geraniums. Seeds are available for reds, pinks, salmons, and white. Plants grow 10 to 18 inches high, and in well-composted soil, form full, round flower heads. The geranium was one of my first seed-starting ventures. I planted seeds indoors in small pots kept on a glass window shelf 25 years ago, and I've never forgotten the thrill of watching those little scalloped-leaved seedlings develop into blooming plants.

Days to Germination: 14 days at 70°F to 80°F (21°C to 27°C)

When to Plant: Seeds started in early February will produce blooming plants in June.

How to Plant: Geranium seeds can be erratic in germination. Try rolling the seeds in damp paper towels and letting them "presoak" in this way for two days before planting in flats. Space first-year plants about a foot apart.

Growing Conditions: Plant geraniums in good, rich soil. Fertilize if leaves turn yellow or seem smaller than normal. My four-year-old geranium lives in a 16-inch clay pot, which I top off with 2 inches of compost each spring. I cut the plant back twice a year to promote bushy form and new bloom.

Remarks: Geraniums are frost-tender, but they are perennials in their native land, so they can be kept going for years if they're given winter protection, or if they are dug up after frost and hung bare root in a cool place till replanting time in February.

Impatiens
Impatiens walleriana

Count on impatiens for wonderful color in shady places. You can buy seeds for dwarf 6- to 8-inch plants or for the regular 1- to 2-foot varieties, in single or double flowers and in colors ranging from white to pink, fuchsia, orange, coral, orchid, and scarlet.

Days to Germination: 15 to 20 days at 70°F to 75°F (21°C to 24°C)

When to Plant: It's best to start impatiens seeds indoors 2½ months before your last frost for outdoor plants. Start them anytime if you will use them as houseplants.

How to Plant: Leave seeds uncovered, because they need light to germinate. Mist planted seeds or water from the bottom. If you cover the flat with plastic, don't seal it tightly. Impatiens is very susceptible to damping-off disease, so provide good air circulation and be sure to use new potting soil and clean equipment. The plant is also very sensitive to frost. Set out transplants a foot apart, a week or so after your last frost date.

Growing Conditions: Impatiens thrives in shade and likes soil that's rich in organic matter.

Lobelia
Lobelia erinus

Another treasure for cool weather and part shade, lobelia is a sprawling 5-inch-high plant with abundant deep blue, light blue, white, or wine-colored flowers. It looks lovely in window boxes or hanging baskets. My sapphire-blue lobelia, planted at the base of a red geranium, lived over winter in the greenhouse, blooming the whole time. The seedlings grow very slowly, and the young plants are wispy-thin, but the impact of a tumbling clump of lobelia, especially the deep blue, is a striking addition to the patio garden.

Days to Germination: 21 days at 75°F (24°C)

When to Plant: Start lobelia seeds indoors in late winter.

How to Plant: Just press the seeds gently into the soil; they need light for germination. In the garden, space plants 6 inches apart and mulch to keep the root zone cool.
Growing Conditions: An eastern exposure is perfect for lobelia seedlings. Keep them well supplied with water in hot weather and mulch to keep the roots cool.

Marigold
Tagetes spp.

If you grow any flowers at all, marigolds are probably among them, but did you know how many marigold choices you have, from the dwarf single-flowered signets to the bushy mid-sized French types to the stately 3-foot, late-blooming American marigolds? Their sunny Aztec colors range from mahogany through orange, yellow, cream, and now even white. A fourth marigold type, the hybrid triploids, are a cross between the American and the French marigolds with intermediate size, wider color range than Americans, and extra-generous blooming habit.
Days to Germination: 7 days at 75°F to 80°F (24°C to 27°C)
When to Plant: Triploid plants are sterile, and seeds for triploids are more expensive and have a lower germination rate than other marigolds, so these types should always be sown under cover, starting in early spring, and set out as transplants after frost danger has passed. American marigolds should also be started early because they are so late to bloom. French and signet marigolds may be either direct-seeded after frost or grown as transplants.
How to Plant: Spacing varies from 8 to 10 inches for shorter varieties up to 18 inches for taller kinds.
Growing Conditions: Marigolds need full sun but not rich soil, and should not be overwatered. They appreciate good drainage.
Remarks: You can save seeds of any nonhybrid marigold.

Morning-Glory
Ipomoea purpurea

Vines grow 8 to 10 feet long and produce 3- to 4-inch-wide trumpet-shaped flowers in rose, white, and blue shades. To my mind, no one has ever improved on the traditional sky blue color of the Heavenly Blue variety.
Days to Germination: 7 days at 75°F to 80°F (24°C to 27°C)
When to Plant: Plant presoaked seeds, either in individual pots a month before your frost-free date or in the ground when the soil warms up, after the last frost.
How to Plant: Morning-glory seeds have tough seed coats. To help things along, nick each black seed with a file, soak the seeds overnight in lukewarm water, and then presprout them on damp paper towels kept in a warm place. I used the top of our insulated electric water heater.
Growing Conditions: Morning-glories flower more generously if grown in lean soil. They need support—a fence, trellis, string, or netting. Both flowers and leaves are edible.

Nasturtium
Tropaeolum majus

Gold, orange, peach, scarlet, or burgundy single or double flowers adorn 6- to 12-inch plants. Vining varieties are also available. Both flowers and leaves are edible.

Days to Germination: 14 days at 65°F to 70°F (18°C to 21°C)

When to Plant: Plant the pea-sized seeds right in the garden around the time of your last frost.

How to Plant: Thin plants to stand 1 foot apart.

Growing Conditions: Nasturtiums bloom better on lean soil and withstand drought. Summer's heat slows them down, but when days turn cool they bloom with great abandon. They need full sun.

Pests and Diseases: Aphids are attracted to nasturtiums.

Nicotiana
Nicotiana alata

Plants grow 1½ to 3 feet tall and produce tubular-based rose, white, and lavender blossoms. The white flowers are delightfully fragrant at night. Newer hybrid strains are more compact, but you can't beat the fragrance of the old-fashioned white nicotiana.

Days to Germination: 14 to 21 days at 70°F to 75°F (21°C to 24°C)

When to Plant: Start nicotiana indoors a month or two before your last frost and set out plants when frost danger is over. Seeds may also be sown directly in the ground after frost.

How to Plant: Seeds are very fine and should be lightly dusted over the soil surface without covering. Space plants about 1 foot apart.

Growing Conditions: Nicotiana prefers good soil and full sun. It will also bloom in part shade.

Pansy
Viola tricolor

What would spring be without pansies? Garden centers often seem to sell limited color selections. I like the grab-bag effect of planting my own seeds from a packet of mixed colors. Hybrid pansies tend to bloom earlier than the species. All pansies have glorious colors—blues, purples, yellows, apricots, burgundies, rusts—in single, bi-color, and tricolor patterns.

Days to Germination: 14 days at 70°F to 75°F (21°C to 24°C)

When to Plant: If you're good at planning ahead and want extra-early bloom, sow the seeds in August and shelter the plants over winter in a cold frame or, if your winters are mild, under straw mulch. A more common practice is to sow seeds indoors in December.

How to Plant: Be sure they're covered; pansy seeds need darkness to germinate. You can set out hardened transplants a good month before your frost-free date. I often stretch that to six to seven weeks in the sheltered east patio exposure where I grow my pansies. Space bedding plants 6 to 8 inches apart.

Growing Conditions: Technically hardy biennials, pansies are grown as annuals. They like cool weather and don't mind light shade. Shear the plants to a 2-inch stub in July for repeated early fall bloom.

Petunia
Petunia ✕hybrida

These include *P. grandiflora, P. multiflora,* and others. A common mainstay in the summer garden, and with good reason. Petunias flower abundantly, cascade charmingly over walls and flower boxes, and appear in a full array of colors, from white to pink, red, lavender, purple, coral, and even yellow. They are available in either single or double forms, in sizes from 8-inch dwarf plants to 18-inch vining mounds.

Days to Germination: 7 days at 80°F (27°C)

When to Plant: Start seeds indoors and plant out after danger of frost has passed.

How to Plant: Seeds are very tiny and need exposure to light in order to germinate, so just pat them into the damp soil surface in a seedling flat. Keep them warm and well supplied with water, and transplant to 2-inch spacing in the flat as soon as they're large enough to handle.

Growing Conditions: Petunias thrive in rich, well-drained soil in full sun.

Portulaca
Portulaca grandiflora

Portulaca is available in every color except blue, and in both single and double-flowered forms. A fleshy-stemmed, spreading plant, it grows 4 to 6 inches high.

Days to Germination: 10 to 15 days at 68°F to 86°F (20°C to 30°C)

When to Plant: After your last frost, plant seeds directly in the ground.

How to Plant: Portulaca flowers quickly from seeds, so it's seldom worth the bother to start seeds indoors, but if you do, plant them in cells or pots because the fleshy stems can be tricky to transplant. Thin the plants in the garden to a 6-inch spacing.

Growing Conditions: Grow portulaca in lean soil in a sunny spot. It can take hot, dry places. Avoid rich soil; portulaca blooms more abundantly if not fed too well.

Scarlet Sage
Salvia splendens

Most varieties of scarlet sage grow 10 to 14 inches tall, but several 24- to 30-inch cultivars are also available. The plants present a dramatic contrast; brilliant red flower spikes rise above deep green foliage.

Days to Germination: 14 days at 75°F (24°C)
When to Plant: Start seeds indoors and set out transplants after frost.
How to Plant: Cover the seeds very lightly, if at all, because they need some light to trigger germination. When transplanting them outdoors, space them a foot apart. Use clean new potting soil for scarlet sage and try to provide good air circulation at the soil surface, because the seedlings are sensitive to the damping-off fungus.
Growing Conditions: Give the plants well-drained, rich soil, and mulch to keep roots cool in midsummer heat. Water well in hot dry weather.

Schizanthus
Schizanthus ×wisetonensis

Also called butterfly flower or poor man's orchid, this plant grows 1 to 2 feet high and has ferny leaves. It produces masses of 1½-inch orchid, white, pink, yellow, or rose flowers, which have yellow throats dappled with dark-hued spots.
Days to Germination: 14 to 21 days at 60°F to 70°F (16°C to 21°C)
When to Plant: For plants that will bloom before the weather turns too hot, start seeds indoors about three months before your last frost date.
How to Plant: Seeds are tiny but need darkness to germinate, so cover them lightly with fine soil, and then keep the flats covered or in a dark place. Plant seedlings a foot apart in the garden after the last frost.
Growing Conditions: Schizanthus prefers good garden soil in full sun. Keep it well watered. It blooms best in cool weather.
Remarks: Schizanthus makes a good pot plant, but sometimes needs a bit of support.

Snapdragon
Antirrhinum majus

Snapdragons are the backbone of my early summer cutting garden, with their delicious colors: apricot, bronze, pink, wine, rose, yellow, and orchid. You can buy seeds for plants ranging from 6-inch dwarfs to 3-foot background beauties, which usually need staking. More open-faced blossoms, recently developed, vie with the traditional "snapping" pouch form in catalogs. I love both.
Days to Germination: 7 to 14 days at 75°F (24°C)
When to Plant: For June bloom, sow seeds indoors in flats in February. Plant seedlings out around the time of your last frost.
How to Plant: Try keeping seeds in the freezer for two days before planting to hasten germination. Snaps need light to sprout, so just press the salt-grain–sized seeds into the soil surface.
Growing Conditions: Snaps do best in full sun and fairly good soil, but can take partial shade. Snapdragons are hardy flowers and, once they become well established, can stand light frosts. Technically they're tender perennials, but most gardeners treat them as annuals.

Spider Flower
Cleome hasslerana

Good for the background of an informal flower border, spider flowers, also called cleome, grow 3 to 6 feet tall and produce 6-inch heads of pink, white, or lavender blossoms.

Days to Germination: 14 days at 70°F (21°C)

When to Plant: A strong self-seeder, cleome is usually planted directly in the ground around the time of your last frost, but may be started early indoors too.

How to Plant: Try chilling the seeds at 40°F (4°C) for five days before planting to encourage earlier germination. When sowing seeds, leave them uncovered, because light improves germination. Warm days and cool nights seem to promote germination of the planted seeds. In the garden, space plants 18 to 24 inches apart.

Growing Conditions: Cleome withstands dry weather and appreciates rich, well-drained soil.

Stock
Matthiola incana

Stocks produce densely flowered wands of pastel pink, yellow, rose, or violet flowers on 15- to 36-inch stems, depending on variety. Their fragrance spiced the air of my cool winter greenhouse, where they were happier than in the hot summer garden. In fact, stocks will bloom only in cool weather, up to about 65°F (18°C).

Days to Germination: 10 days at 65°F to 75°F (18°C to 24°C)

When to Plant: You can plant the seeds anytime for greenhouse or houseplants, but for bedding plants, sow seeds indoors six weeks before your last frost-free date.

How to Plant: Sow seeds uncovered. Set out seedlings, 10 to 12 inches apart, around the time of the last frost, perhaps a week sooner.

Growing Conditions: Although stocks can't take heat, they last a long time in cool weather. Give them rich soil and an occasional side-dressing of fertilizer.

Sweet Pea
Lathyrus odoratus

Another fragrant annual that blooms in cool weather, the sweet pea is available in three forms: dwarf, 15-inch; knee-high, 2½ feet; or vining, 6 feet. Some newer varieties bloom better in warm weather than the older kinds. Colors range from soft blue to rose, pink, maroon, and white. Probably because the cool summers of Great Britain are more to the sweet pea's liking than our hot American weather, many English seed companies carry an excellent selection of the wonderfully fragrant sweet pea cultivars that have been largely eliminated from American seed catalogs.

Days to Germination: 14 days at 55°F to 65°F (13°C to 18°C)

When to Plant: Plant seeds outdoors in early spring as soon as the garden can be dug.

If your winters aren't severe, you can plant the seeds in fall. Sweet peas may also be grown in a cool greenhouse at any time of year.

How to Plant: Cover seeds with ½ inch of soil; they need cool soil and darkness to germinate. Thin seedlings to 10 to 12 inches apart.

Growing Conditions: Plant sweet peas in rich, composted soil. Provide netting or other support for vining types.

Sweet William
Dianthus barbatus

The annual form of sweet william grows to a height of 4 to 12 inches and produces starry flowers in pleasing shades of red, pink, salmon, or lilac. The biennials reach a height of up to 20 inches. An old-fashioned cottage-garden favorite, sweet william flowers best in cool weather.

Days to Germination: 7 to 14 days at 70°F (21°C)

When to Plant: Plant the annual in flats indoors in February for June bloom, or outside in April for summer flowers. For biennials, plant seeds right in the garden for bloom the following year.

How to Plant: Space dwarf varieties 6 to 8 inches apart and taller kinds 12 inches apart. Sweet william self-sows freely, so an established patch may bloom perennially, even if individual plants are short-lived.

Growing Conditions: Sweet william flowers most abundantly in cool weather. It thrives in well-drained soil that is not highly acid, and doesn't demand rich soil.

Zinnia
Zinnia spp.

The centerpiece of many a summer garden, zinnias come in a great variety of flower forms: pompom, button, dahlia-flowered, and cactus-flowered. Sizes range from 6 to 36 inches, and colors include both brassy golds, reds, bronzes, and oranges, and gentle pastel pinks, lavenders, yellows, and white. A magenta-colored flower appears in some inexpensive mixtures, but no blue zinnias exist. A well-grown bed of zinnias is a delight for both display and cutting.

Days to Germination: 5 to 7 days at 70°F to 75°F (21°C to 24°C)

When to Plant: Start the seeds about a month before your frost-free date, or plant seeds directly in the garden after frost.

How to Plant: If you're planting double-flowered zinnias indoors, sow the seeds in individual pots or cells, because transplanting can make them revert to singles. Cover the seeds lightly with fine soil and keep them warm. Thin dwarf zinnias to 6 inches, 15-inch varieties to about a foot apart, and 3-foot zinnias to 15 inches.

Growing Conditions: Zinnias like fertile soil in full sun.

Diseases: In my garden, zinnias often get a disfiguring fungus disease in the summer, especially in a wet season. To prevent disease, grow resistant varieties and avoid splashing water on the plants.

PERENNIALS

Perennials reward your patience by living longer. Except for a few quick starters like flax and blanket flower, which will bloom their first season if seeds are planted early, most perennials begin to flower the year after they're planted. Some—peonies and baby's breath, for example—live for many years. Others, like columbines, may last only four to five years, but like the biennial sweet william, they may reseed themselves so that the colony lives on even though individual plants die out. This is often but not always a desirable trait. Self-seeded phlox plants usually revert to the less desirable generic magenta shade, no matter how delicate the pink of the parents.

How to Plant. To start your seed-grown perennials, you can choose from three methods:

1. My usual procedure is to plant seeds in flats indoors or in the greenhouse in February or March (two to three months before our frost-free date) and then transplant the seedlings to the garden in May.

2. If you have neither greenhouse nor fluorescent lights, you can plant the seeds directly in the garden in May and thin or transplant the seedlings when they're 1 to 2 inches tall. A separate nursery row or raised nursery bed can save many young plants that might otherwise get lost in the garden jungle.

3. A third planting method takes into account the fact that many hardy perennials drop seeds in the fall that, under natural conditions, will endure winter's cold before germinating in spring. So you can plant the seeds in the fall, either in a cold frame or a lightly mulched nursery row, and watch for the tiny plants to appear the following spring. Keep the planting moist but not soggy until seedlings appear.

Except for shallow-rooted perennials like shasta daisies and chrysanthemums, most perennial flowers will winter well without protection, but while they're young and still relatively skimpy-rooted, it's a good idea to cover all of your perennials with a light mulch of straw or evergreen branches. Don't use tightly packed leaves, though, or they may suffocate your flowers.

Here are some favorite perennials to start from seed. Unless otherwise noted, they need full sun.

Anemone
Anemone spp.

These have open, cup-shaped flowers with bright yellow centers, which are followed by attractive seedpods. Snowdrop, which is white, is the hardiest anemone. It prefers a partly shady location. Pulsatilla, which is purple and 6 to 12 inches tall, is slightly less hardy. It has bell-shaped flowers and downy, fringed leaves.

Days to Germination: 21 to 28 days at 70°F to 75°F (21°C to 24°C)
Prechilling aids germination.
Growing Conditions: Both types of anemones need shelter from strong winds, and

fertile, humus-rich soil. They like afternoon shade in warm climates. Space plants 8 to 12 inches apart.

Astilbe
Astilbe ✕arendsii

Also called spirea, these perennials produce spire-shaped blooms. Varieties of these summer bloomers are available in white, pink, coral, lilac, and carmine, in heights ranging from 8 inches to 4 feet.

Days to Germination: 42 to 56 days at 65°F (18°C)

Growing Conditions: Astilbes do not mind some light shade. They *do* mind poor, exceptionally dry or alkaline soil, baking sun, and high winds. Waterlogged soil over winter isn't good for them either, but they are not difficult plants to grow when given the rich, somewhat acid soil they prefer, and they are long-lived perennials. Space plants 15 to 24 inches apart.

Baby's Breath
Gypsophila spp.

A group of long-lived plants, these perennials fill in between more substantial foliage with their airy network of dainty blossoms, which are good for drying. Baby's breath comes in white and pink and grows 30 to 48 inches tall.

Days to Germination: 10 to 15 days at 70°F (21°C)

Growing Conditions: It likes well-drained limestone soil. Allow 1½ to 2 feet between plants.

Remarks: Seedlings started in pots should be set in permanent garden spots while still small, because the plants' deep taproot will make later transplanting chancy.

Balloonflower
Platycodon grandiflorus

The buds of this 1½- to 3-foot plant resemble balloons. Flowers may be white, pink, or violet-blue. Shoots appear late, not until May, and blooms appear in July or August.

Days to Germination: 10 to 15 days at 70°F (21°C)

Growing Conditions: Light, porous, acid soil suits them best. Set plants 1 to 1½ feet apart. Balloonflower seeds need light to germinate.

Remarks: The plants spread slowly in heavy clay soil and form deep taproots, so they should be moved only with care, if at all, and preferably when young.

Basket-of-Gold
Aurinia saxatilis

This hardy, long-lived spring bloomer produces gray-green leaves and masses of small gold flowers on 10- to 12-inch plants that can be trained to tumble engagingly over rock walls.

Days to Germination: 10 days at 55°F to 75°F (13°C to 24°C)

Growing Conditions: Plants prefer well-drained soil. Set them about 8 inches apart.

Remarks: After blooming, cut back each plant by about one-third. Cuttings root readily.

Blanket Flower
Gaillardia ✕*grandiflora*

This dependable summer bloomer is easy to grow and is available in vibrant shades of yellow, red, maroon, and orange. The flowers are followed by attractive seed heads.

Days to Germination: 15 to 20 days at 70°F (21°C)

Growing Conditions: Blanket flower likes good drainage. Space dwarf forms 12 inches apart; taller ones 15 to 18 inches. Seeds germinate better in light.

Remarks: The 12-inch dwarf forms are slightly less tolerant of poor soil and drought than the older, taller (24- to 30-inch) strains.

Candytuft
Iberis sempervirens

Their evergreen foliage makes these 8- to 12-inch-tall spring bloomers good plants to use at an entrance.

Days to Germination: 14 to 21 days at 55°F to 65°F (13°C to 18°C)

Growing Conditions: Give the plants rich, well-drained soil that is high in humus. Set plants every 6 to 8 inches.

Remarks: The mounds of perky white or pink flowers should be cut back after blooming for a neater appearance.

Chrysanthemum
Chrysanthemum ✕*morifolium*

Chrysanthemums come in a wide range of colors, blossom forms, and heights. You can buy seeds for dwarf bedding mums, hardy border mums, and giant football mums, among others. Most mum varieties will flower the first fall, from seeds sown in midwinter. Mums come in white, gold, pink, bronze, rust—almost any color but blue.

Days to Germination: 7 to 28 days at 70°F to 75°F (21°C to 24°C)

Growing Conditions: They prefer rich soil and need water during a dry season. Space dwarf or cushion varieties about 8 inches apart; mid-sized, 12 to 14 inches; and tall varieties, 18 to 20 inches.

Remarks: Pinch back the leaf tips of blooming-age plants until July 1 to encourage the formation of more flower buds. Long nights (actually short days) promote blossoming, which is why commercial growers drape black cloth over their outdoor mums.

Columbine
Aquilegia spp.

These perennials produce spurred blooms in lively colors that blend gracefully with their attractive sage-green foliage. Some species have short spurs, others have long. Some are bicolored as well. Colors range from pastels through deep red, purple, and blue, and flowers appear in spring and summer. Plants may be 8 to 20 inches tall depending on the variety. Dwarf kinds are good for rock gardens and edgings.

Days to Germination: 21 to 25 days at 70°F to 75°F (21°C to 24°C)

Growing Conditions: Columbines like moist, rich soil and need water in a dry season. Allow 8 to 20 inches between plants, according to plant size. Plants do well in light shade. Light promotes germination of the seeds. Prechilling helps also.

Remarks: Individual plants sometimes die out after four to five years, but they self-sow readily and may spread charmingly all over the garden. Columbines have long taproots and should be transplanted only when young.

Coral Bells
Heuchera sanguinea

These grow easily from seed. The tiny pink, white, or coral flowers nod on stems 15 to 24 inches high over rosettes of geraniumlike leaves. Blooming time lasts from May till July.

Days to Germination: 15 to 25 days at 55°F (13°C)

Growing Conditions: These plants are tolerant of some light shade. They need good drainage and should be divided every three years or so, because the central crown grows high and woody. Allow about a foot of space between plants. Prechilling aids germination.

Coreopsis
Coreopsis grandiflora

The perennials (there is also an annual form) make a fine bright spot in the garden and an excellent cut flower. They grow to a height of 1 to 4 feet and produce golden-yellow blooms in the summer. Dwarf varieties are available as well.

Days to Germination: 10 to 12 days at 55°F to 70°F (13°C to 21°C)

Growing Conditions: Coreopsis doesn't mind relatively poor soil. Space dwarf plants 8 to 10 inches apart; tall ones about 15 inches. The seeds need light to germinate.

Remarks: Coreopsis will produce for a long time if the weather is not too hot, and the plants will multiply rapidly and live longer if divided every year. If coreopsis seeds are sown early, the plants often bloom the first year.

Daylily
Hemerocallis spp.

These long-lived perennials are delightful and dependable. Each blossom lasts a day and is succeeded the next day by another. Both buds and spent flowers are edible. Flowers appear in summer on 2- to 3-foot stalks and in gorgeous shades of yellow, gold, peach, salmon, crimson, orange, lavender, or mahogany. Most seed-grown plants will flower in their third year, a few at age two.

Days to Germination: 3 to 7 weeks at 60°F to 70°F (16°C to 21°C)

Growing Conditions: Daylilies prefer neutral to slightly acid soil. Space plants 1½ to 3 feet apart.

Remarks: Hybrid varieties, especially, should be divided regularly.

Delphinium
Delphinium elatum

Delphiniums produce elegant spires of spurred blossoms in blue, white, lavender, pink, yellow, or purple. The flowers appear in June and July, on stems 2 to 7 feet tall. They are said to be short-lived, but the plants I've grown from seed are now 12 years old.

Days to Germination: 8 to 15 days at 60°F to 75°F (16°C to 24°C)

Growing Conditions: Plant only fresh seeds because they quickly lose viability with age. Keep the planted seeds dark. Delphiniums do best in very rich soil that is alkaline to slightly acid. Feed the plants twice a year. Space plants about 2 feet apart.

Remarks: If the stalks are cut back after the early summer blossoming, they usually bloom again in fall. Tall plants will need staking.

Evening Primrose
Oenothera spp.

Large, fragrant, yellow flowers bloom at dusk. In the Tina James strain (*O. glaziovinia*), the flowers pop open suddenly—great fun to watch.

Days to Germination: About 2 weeks at 70°F to 80°F (21°C to 27°C)

Growing Conditions: Any ordinary soil, even poor soil, suits these hardy, drought-tolerant plants. The seeds are tiny. Cover them lightly, if at all. Some species are considered biennials, but all reseed freely.

Flax
Linum perenne

The blue or yellow flowers appear from May through August on 6- to 12-inch-high plants. Flax blooms longer during cool summers. Individual plants live only three to four years, but they multiply generously from seed.

Days to Germination: 20 to 25 days at 70°F (21°C)

Growing Conditions: Flax needs well-drained soil. Thin plants to 10 to 12 inches apart.

Remarks: Try scattering flax seeds among your spring-flowering bulbs. Transplant them only when they are tiny.

Globe Thistle
Echinops exaltatus

These perennials have violet-blue spherical blossoms and prickly leaves. They grow to a height of 3 to 5 feet and bloom in July.

Days to Germination: 15 to 20 days at 65°F to 75°F (18°C to 24°C)

Growing Conditions: They favor light soil that is well drained. Avoid overfertilizing globe thistle plants, or they may grow too rank and need staking. Allow 2 feet between plants.

Remarks: The roots grow deep into the soil, and the plants seldom need to be divided.

Hollyhock
Alcea rosea

Hollyhocks are now bred for double blossoms, but sentimental gardeners like me still prefer the singles. They produce flowers in pink, cream, yellow, rose, peach, or wine on 4- to 6-inch-tall stems in June and July.

Days to Germination: 10 to 14 days at 70°F (21°C)

Growing Conditions: Give hollyhocks rich soil and plenty of water while they're growing. Thin plants to stand 2 to 3 feet apart.

Remarks: Hollyhocks self-sow readily. Seedlings should be moved early to their permanent locations because they develop deep taproots.

Lupine
Lupinus spp.

Lupines have gorgeous, stately columns of pealike flowers in marvelous colors: blue, cream, rose, purple, yellow, and some bicolors. The stalks are 3 to 4 feet high with attractive palmlike foliage.

Days to Germination: 14 to 21 days at 55°F (13°C)

Growing Conditions: To improve germination, keep the seeds in the freezer for two days and then roll them in a damp paper towel. After the seeds have been preconditioned for a day in the damp towel, plant them in individual pots. Lupines like rich, well-drained soil, some shelter from wind, and a steady supply of moisture. Mulch to keep roots cool. Space plants 18 inches apart. Allow 1½ to 2 feet between plants.

Remarks: Both seed size and germination time vary according to flower color. You may find that the weakest lupine seedlings produce the most desirable flower colors.

Phlox
Phlox paniculata

Border phlox, *P. paniculata,* is a very hardy plant, growing to a height of 18 to 36 inches with abundant summer flowers in pink, rose, salmon, white, or lavender. Moss pinks, *P. subulata,* grow 3 inches high and produce white, pink, or lavender flowers in spring.

Days to Germination: 25 to 30 days at 70°F (21°C)

Growing Conditions: Border phlox needs rich, well-drained soil well supplied with moisture. Good air circulation is important to prevent mildew. Moss pinks tolerate dry conditions and less fertile soil. Space plants 2 feet apart where air drainage is good; 3 feet apart in damp climates or near buildings. Cover the seeds when planting; they need darkness to germinate.

Remarks: Keep old flower heads of border phlox clipped off so the plants won't scatter a host of ordinary purplish pink seedlings. Some white varieties of border phlox have a delightful, spicy fragrance in the evening.

Poppy, Oriental
Papaver orientale

Poppies are glorious during their short period of bloom in May or June, but they die back over summer. They like cool weather and are very hardy, yet should be transplanted only with considerable care. They grow 1 to 3 feet tall and produce flowers in pink, red, orange, or white.

Days to Germination: 10 to 15 days at 55°F (13°)

Growing Conditions: Poppies need good drainage. Rich soil is fine but not absolutely necessary. Thin plants to 15 to 18 inches apart. Leave seeds uncovered; light promotes germination.

Primrose
Primula spp.

Primroses grow to a height of 6 to 12 inches. Flower colors may be clear, bright, and intense, or soft, subtle, and pastel. They are available in pink, orange, cream, red, purple, white, and yellow.

Days to Germination: 21 to 40 days at 70°F (21°C)

Growing Conditions: Give these spring-blooming beauties dappled shade and don't be alarmed when they disappear in hot weather; they'll be back next spring. Divide them every third year and water them when the weather is dry. Space the plants about 1 foot apart in rich, moist, slightly acid soil with plenty of humus. The seeds need light to germinate and will appear in desultory fashion for several weeks after sowing, so keep the flats moist. Prechilling aids germination.

Shasta Daisy
Chrysanthemum ✕superbum

This perennial grows easily from seed and is the backbone of any summer bouquet. The large, handsome white daisies bloom on 1- to 4-foot stems. Both tall and compact forms are available. Single-flowered varieties live longer in the garden.

Days to Germination: 10 to 14 days at 70°F to 75°F (21°C to 24°C)

Growing Conditions: Shasta daisies respond well to rich, moist soil. They are summer bloomers, vigorous growers, and should be divided every three years. Allow 1 to 2 feet between plants. The seeds must have light to germinate.

Spiderwort
Tradescantia virginiana

Also called trinity flower, this perennial is an old dooryard favorite that now has hybrid cousins in violet, white, rose, and lilac in addition to the traditional purple. They grow 2 to 3 feet tall

Days to Germination: 30 days at 70°F (21°C)

Growing Conditions: These undemanding perennials will flower in light shade and even in poorly drained, lean soil. Space them about 1 foot apart.

Growing Wildflowers from Seed

Whether you're raising a few specimen plants to soften a rocky corner, introducing new prairie plants to your yard or meadow, or establishing wildflower colonies in your woods, you'll find your patience well rewarded by the delicate beauty and dependable hardiness of native plants like the following. This list is not by any means exclusive, but I had to stop somewhere! Consult one of the excellent wildflower books listed in the Bibliography for more wildflower suggestions. With several exceptions, which are noted in the text, all plants described are perennials. Most of them will bloom the second or third year after seeds are sown. Unless directions specify otherwise, use the usual rule of thumb for planting seeds: Cover with an amount of soil equal to three times the diameter of the seed.

The recommended temperature range for stratifying seeds is 34°F to 44°F (1°C to 7°C). Specified germination temperatures refer to the temperature of the soil or growing medium.

When planting colonies of wildflower seedlings, set small varieties (6 to 8 inches tall) 8 to 10 inches apart. Those that will form medium-sized plants (10 to 15 inches) should be set 12 to 18 inches apart, and taller plants 20 to 36 inches apart. Wider spacing may be used, of course, if you prefer.

Bee Balm
Monarda didyma

Bee balm is a 2- to 4-foot-high, short-lived perennial with shaggy, rosy red blossoms that attract birds.

Collecting Seeds: Seeds mature one to three weeks after flowering. They will fall readily from the blossom head when you bend it over and shake it. Air-dry the seed heads on newspaper and then shake the seeds out. Store them in the refrigerator.

When to Plant: Seeds may be sown in an outdoor seedbed as soon as they ripen. When sown indoors in January, they usually germinate in one to two weeks. Young seedlings grow slowly.

How to Plant: Harry Phillips, author of *Growing and Propagating Wildflowers*, recommends transplanting seedlings to a 3-inch pot after six to seven weeks in the starting flat, and allowing the potted plants to develop sufficient roots to fill the container before planting in the garden. A few plants may even bloom the first fall from seeds sown in February.

Growing Conditions: Bee balm grows well in either sun or partial shade and accepts average garden soil, but it is particular about receiving a good supply of moisture. In the wild, you'll find it in damp meadows and along stream edges. Pinch back the seedlings for a bushier plant.

Black-Eyed Susan
Rudbeckia hirta

The very essence of summer when seen by the wayside, these 1- to 2-foot-high plants bear yellow daisylike flowers with brown centers. Blooming from June to August, they are short-lived perennials that self-sow readily. Sometimes they behave like biennials.

Collecting Seeds: When the central cone fades somewhat and looks a bit shaggy, shake out the seeds and either store them in the refrigerator or a late-winter indoor or cold frame planting, or plant them in a nursery bed in the fall.

When to Plant: Sow seeds indoors or in a cold frame in late winter, or plant them in a nursery bed in the fall.

How to Plant: Pot up started seedlings after they have developed several true leaves. Once potted plants recover from transplanting, let the soil in the pots dry before watering again.

Growing Conditions: Give them average, well-drained soil and full sun. The plant will be more compact in soil that is not too rich.

Bloodroot
Sanguinaria canadensis

In March and April, the lovely starlike, white flowers appear on 6- to 8-inch stems, with the developing leaf furled around the stem. The attractive lobed leaves usually disappear in summer.

Collecting Seeds: Seeds develop quickly. Capsules split while still green and are often hidden by the leaves. Start checking the seed capsules a month after flowering. When seeds start to turn brown, collect some capsules. Put them in a plastic bag with ¼ cup of moist soil, and keep them cool until the capsules split and release their seeds.

When to Plant: Sow seeds immediately after they have been released from the capsule. Don't store bloodroot seeds; they rot if stratified and lose viability if the fleshy aril (a small white appendage on the seed) dries out.

How to Plant: Sow seeds in a seedbed enriched with a good supply of humus.

Growing Conditions: Bloodroot likes light to moderate shade, especially under hardwoods, and well-drained soil.

Blue-Eyed Grass
Sisyrinchium angustifolium

A short-lived, 12-inch-tall perennial, blue-eyed grass blooms at the edges of our meadows in May and June. Its perky blue flowers dotted on grasslike stems are often nearly hidden by taller competing grasses. In cultivation, the plant is larger and more handsome.

Collecting Seeds: The pealike seed capsules start to shrivel about a month after flowering. Crack them open then and remove the seeds.

When to Plant: Plant seeds in midsummer.

How to Plant: Sow the tiny black seeds in an outdoor bed or cold frame with some protection from full sun. A short period of stratification—a month or so—improves germination. The plant self-sows readily.

Growing Conditions: It likes average soil, not too rich, or it may grow unattractively rank. Full or dappled sun, steady moisture, and good drainage complete blue-eyed grass's modest requirements. Avoid a heavy mulch.

Bottle Gentian
Gentiana andrewsii

A few of these elusive 1- to 2-foot perennials bloom at the edge of our wet meadow each August. The blue bottle-shaped blossoms, borne in the axils of the upper leaves, are often partly hidden by taller grasses. We know the gentians are there, but we must seek them out—a good excuse on a late summer day to let the tomatoes pile up and the garden weeds continue to flourish, and declare a free morning for wildflower rambling.

Collecting Seeds: Seeds should be ready after frost, when capsules begin to open. Shake them out.

When to Plant: If you don't plant them in flats in the fall, keep them refrigerated for spring sowing.

How to Plant: It's a good idea to plant bottle gentian seeds in plant bands or small milk cartons, so the roots needn't be disturbed when you set them in their permanent places. If your seedlings are in flats, transplant them to pots in clumps, not individually. When planting out, set the crown 1 inch deep.

Growing Conditions: Bottle gentian grows well in average soil, preferably slightly acid, with plenty of moisture. In central Pennsylvania, the plants thrive in full or part sun; farther south, they'd prefer a bit more shade.

Bunchberry
Cornus canadensis

This diminutive dogwood relative, a favorite of mine from childhood summers spent in Maine, needs special conditions, so it isn't for everyone. The white, yellow-centered, open flowers (technically bracts) bloom above a whorl of quilted-veined leaves from May until July. They turn to clusters of bright red berries by summer's end. Carpets of these 4- to 8-inch-tall wildflowers enrich shady, cool northern woods.

Collecting Seeds: Pick berries at the red ripe stage and remove the seeds.

When to Plant: Marie Sperka, author of *Growing Wildflowers,* keeps her seedlings in pots and plants out the whole potted clump without separating individual plants. Because the roots are scarcely disturbed by the move from pot to woodland ground, bunchberry clumps may be set out whenever the ground can be dug. Transplants dug from a nursery bed should be moved only when dormant.

How to Plant: Sperka plants six to eight seeds in a pot and keeps them moist. She's found that bunchberries sometimes germinate late—two to three years after planting—and that they thrive best if planted in groups. Single plants seem to languish.

Growing Conditions: Bunchberries need strongly acid soil, shade, abundant humus, and a steady supply of moisture with good drainage. Sperka reports that she has seen bunchberry colonies growing in soil with a relatively high pH of 6, leading her to suspect that the other conditions—shade, humus, moisture, and drainage—are more vital to bunchberries than highly acid soil. You'd need to be very determined, though, to grow these beauties in any but their ideal conditions, which also include coolish summers and what horticulturists call a cool root run—mulch, stone, or other surface insulation to keep the roots cool.

Butterfly Weed
Asclepias tuberosa

One of those wild beauties that seldom survives transplanting because of its deep taproot, butterfly weed is easy to raise from seed. The clusters of brilliant reddish orange (occasionally yellow, rarely red) flowers resemble those of its weedier relative, milkweed. The 2- to 3-foot-tall plants bloom in June and July. Butterfly weed looks glorious in a flower border.

Collecting Seeds: Collect seeds when the pods start to split, and remove the silks before sowing the seeds. Butterfly weed seeds need no special treatment; just keep them cold until planting time.

When to Plant: Start them indoors or outdoors in spring, as you would any other perennial plant; or you can sow seeds as soon as you collect them in fall.

How to Plant: Pot up the seedlings when they develop their true leaves and plant them out in the garden as soon as possible so the root can go the way it wants. Marie Sperka, owner of a wildflower nursery in Wisconsin and author of *Growing Wildflowers,* likes to trim the seedling roots to 4 inches when setting them out, to encourage new feeder roots.

Growing Conditions: They prosper in full sun and average soil. They accept dry locations but cannot tolerate standing water, so plant them in a well-drained place. Avoid mulching, too, or the crown may rot. In the South, you might give them some light shade.

Remarks: Tubers of young plants sometimes heave in winter thaws until they grow enough roots to anchor themselves. If this happens, simply push them back into the ground. Then, in summer, sit back and watch for the butterflies that will multiply your pleasure from this easy and decorative plant.

Cardinal Lobelia
Lobelia cardinalis

Also called cardinal flower, these stately 2- to 5-foot plants, with their vibrant red flowers, are startling in their beauty. The flowers bloom for a long time during August and September.

Collecting Seeds: Seeds ripen in October. Watch for the capsules to open. When seeds are dry and shake out readily, put cut stalks head down in an open paper bag and jostle the bag to loosen the seeds. Seeds should remain viable for three years.

When to Plant: Seeds sown right after collection will often germinate promptly, but for more complete germination, stratify the seeds for two months. Seeds sown indoors in February may bloom by early fall. At 70°F (21°C), seeds germinate within two weeks.

How to Plant: Space seeds ¼ inch apart and cover them very lightly with fine soil.

Growing Conditions: Although cardinal lobelia is found in very moist soil in the wild, it is an adaptable plant that does well in a perennial border in average garden soil that is reasonably well supplied with water. Light shade or morning sun is ideal; in full sun, flower color isn't as rich and the plants need more water. To prevent loss of plants from rotting, keep the plant crown free of mulch and fallen leaves.

Remarks: Cardinal lobelia is a relatively short-lived perennial, so you might want to raise new batches of seedlings every few years.

Columbine
Aquilegia canadensis

In April and May, airy yellow and red blooms nod at the top of 2-foot wiry stems above attractive gray-green foliage. In the wild, columbines often perch on improbable rocky slopes or ledges.

Collecting Seeds: The seeds mature quickly, and can be collected as soon as they turn black, which often happens before the seed capsule dries and splits. Cut open a few capsules to check. When seeds are black, cut stalks and put them in a paper bag. Seeds will ripen in a week and collect in the bag. Shake the bag to release them.

When to Plant: You can either plant collected seeds in the garden in fall or stratify them over winter. This is one wild plant for which it does make some sense to scatter extra seeds in the woods, because seeds are relatively easy to obtain and needn't be covered in order to germinate well.

How to Plant: They need light in order to germinate and therefore should not be covered.

Growing Conditions: Columbines like well-drained soil, not too rich or too heavy, and light shade.

Foamflower
Tiarella cordifolia

Starry wands of small white flowers nod 6 to 12 inches above basal foliage shaped like maple leaves. In protected locations, the foliage may be evergreen or actually bronze-tinged in the cold.

Collecting Seeds: Wisconsin wildflower grower Marie Sperka says that foamflowers rarely set seed there, but farther south you can collect seeds a month after the first flowers form. The first seeds to ripen will be those on the lower part of the stalk. Air-dry cut stalks of foamflowers in open paper bags for several days, then shake out the seeds into the bag. Keep seeds cool till planting.

When to Plant: Plant the seeds in early spring.

How to Plant: The tiny seeds are best sown in a flat, either indoors or in a cold frame. It isn't necessary to stratify them.

Growing Conditions: These shade-loving May-blooming ground covers like humus-rich soil, a steady supply of moisture, and good drainage.

Forget-me-not
Myosotis scorpioides

Matted drifts of these wonderful sky blue flowers can be found in the wild at the edges of small streams, often in full sun. Our city neighbors planted them around the perimeter of a sunken bathtub pool in their yard, where they tumbled engagingly over the edge and hid the mechanical appearance of the contrived pool. The sprawling plants grow 6 to 8 inches high.

Collecting Seeds: Fasten cloth or paper bags around a few fading flower heads to collect the seeds.

When to Plant: Sow seeds in flats in spring.

How to Plant: Forget-me-not seeds need no special treatment. I purchased packets from a mail-order seed company and planted them in flats in the spring like any other perennial flower. Later in spring, I potted up the seedlings and, when they grew to a height of 3 inches or so, tucked them into rocky pockets at the edge of the pond. If you're planting out seedlings along a stream bed, protect the roots with stones or gravel so the water won't wash the plants away. Plant them near the water, not in it.

Growing Conditions: Apart from their need for abundant moisture, forget-me-nots are adaptable plants that thrive on slightly acid or neutral soil and in sun or light shade.

Goldenrod
Solidago spp.

Arching sprays of these rich gold flower heads bloom in roadside fields from August until October. For those who love the fall, goldenrod evokes all our feelings about this bittersweet season of abundance and finality. Too often taken for granted here, goldenrod is valued and cultivated in Europe, where named varieties are sold. Wayside goldenrods (usually *S. altissima* or *S. canadensis*) grow to a height of 3 to 4 feet.

Collecting Seeds: In October, when seeds develop a grayish cast, they are ripe enough to collect. White seeds are immature. Shake the seeds into a bag and either plant them right away or refrigerate them until spring.

When to Plant: Seeds may be planted in the fall or spring.

How to Plant: In any case, sow the seeds thickly, because goldenrod seeds often have low vitality. Harry Phillips suggests mounding the soil around the stem of potted goldenrod seedlings so their leaves won't rot from contact with the soil.

Growing Conditions: Goldenrod prefers full sun and will grow in an ordinary well-drained soil.

Varieties: If you like goldenrod, you might also want to grow seaside goldenrod (*S. sempervirens*) and sweet goldenrod (*S. odora*), which has anise-flavored leaves that make good tea.

Hepatica
Hepatica americana

In March and April, soon after the snow melts, these ground-hugging jewels of lavender, white, or pink blossoms appear in our woods. Flowers on 3- to 4-inch stems nestle among droopy three-lobed leaves left from the previous year.

Collecting Seeds: Chipmunks often pirate the seeds, so to save some for yourself you might need to cage the plants.

When to Plant: Sow seeds as soon as you collect them.

How to Plant: Plant seeds either in a nursery bed or in flats in the cold frame. Treat seedlings gently to avoid root injury when transplanting. It's all right to leave two or three seedlings in a small clump.

Growing Conditions: Light morning sun is fine. Hepaticas grow slowly. *H. americana* thrives in oak wooks. *H. acutiloba,* which has pointed leaves, prefers neutral soil and favors well-drained slopes.

Jack-in-the-Pulpit
Arisaema triphyllum

A universal childhood favorite for its distinctive shape with three-pointed leaflets and the hooded sheath curving protectively over the spathe or "Jack." Jacks appear in shady, moist woods in April and May, growing 1 to 2 feet tall. By September, the spathe has turned to an upright club of bright red berries, and the remainder of the plant has often rotted or at least wilted on the ground.

Collecting Seeds: Break up the berry cluster, soak berries in water to soften them, and rub out the seeds. Put them in a strainer and wash the pulp off.

When to Plant: Stratify the seeds for eight weeks or more, and plant them indoors.

How to Plant: When sown in a flat and kept moist, they will sometimes germinate in a few weeks, but some northern gardeners find that germination is delayed until the second spring after planting. Avoid fertilizing the seedlings during the growing season, or they go into early dormancy.

Growing Conditions: Plant them in a shady spot in rich soil with a good supply of moisture.

Remarks: They combine well with ferns, which are still unfurling when the jack is at its best, and then fill in with their airy fronds when the jack leaves go dormant, leaving their red-berried stalks to punctuate the ferns.

Jacob's Ladder
Polemonium reptans

These compact charmers were one of the first wildflowers I adopted for cultivation, and they looked wonderful alternated with coral bells on a stone wall in the garden of our old Philadelphia house. Their neat 8- to 12-inch-high mounds of small, rounded, opposite leaves and blue, bell-like, May-blooming flowers make them a fine addition to any garden, wild or tame.

Collecting Seeds: Collect the dried seed heads in early summer and shake the seeds out into a bag.

When to Plant: You can sow the seeds in a nursery bed or cold frame soon after you collect them, or keep them cold and plant them indoors in late winter. Outdoors, plant seeds in either regular garden soil or a bed enriched with woods soil. Indoors, use potting soil or a seed-starting medium.

How to Plant: Move seedlings to their garden or woodland spots in either spring or fall. Plant them about 1 foot apart.

Growing Conditions: Mine have done well in light shade or part sun and average garden soil.

Varieties: A white form, *P. mellitum,* which is a somewhat smaller plant, is also available.

Remarks: Jacob's ladder self-sows readily in the rich woods where it is usually found.

Lady's Slipper
Cypripedium spp.

This genus includes *C. calceolus* var. *pubescens,* large yellow; *C. acaule,* pink; and *C. calceolus* var. *parviflorum,* small yellow. The showy, pouch-shaped flowers appear in May on 10- to 20-inch-tall plants.

Collecting Seeds: Collect lady's slipper seeds *only* if you are experienced in seed starting and serious about caring for the seedlings. Even then, collect sparingly, and only from a well-established colony, never from a small or solitary stand of the plant. A single seed capsule contains hundreds if not thousands of seeds, and that should be plenty to give you a start. If you're inexperienced but determined, purchase seeds from a wildflower specialist.

When to Plant: Plant collected seeds soon after you gather them, or refrigerate them for spring planting. Purchased seeds may be planted in spring or fall.

How to Plant: Elda Haring, an expert gardener, suggests planting lady's slipper seeds in a sterile medium under fluorescent lights, but you could also plant them in a cold frame. To prevent transplanting shock, sow seeds in pots. Established plants often spread into handsome colonies.

Growing Conditions: Of the three difficult wildflowers mentioned here, large yellow lady's slipper is the most adaptable. Unlike the more demanding pink lady's slipper, which must have highly acid soil, the yellow thrives in slightly acid or neutral soil, or soil well supplied with moisture and humus. Lady's slipper appreciates dappled to partial shade and hardwood-leaf mulch. Mycorrhizal fungus from woods soil helps the plants to absorb the soil nutrients they need. The variety *parviflorum* likes cooler weather and somewhat more moisture.

Purple Coneflower
Echinacea purpurea

A handsome plant, purple coneflower grows 2 to 4 feet tall. It is often tamed for the flower border and grown as a healing herb. Purple coneflower has 4-inch-wide, pinkish lavender, daisylike flowers with a deep purple central cone.

Collecting Seeds: Seeds ripen four to five weeks after flowering and, unfortunately, tend to have low viability. When they can be readily shaken from the dry cone, they're ripe and ready to harvest. The seed heads attract goldfinches, so you might need to bag a few to be sure of having seeds for yourself.

When to Plant: Sow the seeds in an outdoor nursery bed after collecting, or store them cold and sow them in spring, either indoors or out. Stratifying the seeds for four weeks promotes better germination.

How to Plant: Cover the seeds with ⅛ inch of soil or starting medium.

Growing Conditions: Purple coneflower likes rich, well-drained soil that is not too acid; a pH of 6 to 7 is fine. It thrives in either full sun or light open shade. The plant will be more compact in full sun.

Rue Anemone
Anemonella thalictroides

Look on wooded slopes under high open shade for these white-flowered, 4- to 6-inch-high charmers with wiry stems, bearing alert, trembling leaves. For nearly a month in April and May, our walk to the mailbox is enriched by their bloom, which lasts a long time for plants of such fragile appearance.

Collecting Seeds: Use a stake to mark plants from which you want to collect seeds, because they go dormant after they bloom and set seed. Make a note on your calendar to watch carefully for dry seed heads after the flowers fade.

When to Plant: Sow seeds soon after you collect them.

How to Plant: Plant the seeds in flats in a shaded location. They form small tubers, which can be planted in their second year.

Growing Conditions: Rue anemones prefer a sheltered spot with dappled shade, in well-drained neutral to slightly acid soil that contains plenty of humus.

Varieties: A double pink rue anemone (Schaaf's Double Pink) is sometimes cultivated, but like many other double flowers, it is sterile.

Solomon's Seal
Polygonatum commutatum

Graceful arching 2- to 3-foot stalks make these woodland plants an attractive accent on banks or under trees. The small, pale yellow, pendent bells that dangle from the leaf axils turn to dark blue berries in the fall.

Collecting Seeds: Collect berries when they turn blue. Berry flesh of some wildflowers contains germination inhibitors. Although it's not certain that this is true of solomon's seal, you might want to remove the seeds from the berries before planting.

When to Plant: Plant them as soon as the seeds are ripe, and keep them evenly moist. Even under ideal conditions, germination may be irregular, and the seedlings may develop slowly. Purchased seeds, which have been stored, tend to go dormant, and may take two years to germinate.

Growing Conditions: Plant young solomon's seal in a shady, moist spot with a good supply of humus. If soil moisture is adequate, a morning sun exposure is fine.

Trailing Arbutus
Epigaea repens

The oval, hairy leaves and woody stems of this creeping, matlike evergreen plant are often rather scruffy in appearance, but its tubular white or pale pink flowers, which bloom in April and May, possess a haunting fragrance. The delightful fragrance might be an aid to pollination, but hasn't done much for the survival of arbutus plants, which have been plundered for years to bring that delicate aroma into our living rooms, where it doesn't even make a decent bouquet. In the wild, the small, five-lobed, seed-containing berries are often pilfered by squirrels and insects. In many states, trailing arbutus is now on the list of endangered or protected plants.

How to Plant: Tackle this one only if you enjoy a challenge and if you have the right growing conditions in your area. I've been in touch with two people who have grown this difficult plant from seed. Elda Haring, an expert gardener, purchased seeds from a wildflower nursery and planted them in a sterile medium under fluorescent lights at 80°F (27°C). Ralph Reitz, who is plant propagator at Bowman's Hill Wildflower Preserve in Washington's Crossing, Pennsylvania, showed me several flats of trailing arbutus plants that he had raised from seed. The plants were two years old and still tiny. "The seeds are as fine as powder," Reitz says. "We use a sterile acid medium. These flats were planted in October and exposed to cold winter temperatures. Trailing arbutus is not an easy plant to grow; we use a mist house for ours. They take three years to reach flowering size." Seeds may be purchased from Bowman's Hill Wildflower Preserve Association. The address is found under the list of suppliers at the back of this book.

Growing Conditions: Trailing arbutus needs a location in acid-soiled woods with open shade and some filtered spring sunlight. Colonies of trailing arbutus often appear under pines and hemlocks in somewhat poor soil with good drainage. Started plants may be set out in spring or fall. Water them regularly until new growth has started, and spread a light mulch around them. Wildflower expert Marie Sperka recommends a mixed mulch of evergreen needles and birch and maple leaves.

Remarks: Where trailing arbutus thrives, its roots are usually colonized by soil fungi that help it to absorb nourishment from the soil—a fact which helps to explain the common failure of established plants to survive transplanting.

Trillium
Trillium grandiflorum

The long-lasting, three-petaled white flowers grow on sturdy 12- to 18-inch stems and appear in April and May three to five years after sowing. Their dramatic beauty makes them worth the wait.

Collecting Seeds: Trillium fruit is a ½- to ¾-inch white, oval berry. Check seed color in the berry five to six weeks after flowering. If the seeds have turned from white to reddish brown, pick the berries and let the seeds ripen a bit more inside them.

When to Plant: It's best to sow the seeds right after collecting them, or as soon as they ripen, but you can also stratify them until you're able to plant them. Don't let the white appendages on the seeds dry out.

How to Plant: Sow seeds in a shaded nursery bed of humus-rich soil or in pots filled with a mixture of equal parts of peat moss, sharp sand, and soil dug from a thriving natural colony. Be sure to keep the soil moist.

Growing Conditions: Trilliums grow best in a shaded location and well-drained, humus-rich, neutral or slightly acid soil. They need a good supply of moisture. Transplanting often kills them, and careless digging has sadly decimated many native stands, so every seed-grown trillium plant is a being of great value. Some trilliums are difficult to grow, but this large white form is not. All you need is patience. The plants may need to be watered in dry weather and they often thrive in a mulch of leaf mold. Our Philadelphia neighbor had an impressive border of trilliums growing on the north side of his house, and he'd count and record the number of blossoms (as many as 40) each spring.

Virginia Bluebell
Mertensia virginica

Virginia bluebells soften low places and stream edges in April and May with enchanting 1- to 2-foot-high sprays of China-blue, bell-shaped flowers that bloom above light green, oval leaves.

Collecting Seeds: After bloom, the leaves yellow and disappear, and the stems collapse, so finding the seeds can take some persistence. Mark the site with a stake during the plant's showy phase and look on the ground for seed capsules.

When to Plant: Sow the seeds soon after collecting them.

How to Plant: Plant seeds in a shaded seedbed, or stratify them in a flat in the refrigerator for six weeks, and then set the flat outdoors in a sheltered spot. Keep the seedbed moist, and plant the seedlings in rich damp soil that receives some shade.

Growing Conditions: They like light to moderate shade and rich, moist soil with plenty of humus.

Wild Blue Indigo
Baptisia australis

This wildflower makes a gorgeous addition to a flower border, or, as wildflower expert Bebe Miles has done, you can plant it as a hedge. The blue pealike flowers grow on 3- to 4-foot-high stalks, and even after bloom, which occurs in May and June in the Northeast, the attractive mounds of gray-green leaves add interest to the garden. A well-established blue indigo plant can live for many years.

Collecting Seeds: The seed stalks extend above the leaves and bear 1½-inch-long pods that contain ⅛-inch seeds. Pick the pods when they turn black or start to split.

When to Plant: Sow seeds outdoors in spring.

How to Plant: Before planting, soak the seeds in water overnight. You might want to scarify the seeds, too. Either sow the seeds outdoors in a cold frame, or in pots or a seedbed. Warm soil, 70°F (21°C), encourages germination, which will be irregular over two to seven weeks after spring planting.

Growing Conditions: Avoid overwatering the seedlings, but don't let the soil dry out either. Baptisias do well in ordinary garden soil that is alkaline or neutral, in either sun or light shade.

Wild Geranium
Geranium maculatum

One of the first wildflowers we adapted to our backyard garden, wild geraniums came to us on a returned pie plate, part of a wildflower sampler that could have been cut from the corner of a Botticelli painting; a miniature "wildscape" tufted with a small hummock of moss surrounded by a tiny yellow violet, a downy-leaved hepatica, an unfurling mayapple, silky heart-shaped wild ginger leaves, a still-sleeping clump of Jacob's ladder, and a soon-to-bloom wild geranium. We planted the wild geranium under the high shade of several young trees, where it became a thriving colony. The 12- to 18-inch plants bear pleasant open-faced pink flowers in April and May, above deeply-cut attractive leaves. Wild geraniums self-sow readily.

Collecting Seeds: The seeds can be tricky to collect because they're forcibly dispersed when the seed capsule springs open, and the capsule can ripen quickly when you're not looking. To be sure of getting some seeds, you can collect the capsules when they turn from light to dark green and let them ripen further in a paper bag; then store them in the refrigerator until you're ready to plant them.

When to Plant: You can either sow them as soon as you collect them, or plant them out in spring.

How to Plant: Give them a shaded seedbed and don't give up if seedlings don't appear soon. Sometimes they seem to need a second winter of cold before they're ready to germinate.

Growing Conditions: Set plants in moist, fertile soil in light shade. Morning sun is all right. In deeper shade, you'll get fewer flowers.

Wild Ginger
Asarum canadense

We learned to love this ground-hugging plant from our Philadelphia neighbors, who gave us a start of it, so you see that it can thrive in a shaded city backyard as well as in the woods. The long-lasting, maroon-colored blossoms, which appear in April and May, are hidden under silky-sheened, heart-shaped leaves.

Collecting Seeds: You may have to get down on hands and knees to collect the seeds; the capsule is well hidden, and chipmunks often get there first. Start checking about a month after the first flowers appear. Pluck off the capsules and tap out the seeds.

When to Plant: Wild ginger germinates best when sown immediately after collection, but it may be stratified in the refrigerator for later planting. Just don't let the seeds dry out before planting them.

How to Plant: Use some humus-rich woods soil in your seed-starting mix, if possible.

Growing Conditions: Wild ginger appreciates fertile soil with plenty of humus, and a shaded location. In moist ground, it self-sows readily.

Growing Trees and Shrubs from Seed

Here are specific directions for starting seeds of a good many favorite trees and some shrubs. Descriptions of any trees with which you're not familiar can be found in any tree identification book. Trees take time to grow from seed, but they last a long time, too.

For general information on starting trees and shrubs from seed, and for details on stratification and scarification, see chapter 26. Information that is specific to individual species is given below. Also note that specified temperatures here refer to soil or seed-starting media rather than to air temperatures. Unless otherwise indicated, cover seeds with an amount of soil equal to three times their diameter.

TREES

Ailanthus
Ailanthus altissima

Also called the tree of heaven, this prolific seeder is often found in cities. Male and female flowers are borne on separate trees. When in bloom, the male flowers have an unpleasant odor. Ailanthus trees begin to bear at 15 to 20 years.

When to Plant: The small winged fruits should be planted as soon as they're ripe, if you are able to collect them directly from the tree.

Breaking Dormancy: Stratify purchased seeds for eight weeks at 40°F (4°C) before planting.

Alder
Alnus spp.

The small-winged fruits of the alder are borne in ½-inch cones. The trees produce good crops of cones at least every four years, beginning at about age ten.

Collecting Seeds: Collect the cones when they begin to dry and fan open, dry them on screens for a few days, and then shake out the seeds. You can also cut branches,

334

leaving the cones on, and shake the seeds onto a sheet.
When to Plant: Plant seeds in fall or stratify them and then plant in spring.
Breaking Dormancy: Stratify seeds for three months at 40°F (4°C).
How to Plant: Alder seeds are prone to rotting and have a short lifespan, so plant them thickly in finely screened soil.
Remarks: Alders are good for erosion control and reclaiming spoiled banks.

Apple
Malus spp.

My favorite apple seed story is the true account of the woman who noticed a bee pollinating the flower on one of her apple trees. On a whim, she marked the fruiting spur, then she picked the resulting apple and saved and planted its seeds. One grew into a tree that bore delicious fruit—now named Cox Orange Pippin.

Crabapple trees may bloom as early as 4 years. Full-sized apples will bloom in 10 to 12 years.
Collecting Seeds: Cut the apple open and remove the seeds, which will be brown when fully ripe. Select seeds from fruits with good flavor.
When to Plant: Don't let apple seeds dry out too much before you plant them. Collected seeds should be planted in the fall.
Breaking Dormancy: If you're stratifying purchased seeds or traded seeds, chill them for at least three months, adding 1½ to 2 ounces of water to each ounce of dry moss.
Pests and Diseases: The seedlings are sometimes susceptible to powdery mildew and aphids.
Remarks: Once they reach a height of 9 inches, apple seedlings may be used for grafting.

Ash
Fraxinus spp.

Most ash seeds are winged fruits and all exhibit dormancy. Normally the winged fruits remain on the tree until late winter or early spring, when they are scattered by the wind.
Collecting Seeds: Collect seeds in late fall before they blow away.
When to Plant: The seeds lose viability in storage and should be planted as soon as possible after collecting. A fall planting might germinate the next spring.
Breaking Dormancy: Some species of ash may take two years to germinate. Many ash seeds germinate more completely if given at least a month of warm stratification before chilling them to help break down the seed coats. Flowering ashes, *F. americana* and *F. pennsylvanica,* have less stubbornly dormant seeds and will often germinate well the spring following fall planting. If you can't get them planted promptly, or if you're using seeds that have been stored, stratify the seeds for 2½ to 3 months at 40°F (4°C).
Growing Conditions: Shade the seedbeds lightly for a few weeks after germination.

Ash, Mountain
Sorbus americana

Clusters of small, reddish orange fruits, each containing two to three seeds, adorn the mountain ash from August to October. Trees bear good annual crops, starting at 15 years.

Collecting Seeds: You can collect fruit, either ripe or a bit on the green side, and ferment it for a few days to extract the seeds.

When to Plant: Allow seeds to afterripen and stratify them before planting.

Breaking Dormancy: The seeds seem to need a period of afterripening, and they develop delayed dormancy when allowed to dry. To help the embryo mature, give mountain ash seeds a three- to four-week warm stratification at 68°F to 86°F (20°C to 30°C) in a warm greenhouse or enclosed porch, followed by outside planting, or cold stratification for at least two months.

How to Plant: When planting, cover seeds very lightly with fine soil, or mix them with sand and spread on the surface of the seedbed.

Remarks: The seeds stay viable for several years if kept cool and dry. Seedlings are quite hardy.

Beech
Fagus spp.

Beech nuts, each containing two seeds, are enclosed by prickly burrs that split open when the seeds are ripe. Squirrels and birds love them. Beech trees produce good crops in cycles. On "off" years, the nuts usually don't contain any seeds.

When to Plant: The seeds lose viability in storage and should not be allowed to dry out. Plant seeds in the fall or stratify them and sow them in the spring.

Breaking Dormancy: Stratify seeds for three months at 40°F (4°C).

How to Plant: The trees develop a taproot, so should be transplanted early to their chosen places. Give the seedlings light shade for most of their first summer.

Birch
Betula spp.

Birch seeds are contained in fragile cones, called strobiles, which dry out and disintegrate while hanging on the tree. Black birch (*B. nigra*) trees disperse their seeds in summer, other species in fall.

Collecting Seeds: Pick the cones before they break apart. Rub them to loosen the small oval winged seeds. Separating seeds from scales is an impossible job; just sow the mixture.

When to Plant: Sow in the fall or in spring after one or two months of stratification at 34°F to 40°F (1°C to 4°C).

Breaking Dormancy: Birch is one of the few trees that bear seeds that seem to have some sensitivity to light. Unchilled birch seeds may need exposure to light to germinate. When exposed to light, the seeds may germinate without stratification.

How to Plant: Seeds should be just pressed into the surface of the seedbed. When chilled, as they will be if sown in fall, they may be lightly covered. Give the seedlings light shade during their first summer.

Catalpa
Catalpa bignonioides

This is an easy tree to start from seed. Catalpa seeds are born in narrow, cylindrical capsules 6 to 12 inches long. The thin, papery seeds are ¼ inch wide and 1 to 2 inches long. They don't need to be stratified.

Collecting Seeds: You can collect them in February and March when the capsules begin to split.

When to Plant: Plant them in either the fall or spring. Germination is usually prompt (two weeks in spring), and the germination rate is high.

Cedar, white
Thuja occidentalis

Also called arborvita, this tree produces male and female flowers on the same tree, but on separate twigs. The arborvita produces small cones containing ¼-inch-long winged seeds. Good seed crops occur every three to five years.

Collecting Seeds: Watch as the cones ripen from green to yellow to brown. They open seven to ten days after ripening. Closed cones may be sun-dried until seeds can be shaken out.

When to Plant: Fall planting is probably best, but you can also stratify stored seeds and plant them in spring.

Breaking Dormancy: Stratify the stored seeds for two months at 34°F to 40°F (1°C to 4°C).

How to Plant: Many seeds are empty, so plant about 50 per square foot of ground.

Growing Conditions: Give the seedlings half shade for their first summer.

Chestnut
Castanea spp.

This genus includes *C. mollissima,* the Chinese chestnut, and *C. dentata,* the American chestnut. Chestnut seeds are good to eat, and the squirrels know it. Chestnuts are protected by a wickedly spiny burr, but they usually fall free when the burr hits the ground.

Collecting Seeds: Good nuts should have a smooth surface. If they've deteriorated, the nutshell will be slightly wrinkled and dull. Weevils often infest the nuts. You can float off the damaged ones. Good ones will sink. If you're collecting nuts for a week or more in fall and saving them to plant all at once, keep the first ones in a paper bag in a cool place.

When to Plant: They spoil quickly, so plant gathered chestnuts immediately in the fall.

How to Plant: Bury them 1 inch deep. If chestnuts dry out, they lose viability, but they'll also decay readily if they're too damp. Purchased seeds will probably be dry and should be soaked for 24 hours before planting.

Growing Conditions: Mulch fall-planted beds and remove the mulch in spring.

Dogwood
Cornus florida

The beautiful spring-flowering dogwood produces striking red ¼-inch oval berries (technically, they're called drupes) in the fall. Each contains a small "stone" that shelters one or two seeds. Seeds of the familiar flowering dogwood, *C. florida,* are often empty if the tree is not cross-pollinated by another dogwood. The seeds are slow starters because they have both hard seed coats and dormant embryos. Dogwood volunteers are all over the woods, though, and we recently found at least a dozen seedlings near a parent tree in my aunt's yard, so don't be discouraged.

Collecting Seeds: The fruit flesh contributes to the seed's dormancy, so remove the stones as soon as the berry turns red, and plant them immediately.

When to Plant: Plant immediately in the fall.

Breaking Dormancy: They may not germinate until their second spring in the ground. If you want to hasten things along and don't mind going to some extra trouble, you can give them 60 days of warm stratification followed by three to four months of cold stratification, or treat them with sulfuric acid for one to three hours. You might try nicking or rubbing the stones between sandpaper also. Seeds that have dried should be soaked in water for a day.

How to Plant: Mulch the planted seeds over winter with ½ to 1 inch of sawdust or other fine stuff.

Remarks: Kousa dogwoods (*C. kousa*) have smaller seeds and are even more difficult to start from seed. Dogwoods suffer in a drought, so be sure to keep both seedlings and saplings well watered.

Elm
Ulmus spp.

American elms disperse their short-lived, winged seeds in the fall. Viable seeds have firm, slightly rounded centers.

Collecting Seeds: Collect the flaky winged fruits as soon as they're shed, and air-dry them for a few days.

When to Plant: Sow them right away after they have been air-dried.

Breaking Dormancy: Seeds that have been stored should be stratified for two to three

months, and planted thickly because the germination rate is often lower.
Remarks: Seedlings don't need shade.

Fir
Abies spp.

Starting at age 20 to 30, fir trees produce 3- to 10-inch-long female cones that contain fragile winged seeds with soft, thin seed coats. Good seed crops appear in two- to four-year cycles.

Collecting Seeds: Seeds usually start to blow out of the cones about one month after the cones turn brown. Collect the cones just before they open and dry them on screens. Seeds at the tips of the cones are usually infertile. Don't remove the seed wings. Keep the cones in a damp, cool place.

When to Plant: Plant them in fall if you can, spring if you must.

Breaking Dormancy: The average germination rate of fir seeds is 20 to 50 percent, and drops lower for seeds that are one year old. Unlike most conifers, fir seeds tend to go dormant and benefit from a brief period of stratification, around 2 to 4 weeks. Seeds left in the stratifying medium for longer periods of 15 weeks or so will often germinate at the stratifying temperature of 40°F (4°C).

Disease: Seedlings are susceptible to damping-off.

Ginkgo
Ginkgo biloba

Trees start to bear at 30 to 40 years. To produce seeds, a ginkgo tree bearing female blossoms must be pollinated by one with male blossoms. The resulting yellow fruits, which contain single seeds, have an unpleasant odor when crushed. Ginkgos are easy to raise from seed. Just be sure that the seeds don't dry out.

Collecting Seeds: The fruits usually drop after frost. At that time, the seed embryos are immature. They will continue to develop for six to eight weeks after the fruits fall. Store the fruits in a warm place until their flesh becomes soft enough to wash off.

When to Plant: Sow them in late fall. If you miss the fall planting, stratify the seeds before spring sowing.

Breaking Dormancy: The seeds don't actually go dormant, but chilling seems to increase their germination. Stratify the seeds for one to two months at 34°F to 40°F (1°C to 4°C).

How to Plant: Cover the seeds with 2 inches of soil.

Growing Conditions: Ginkgos thrive in most ordinary soils.

Golden-rain
Koelreuteria paniculata

Also called varnish tree, the fruits of this handsome street tree ripen in September and October. They are brown, triangular husks about 1½ inches long. Each contains three round black seeds. The tree produces good seed crops almost annually and the seeds keep well.

Collecting Seeds: Gather the husks when they fall from the tree in autumn.

When to Plant: If you can obtain fresh seeds, soak them in water for two hours, and then plant them in the fall.

Breaking Dormancy: The seeds have impermeable seed coats, and their embryos go dormant. If you're working with seeds that have been stored and are, therefore, more deeply dormant, try nicking or sanding the seeds, and then give them the boiling water treatment described in chapter 26, or scarify them with sulfuric acid for an hour, and then stratify for three months.

Remarks: Golden-rain trees do well in average soil, but they need full sun to germinate well.

Hawthorn
Crataegus spp.

This slow-growing tree produces white flowers in spring, followed by small red or yellow fruits that contain hard-coated, nutlike seeds.

Collecting Seeds: Pick the seeds out of the fruit pulp and wash them before planting.

When to Plant: Plant seeds right away; they will sometimes germinate the following spring. Stored seeds can be planted in the summer after stratification.

Breaking Dormancy: Before the seeds can imbibe water and start germinating, the seed coats must decompose. In addition, the embryos need chilling in order to break dormancy. If the seeds have been stored, stratify them over winter at room temperature to break the seed coats and then chill them for four to five months in the refrigerator before planting out in summer. If you decide to use an acid treatment on hawthorn seeds, treat only seeds that have been kept at room temperature for several weeks, because the embryos in fresh seeds are damaged by the harsh acid.

How to Plant: Plant seeds thickly, because some will probably be infertile. Some should germinate the first year, but some will take another year to sprout.

Remarks: Hawthorns develop taproots and should be transplanted out of the nursery bed by the time they're a year old.

Hemlock
Tsuga canadensis

Seeds of these short-needled evergreens are borne in small oval cones ½ to ¾ inch long. Seed production begins at age 20 to 30, or a bit later if the tree is in the shade. Most hemlocks bear frequent large crops.

When to Plant: Either plant seeds in fall or stratify them for three months.

Breaking Dormancy: Although some kinds of hemlock seeds don't go dormant, stratification still improves hemlock seed germination. Seeds planted in the South need less stratification than those grown in the North, and recommended temperature is higher for southern seeds (70°F, 21°C) than for northern (55°F, 13°C).

How to Plant: If you can, plant seeds from trees growing in a latitude and elevation similar to yours. Some growers use a sterile medium for hemlock seeds because they are prone to damping-off.

Remarks: The seedlings are small and delicate and should be shaded for their first two summers. They're usually kept in a nursery bed until their second or third season.

Hickory
Carya ovata

C. ovata is the shagbark hickory, *C. laciniosa,* the shellbark hickory. Hickory nuts are enclosed in woody husks that split open when they fall to the ground. Good crops are periodic. Our shagbark hickory tree has had two excellent crops in 14 years, with poor or fair harvests in the remaining years.

Collecting Seeds: Gather the nuts promptly as they drop, because squirrels and mice collect them, too.

When to Plant: As soon as they drop or after stratification.

Breaking Dormancy: If you are planting saved seeds, chill them for one to five months at 33°F to 40°F (1°C to 4°C). Hickory nuts may be stratified in plastic bags without moss or other surrounding medium. Spread mulch over planted nuts.

How to Plant: Plant the nuts ½ inch deep and mulch with 1 or 2 inches of peat moss. Bury several nuts in the spot where you want to have a hickory tree, and then cut off all but the strongest seedlings.

Remarks: Hickories are difficult to transplant successfully.

Hornbeam
Carpinus betulus

Clusters of ribbed nutlets ripen in late summer and fall. Under natural conditions, seeds often germinate late—in their second spring.

Collecting Seeds: Experienced woodsmen recommend collecting the seeds while they're still slightly green.

When to Plant: Fresh seeds should be sown immediately without allowing them to dry. Seeds treated in this way have germinated the spring after planting. If seeds are collected when brown and dry, stratify them before planting.

Breaking Dormancy: Stratify seeds for three to four months at 34°F to 40°F (1°C to 4°C).

How to Plant: Place them in a seedbed of good, friable soil, mulch fall sowings, and provide light shade for the seedlings during their first year.

Larch
Larix laricina

L. laricina is the tamarack, *L. decidua,* the European larch. Larches hybridize readily, and good seed crops are periodic—every three to six years for the tamarack and every three to ten years for the European larch. The seeds should remain viable for at least three years if kept cool and dry.

Collecting Seeds: Collect fresh, dry cones in the fall, picking from the tree if possible because cones may remain on the tree until seeds have blown away.

When to Plant: Either sow the winged, triangular seeds in the fall or stratify them and plant them in spring.

Breaking Dormancy: Most larch seeds germinate well without any pretreatment, but some will enter an easily-broken dormancy. Stratify them for three months at 32°F to 41°F (0°C to 5°C).

Growing Conditions: Shade seedlings for the first two years. Larches like a good supply of moisture.

Linden
Tilia americana

T. americana is the American linden, *T. cordata*, the European linden. The hauntingly fragrant June and July blossoms of the linden are followed in early fall by small seed pellets. These are attached by wiry stems to the leaflike bract that earlier accompanied each cluster of flowers. The pelletlike capsules most commonly contain a single seed, but may contain two or four.

If you're looking for a seed-starting challenge, try the linden. Its seeds are probably the most difficult to start of any familiar tree. In nature, the seeds don't germinate until the second spring, and even then incompletely, with more stragglers following over the next six to seven years. Even methods like acid treatment are not always effective in breaking dormancy. Both the linden seed and its woody covering are very hard, and its embryo is dormant and possibly immature. In order to break its especially deep and complex dormancy, both the seed coat and its pericarp covering must be softened to admit water, and the embryo must undergo a period of afterripening.

Collecting Seeds: Pick the seeds while they're still green, before the seed coats harden. You can either pluck them as soon as the fruit color changes from green to buff or be more daring and pick a few weeks before the color change. For early seed harvests, you should just barely be able to puncture the outer covering of the seed with the thumbnail. If they are too tender, wait a week.

When to Plant: Sow green seeds immediately, or plant stored or dried seeds after stratification.

Breaking Dormancy: Stratify brown, dry seeds at 50°F to 85°F (10°C to 29°C) for four to five months to mature the embryos, then chill for four to five weeks at 34°F to 40°F (1°C to 4°C).

You can also try acid scarification. First use nitric acid to soften the pericarp, then sulfuric acid to etch the seed coats. Soak the acid-treated seeds in water for 24 hours and stratify for four months.

Remarks: The good news about starting lindens from seed is that once seedlings do appear, they are vigorous and grow rapidly.

Locust
Robinia pseudoacacia

Panicles of white, pealike flowers in May turn to brown pods by fall, each containing four to ten black seeds, each about ¼ inch long. Black locusts may bear as young as six years, but years of abundant pod production are periodic.

Collecting Seeds: Collect the pods in fall while they're still whole.

When to Plant: Plant seeds in fall, soon after they drop.

Breaking Dormancy: The impermeable seed coats can cause dormancy, so nick or abrade the seeds and treat them with boiling water before planting, or use acid scarification for about ten minutes to one hour.

How to Plant: Cover them with about ¼ inch of soil, sand, or sand and sawdust. Raised beds make transplanting easier.

Remarks: Locusts volunteer readily in our garden and hedgerows, so you'll probably get plenty of trees if you plant seeds casually in sunny places where you might like to have them growing.

Magnolia
Magnolia spp.

Magnolias may bear ten years after planting, and they usually produce good annual seed crops. The magnolia's showy spring flowers turn to 3- to 5-inch-long cones honeycombed with seed-containing follicles. The fleshy follicles split open to release ½-inch-long red drupes—an exotic-looking arrangement.

Collecting Seeds: As the cones start to split, spread them out to dry a bit so they'll release more seeds, which should *not* be allowed to dry.

When to Plant: Sow the seeds immediately or after stratification.

Breaking Dormancy: Stratify seeds for three to six months at 32°F to 41°F (0°C to 5°C).

How to Plant: Seedlings appreciate half shade during their first summer. Pot them up while they're still small so their extensive root system won't be lost in transplanting.

Maple
Acer spp.

Most trees of this large and diverse genus grow easily from seed if you don't let the seeds dry out. Technically samaras, the winged seeds are often called "maple keys." Maples usually produce abundant annual seed crops.

When to Plant: Plant collected seeds immediately. Stored seeds must be stratified first.

Breaking Dormancy: If allowed to dry, the seeds develop hard coats. A two- to three-month period of warm stratification (68°F to 86°F, 20°C to 30°C) followed by two to three months of cold stratification promotes better germination in maples. These conditions are met by prompt planting of seeds soon after they fall.

How to Plant: You can plant seeds when still "green," before the wings dry completely. In that case, leave the wings on. It's all right to remove wings from dried seeds before planting if you wish. A seedbed enriched with leaf mold will encourage a good root system.

Mulberry
Morus alba

M. alba is the white mulberry, *M. nigra,* the black mulberry. Mulberry trees begin to bear at about five years. Each mulberry is actually a multiple fruit composed of small drupes. The seeds probably have some dormancy, but they're not difficult to start.

Collecting Seeds: Save out some good large berries while you're out there munching under the tree. Wash the pulp from the seeds.

When to Plant: I started seeds of *M. alba* under grow lights and, even without stratifying, got a few seedlings. The germination rate was low, though. Next time, I'd stratify the seeds first, or plant them in the fall after soaking them in cold water for several days.

Breaking Dormancy: Stratify the seeds for one to three months at 34°F to 40°F (1°C to 4°C).

How to Plant: Cover the seeds very lightly with fine soil, and give the nursery bed partial shade for two to three weeks after seedlings germinate.

Oak
Quercus spp.

This large group of trees includes species with a wide range of forms and germinating times. White oaks shed their acorns the same season they're formed; black oaks retain theirs until the following season.

Collecting Seeds: It's best to collect acorns from trees near others of the same species, because self-pollinated oak trees produce seeds of poor quality.

When to Plant: Seeds of the white oak can be planted immediately; they don't go dormant and will often germinate in fall. Black oak acorns need to be chilled to break dormancy. Sow them immediately, or stratify them and plant in spring.

Breaking Dormancy: Stratify black oak acorns for two to three months at 34°F to 40°F (1°C to 4°C).

How to Plant: Plant only sound acorns, without any evident cracking or weevil damage. Don't let the acorn dry. Fall-planted black oak acorns will germinate fairly late in spring. Keep the seedbed lightly shaded.

Remarks: Depending on the local population, you might need to screen your oak seedbed to keep out squirrels.

Pear
Pyrus communis

As fruits go, pear trees can be slow to bear, and some varieties have their "off" years, too. The ripe fruits contain four to ten seeds. A seedling pear we have here on the farm blossomed for the first time in its tenth spring.

Collecting Seeds: Pick seeds out of ripe fruits and rinse them in a strainer.

When to Plant: Sow them in the fall or after a period of stratification.

Breaking Dormancy: Pear seeds can be treated like apple seeds, but they need less chilling. Stratify presoaked seeds at 32°F to 36°F (0°C to 2°C) for eight to ten weeks.

Disease: Powdery mildew can sometimes be a problem. To help prevent it, arrange for good air circulation in the seedling bed.

Pine
Pinus spp.

Pine seeds vary greatly, from the large edible pinyon (*P. edulis*) nuts to the very small flakes of the mugo pine (*P. mugo*). All are borne in cones of varying sizes. Most cones grow fairly high on the tree. They ripen and shed seeds in fall and winter.

Collecting Seeds: If possible, collect the cones just before they open, and dry them to free the seeds. Cones of a few pine species are held together by resin that has a high melting point, and they will germinate best after fire melts the resin and releases the seeds. Some species of pine produce good crops annually, others only periodically. Pine seeds have a long storage life—five to ten years is not unusual, as long as they are kept cool and dry.

When to Plant: Plant them in the fall or after a period of stratification.

Breaking Dormancy: Like those of most conifers, pine seeds will often germinate well without any treatment, but chilling them for six weeks before planting at 34°F to 40°F (1°C to 4°C), or planting them in the fall will result in more complete germination. For seeds grown in the South, a shorter stratification period will be sufficient, but such seedlings will be more frost tender than those raised from northern-grown seeds.

How to Plant: Some growers use a sterile soil mix to start pine seeds, because they are subject to damping-off.

Prunus spp.

This genus includes the stone fruits, such as *P. cerasus,* sour cherry; *P. persica,* peach; *P. armeniaca,* apricot; *P. domestica,* plum; and *P. dulcis,* almond. Seeds of this genus, especially peaches and apricots, volunteer freely around our farm. We often toss the pits around in spots on our land where we think a peach tree would be welcome. Since peach trees are short-lived in our area, often succumbing to borers and disease after bearing for only a few years, this simple strategy ensures that we always have at least some bearing peaches. Apricot blossoms are often killed by a sneak frost in late spring, so it's nice to have plenty of trees in different locations.

Collecting Seeds: For intentional planting, collect sound pits from ripe fruits and clean off the pulp. You can float off empty seeds.

When to Plant: Don't let the seeds dry for more than a few weeks before you plant them. Sow seeds in August or early September, or in spring after a period of stratification.

Breaking Dormancy: Most of these summer-ripening fruit seeds will germinate better if they have a short period, at least two weeks, of warm stratification at 68°F to 86°F (20°C to 30°C) followed by six months or so of cool stratification at 34°F to 40°F (1°C to 4°C). You can approximate these conditions by sowing the seeds in August or early September. If you stratify the seeds in the refrigerator, sow them as early as possible in spring.

Redbud
Cercis canadensis

C. canadensis is the Eastern redbud, *C. occidentalis,* the California redbud. When the redbuds bloom in April, the fringes of our woods blush deep pink. A few of our naturally-seeded trees have lighter, shell-pink blossoms. The flat brown pods are borne in abundance, each containing 7 to 12 seeds.

Collecting Seeds: Collect the pods as soon as they turn dry and brittle.

When to Plant: Freshly collected seeds planted in fall should germinate in spring.

Breaking Dormancy: Stored seeds develop hard seed coats. To soften the seed coats, scarify the seeds by nicking or sanding and then treating them with boiling water, or use acid scarification. Stratifying seeds for 5 to 8 weeks for eastern species or 12 weeks for western species will help to break dormancy.

How to Plant: Pot up seedling trees at the end of their first season, or transplant them to their permanent locations, well marked with stakes. Potted trees may be held for a year or two while roots develop.

Remarks: Older trees don't transplant well.

Spruce
Picea spp.

Mature spruce trees like a good supply of moisture, but tolerate exposure to wind and cold. Their pendent cones ripen in the fall.

Collecting Seeds: If you need a lot of seeds, try to collect cones before they open, but even fallen cones will still contain a few of the small winged seeds for small plantings. Avoid over-drying the seeds. They're somewhat less substantial than those of the firs, and have a lifespan of only about three years.

When to Plant: Sow the seeds in fall as soon as you collect them, or after a period of stratification.

Breaking Dormancy: Dormancy in spruces is mild and probably not universal. You might find that unstratified imbibed seeds will germinate well under lights, but, as with most woody plants, stratifying the seeds encourages more complete germination. Stratify stored seeds for about two months.

Disease: Like the seedlings of other conifers, spruce seedlings are susceptible to damping-off.

Remarks: Seeds of Engelman and blue spruces can germinate at low temperatures. If fall-planted seedlings of these species appear before winter, mulch them and cover lightly with evergreen branches to ward off frost damage.

Sweet Gum
Liquidambar styraciflua

The prickly, round fruiting heads of the sweet gum tree mature in September and October. Most trees produce some seeds every year, with large crops occurring every three years or so.

When to Plant: Shake out the small winged seeds and sow them immediately, so they'll receive a natural three-month chill to overcome the dormancy of their embryos. Stored seeds must be stratified first.

Breaking Dormancy: Stratify stored seeds at 34°F to 40°F (1°C to 4°C) for at least a month before outdoor planting.

How to Plant: Seed viability is low, so sow with a generous hand. Press the seeds lightly into the surface of the soil.

Sycamore
Platanus occidentalis

Plane trees or buttonballs, as sycamores are also called, can produce seeds when they are as young as five years, if grown in the open. Trees growing close together will fruit later. Sycamores usually produce good crops in alternate years. The "buttonball" that falls from the tree in late winter is really a densely-packed mass of thin, slender, pointed seeds. It's usually slightly smaller than a Ping-Pong ball.

Collecting Seeds: Seed-bearing trees are tall, so unless you want to climb, you'll need to wait for the seed balls to fall.

When to Plant: Sow seeds immediately or after stratification if stored.

Breaking Dormancy: Promptly planted seeds usually germinate well without any pretreatment, but seeds that have been stored should be stratified for at least two months before planting.

How to Plant: Crush the dried balls and sow the fibers in a well-prepared seedbed.

Tulip
Liriodendron tulipifera

The straight-trunked tulip tree produces conelike fruits that look like tiny umbrellas when they've lost their seeds. Each long, winged flake that snaps off from the cone's pointed central shaft resembles an exclamation point and contains one or two seeds. Tulip trees start to bear at age 15 to 20 and produce good annual seed crops.

When to Plant: The seeds deteriorate quickly; on average only about 10 percent are fertile, so they should be planted promptly. Stored seeds must be stratified.

Breaking Dormancy: Stored seeds need stratification at 34°F to 40°F (1°C to 4°C) for two to three months to break dormancy. Purchased seeds may not germinate until the second spring.

How to Plant: Because seeds deteriorate rapidly, plant thickly—about 50 to 75 seeds per linear foot. Thin to five to ten seedlings per square foot.

Remarks: Pot up tulip tree seedlings during their first year, before their roots grow deep, because their long taproots make later transplanting difficult.

Walnut
Juglans spp.

J. nigra is the black walnut, *J. regia,* the English, or, more properly, Persian, walnut. Walnuts produce male and female flowers on the same tree, but at different times— probably a safeguard against too much self-pollination. You need at least two trees for good crops. The delicious edible seeds are oily nuts guarded by hard shells encased in husks.

Collecting Seeds: Persian walnuts usually fall free of their husks, but black walnuts hit the ground still surrounded by the fleshy ¼- to ⅜-inch-thick green covers. Black walnut husks can be difficult to remove if allowed to dry on the nut. It is not absolutely necessary to husk the seeds before planting.

When to Plant: The nuts usually germinate naturally in spring after fall planting. Spring planting may be done after a period of stratification.

Breaking Dormancy: Stratify purchased seeds or stored seeds for three to four months at 34°F to 40°F (1°C to 4°C).

How to Plant: Cover the nuts with 2 inches of soil. You might need to screen the seedbed to keep out hungry squirrels. Transplant walnuts to your chosen spot during their first year because their taproots can make later transplanting difficult.

Remarks: Persian walnut seeds are more likely to dry out in storage than those of the black walnut, and the seedlings are very susceptible to frost injury.

SHRUBS

The general principles for starting shrubs from seed are the same as for trees. Shrubs bear their seeds at an earlier age than trees, and fewer of them develop taproots. The most outstanding difference is probably the fact that, because they play a smaller part in our economy, shrubs have been studied much less than trees. Because there is less detailed information available on starting shrubs from seed than for trees, your adventures in this area will be even more experimental, but no less rewarding.

Barberry
Berberis spp.

This large, diverse group of plants bears good annual crops of berries. The familiar Japanese barberry often used for hedges produces red berries; other species produce purple and bluish black berries. The different species of barberry hybridize readily. Each fruit contains several berries.

When to Plant: If you don't get the seeds planted in fall, when they ripen, don't worry. You can plant them as late as March (February in the mid-South), and they will still receive enough chilling to break dormancy.

Breaking Dormancy: Seeds have mild dormancy, easily overcome by brief chilling.

How to Plant: The pulp doesn't seem to contain any germination inhibitors, so you can plant the whole berry without extracting the seeds, although if mold has been a problem in your seedbed, it might be best to remove the seeds from the pulp.

Diseases: Barberry seedlings are susceptible to damping-off, and the plant is also a host to black stem rust of wheat.

Blueberry
Vaccinium spp.

Blueberry fruits vary widely in dormancy and chilling requirements. Some species will germinate without chilling.

Collecting Seeds: Begin by collecting ripe fruits, always a pleasant job when there are plenty to eat, too, and keep them refrigerated for a few days. Then put the berries in the blender with ½ cup of water and blend them briefly just to break up the pulp. Let the blueberry slush stand in a jar until the good seeds settle. Pour off the pulp and the empty seeds. Next, dry the good seeds and refrigerate them. The seeds keep well.

When to Plant: Mix them with the stratifying medium, or try planting them in early spring in a flat of mixed sand and peat.

Breaking Dormancy: In the absence of specific information for each variety— lowbush, highbush, dwarf, box, and early blueberry, and red huckleberry—you might as well plan to stratify blueberry seeds for two to three months.

Growing Conditions: Blueberries need an acid, well-drained soil. The best bushes we've ever grown were in front of a stone wall, where they received morning sun and light open shade in the afternoon. We kept them mulched with peat moss.

Cotoneaster
Cotoneaster spp.

This varied group of plants ranges from ground covers to large shrubs. Different species cross readily. Cotoneasters purchased from nurseries may be hybrids. The red or black fruits, which resemble rose hips, ripen in fall and contain up to five seeds.

Collecting Seeds: The fruits remain on the bush into winter. To extract the seeds, you can use the blender treatment suggested under blueberries.

Breaking Dormancy: Most have a hard seed coat and need warm stratification before chilling to help to soften the seed coat. Collecting green berries might circumvent the hard seed coat problem, but there are no studies to prove it. Either give the seeds four months of warm stratification at 75°F (24°C) followed by about three months of chilling at 34°F to 40°F (1°C to 4°C), or sow acid-scarified seeds in a cold frame or seedbed in midsummer or fall. Don't let stratified seeds dry out before planting.

Elderberry
Sambucus canadensis

Lacy umbels of white flowers turn to flat-topped clusters of purple-black berries. If you've ever made elderberry jam, you know what purple is. The seeds keep well, but they have dormant embryos and hard seed coats.

Breaking Dormancy: A two- to three-month period of warm stratification at 68°F to 86°F (20°C to 30°C) followed by a three-month chill at 34°F to 40°F (1°C to 4°C) should help to break dormancy. You can also cold-stratify the seeds and sow in spring. Either way, germination is often delayed until the second spring.

How to Plant: Plant the whole berry if you have a good supply. If not, you'll find three to five seeds in each berry. To extract the seeds, use the blender treatment described under blueberries. Low places well supplied with moisture make good sites for elderberries.

Gooseberry and Currant
Ribes spp.

Both of these low shrubs and their many cousins produce smooth, round berries that contain many seeds and ripen in early to midsummer. Plants start to bear at 3 to 5 years. The seeds can remain viable for as long as 13 to 17 years.

When to Plant: Either sow seeds in fall or stratify them and plant in the spring.

Breaking Dormancy: Many ripe seeds will germinate in spring after fall planting, but a certain number of seeds also seem to stay dormant for varying periods—an advantage for the plant in adapting to changing growing conditions; some seeds are always there waiting for better weather. Some *Ribes* seeds have hard seed coats, and most need a fairly long period of cold stratification. Stratify the seeds at 34°F to 40°F (1°C to 4°C) for three to four months.

How to Plant: You can either plant whole berries or extract seeds as for blueberries.

Remarks: Seeds of the Sierra gooseberry and possibly some others do not germinate well if planted fresh; they should be dried before sowing.

Hazelnut
Corylus americana

The hazelnut's round, hard-shelled, dark tan nuts are encased in fringed husks from which they readily fall free.

Collecting Seeds: When the husks dry, squirrels often get them first, so we pick the clusters of nuts while the husks are still green.

When to Plant: Plant the nuts as soon as they fall, or keep them cool and plant them in October or November.

Breaking Dormancy: For spring sowing, prechill the nuts for several months.

Holly
Ilex spp.

The familiar round red holly berries are produced only by female plants. Bearing begins at eight to ten years. The berries, which ripen in late fall, contain one to four seeds.

Breaking Dormancy: Hollies don't start easily from seed. In nature, their complex dormancy controls often delay germination until their third or fourth spring. Holly seeds have very tough seed coats, their embryos are immature, and they require chilling to break dormancy—not an easy set of conditions to get around. Acid stratification will soften the seed coat, but it might damage the immature embryo. The best plan is probably to give winter- or spring-gathered seeds a period of warm stratification, 60°F to 90°F (16°C to 32°C), until the following fall. This will help to mature the embryo and soften the seed coats. Then sow the seeds in a seedbed where they will naturally receive the fairly intense winter chill they need in order to sprout.

Lilac
Syringa spp.

Plant seeds only from lilacs that are not hybrids. The seeds keep well. The fruit is a two-celled capsule containing four flat, thin, ½-inch-long seeds.

When to Plant: Sow the seeds in late summer as soon as they are mature, or stratify until spring.

Breaking Dormancy: Lilac seeds have little, if any, tendency to go dormant, and may also be sown in spring without stratifying. Fall-planted seeds usually germinate the following spring.

Privet
Ligustrum spp.

Fruits of the common privet hedge are black drupes ⅓ to ½ inch long. Each contains several seeds.

When to Plant: Plant immediately or after a period of stratification.

Breaking Dormancy: Extracted seeds that have not been allowed to dry should germinate well without stratifying. Stored dry seeds need two to three months of stratification at 40°F (4°C).

Rhododendron and Azalea
Rhododendron spp.

Collecting Seeds: Collect the tiny seeds as soon as the oblong capsules start to turn brown, and before they open.

When to Plant: You can start seeds indoors under fluorescent lights, in a cold frame in spring, or in a cool greenhouse in winter.

Breaking Dormancy: Rhododendron seeds need no pretreatment to promote germination.

How to Plant: Use an acid seed-starting medium, a mixture of vermiculite and peat or sphagnum moss, and incorporate some leaf mold in the nursery bed. Press the seeds into the surface of the flat or bed, and cover them lightly with finely sifted sphagnum moss. Don't bury the seeds; light aids germination. Give the fragile seedlings partial shade for their first year, and be sure to provide acid soil, plenty of moisture, and good drainage.

Rose
Rosa spp.

Collecting Seeds: Collect rose hips when their color is changing from green to orange. At this stage, the seeds pop out readily when you squeeze the fruit. Later, when rose hips are fully ripe, it's more difficult to separate seeds from pulp.

When to Plant: Sow seeds early outside while soil is still warm. Stored seeds can be sown in spring after a period of stratification.

Breaking Dormancy: Rose seeds have tough seed coats. Planting them just before cool temperatures arrive provides the two to three weeks of warm stratification that seem to help to break down those hard seed coats.

If you're planting stored seeds, first stratify the seeds at 81°F to 90°F (27°C to 32°C) for two to three weeks; then chill them and try to sow them in spring when the soil temperature will be rising. If you do use acid to treat rose seeds, make sure that they're dry.

Rubus spp.

This genus includes the various brambles *Rubus,* blackberry; *R. macropetalus,* dewberry; and *R. idaeus,* raspberry. Whether black or red, these berries (actually aggregates of small single-seeded drupes) are delicious and well worth propagating.

When to Plant: Sowing seeds immediately in midsummer when they ripen naturally provides the necessary warm and cool stratification needed to break down the seed coats and overcome dormancy.

Breaking Dormancy: *Rubus* seeds will often have both impermeable seed coats and dormant embryos. You can acid-scarify them for 20 to 60 minutes and then stratify for one to three months at 34°F to 40°F (1°C to 4°C).

Remarks: In the wild, seeds that have fallen to the ground retain viability for several years.

Russian Olive
Elaeagnus angustifolia

E. umbellata is the autumn olive.

Collecting Seeds: Collect the silver-gray fruits in the fall.

When to Plant: Plant seeds right away, or store them and stratify them for later planting.

Breaking Dormancy: Stratify stored seeds for two to three months at 34°F to 50°F (1°C to 10°C). Russian olive seeds sometimes have hard seed coats, which can be softened by 30 minutes of acid scarification.

Serviceberry
Amelianchier spp.

These shrubs produce small applelike fruits in midsummer.

When to Plant: Sow seeds in the fall or in the spring after stratification.

Breaking Dormancy: In nature, germination usually occurs in early spring of the second year after fruits have fallen, but if they're not allowed to dry, a few fall-planted seeds sometimes germinate in spring. Purchased seeds, which will have developed hard seed coats in addition to the embryo dormancy exhibited even by early-collected seeds, will benefit from warm stratification at 68°F to 86°F (20°C to 30°C) followed by chilling for two to six months at 34°F to 40°F (1°C to 4°C).

Viburnum spp.

This genus of ornamental shrubs includes *V. dentatum,* arrowwood; *V. lentago,* blackhaw; and *V. opulus,* cranberry bush. Except for the reddish orange cranberry bush, these decorative shrubs bear bluish black, soft-pulped fruits, which often remain on the plant all winter. The seeds have both impermeable seed coats and dormant embryos.

Collecting Seeds: Collect fruits while they are still red, and remove the seeds before the fruit turns black.

When to Plant: Sow seeds immediately. Seeds that have been stored do not germinate as readily. Plant them in spring in nursery beds, and look for seedlings the following spring.

Breaking Dormancy: Acid pretreatment is too harsh for the viburnum seed's hard but thin seed coat. The spring-planted seeds will undergo a natural warm stratification over summer, followed by a cold stratification in winter that should help them break dormancy in their second spring.

Remarks: Most viburnum species tolerate some shade.

Witch Hazel
Hamamelis virginiana

The fall-flowering witch hazel ripens its fruits the following fall. Look for reddish brown capsules, which contain two shiny black seeds.

Collecting Seeds: If you can collect the fruits in late summer before they're completely dry, the seeds will have a more permeable coat.

When to Plant: Plant seeds right away. In nature, seeds usually germinate the second spring after they fall from the bush, but germination of dried commercial seeds is slower and often spread over several years.

Breaking Dormancy: For stored seeds, stratify for three months at 60°F (16°C) or higher and then chill at 34°F to 40°F (1°C to 4°C) for two months before planting.

How to Plant: Keep the seedbed moist so the seeds won't dry out and develop a hard surface. Lots of leaf mold in the nursery bed will help to hold moisture.

Yew
Taxus spp.

The fleshy red pulp of the bell-shaped fruit partly surrounds a hard, single seed.

Collecting Seeds: Seeds ripen in late summer or early fall.

When to Plant: Sow seeds as soon as possible after collection.

Breaking Dormancy: The seeds are difficult to start. Neither acid pretreatment nor early collection helps here. In nature, the seeds germinate in their second year, often after being eaten and eliminated by birds. Don't let the collected seeds dry out. Pretreat them with boiling water before planting. You can also stratify seeds at room temperature for five months, followed by a three-month 40°F (4°C) chill.

How to Plant: Plant the seeds as soon as possible in soil that is well supplied with leaf mold to help decompose the seed coats.

List of Seed Suppliers

Catalogs are free unless otherwise noted. Sources marked with an asterisk also offer flower seeds.

General Seed Sources

Abundant Life Seed Foundation*
P.O. Box 772
Port Townsend, WA 98368
Sample catalog $1. ($5-15 membership brings you the catalog, book list, and periodic newsletters.)

Becker's Seed Potatoes
R.R. #1
Trout Creek, Ontario
Canada P0H 2L0

Burgess Seed and Plant Company*
905 Four Seasons Rd.
Bloomington, IL 61701

W. Atlee Burpee*
300 Park Ave.
Warminster, PA 18974
and
W. Atlee Burpee Company*
Research & Development
335 S. Briggs Rd.
Santa Paula, CA 93060

D. V. Burrell Seed Growers Company*
P.O. Box 150
Rocky Ford, CO 81067

Comstock, Ferre, and Company*
263 Main St.
Wethersfield, CT 06109

The Cook's Garden
Box 65
Londonderry, VT 05148
Catalog $1

William Dam Seeds*
P.O. Box 8400
Dundas, Ontario
Canada L9H 6M1
Catalog $1 (to U.S. residents, free to Canadians).

De Giorgi Company, Inc.*
P.O. Box 413
Council Bluffs, IA 51502
Catalog $1

Farmer Seed and Nursery Company*
P.O. Box 129
Faribault, MN 55021

Henry Field Seed and Nursery Company*
2176 Oak St.
Shenandoah, IA 51602

Glecklers Seedman
I-4800 Sh 120, Box 189
Metamora, OH 43540
(Send two first-class stamps for catalog.)

Good Seed Company
P.O. Box 702
Tonasket, WA 98855
Catalog $1

Gurney Seed and Nursery Company*
Second and Capital
Yankton, SD 57078

Harris Seeds*
3670 Buffalo Rd.
Rochester, NY 14624

The Charles C. Hart Seed Company*
304 Main St.
Wethersfield, CT 06109

H. G. Hastings Company*
P.O. Box 4274
Atlanta, GA 30302

J. L. Hudson, Seedsman*
P.O. Box 1058
Redwood City, CA 94064
Catalog $1

Johnny's Selected Seeds*
299 Foss Hill Rd.
Albion, ME 04910

J. W. Jung Seed Company*
335 S. High St.
Randolph, WI 53957

Kitazawa Seed Company
1748 Laine Ave.
Santa Clara, CA 95051
(Oriental seeds)

Liberty Seed Company*
Box 806
New Philadelphia, OH 44663

Earl May Seed and Nursery Company*
208 N. Elm St.
Shenandoah, IA 51603

Mellinger's Inc.*
2310 W. South Range Rd.
North Lima, OH 44452

Native Seeds/Search
3950 W. New York Dr.
Tucson, AZ 85745
Catalog $1

Nichols Garden Nursery*
1190 N. Pacific Hwy.
Albany, OR 97321

Park Seed Company*
Cokesbury Rd.
Greenwood, SC 29647

Pinetree Garden Seeds*
Rt. 100
Gloucester, ME 04260

Redwood City Seed Company*
P.O. Box 361
Redwood City, CA 94064
Catalog $1

Seeds Blum*
Idaho City Stage
Boise, ID 83706
Catalog $2

Seedway, Inc.*
Hall, NJ 14463

Siberia Seeds
Box 2528
Olds, Alberta
Canada T0M 1P0
(Send SASE for price list.)

Southern Exposure Seed Exchange*
P.O. Box 158
North Garden, VA 22959
Catalog $3

Stokes Seeds, Inc.*
P.O. Box 548
Buffalo, NY 14240

Sunrise Enterprises*
P.O. Box 10058
Elmwood, CT 06110
(Oriental seeds)
Catalog $1

Thompson and Morgan*
P.O. Box 1308
Jackson, NJ 08527

Tsang and Ma International
P.O. Box 294
Belmont, CA 94002
(Oriental vegetables)

Otis S. Twilley Seed Company Inc.*
P.O. Box 65
Trevose, PA 19047

Vermont Bean Seed Company*
Garden Ln.
Fair Haven, VT 05743

Vesey's Seeds Ltd.*
York, Prince Edward Island
Canada C0A 1P0

Garden Flower Seed Sources

The Country Garden
Rt. 2, Box 455A
Crivitz, WI 54114
Catalog $2

Far North Gardens
16785 Harrison, Dept. RD
Livonia, MI 48154
Catalog $2 for 3 years

J. L. Hudson, Seedsman
P.O. Box 1058
Redwood City, CA 94064
Catalog $1

Herb Seed Sources

Catnip Acres Herb Farm
Christian St.
Oxford, CT 06483
Catalog $2

Fox Hill Farm
444 W. Michigan Ave.
Parma, MI 49269
Catalog $1

Mellinger's Inc.
2310 W. South Range Rd.
North Lima, OH 44452

Sandy Mush Herb Nursery
Rt. 2, Surrett Cove Rd.
Leicester, NC 28748
Catalog $4

Richters
Goodwood, Ontario
Canada L0C 1A0
Catalog $2.50

The Rosemary House
120 S. Market St.
Mechanicsburg, PA 17055
Catalog $2

Well-Sweep Herb Farm
317 Mt. Bethel Rd.
Port Murray, NJ 07865
Catalog $1

Tree and Shrub Seed Sources

Abundant Life Seed Foundation
P.O. Box 772
Port Townsend, WA 98368
(Membership, $5 to $15 donation.)

Girard Nurseries
P.O. Box 428
Geneva, OH 44041

Maver Seeds
Rt. 2, Box 265B
Asheville, NC 28805
(Also for perennials, grasses, and woody
plants. Complete list $5; wildflower
list only, $1.)

Bruce J. Miller International
P.O. Box 66
Germantown, WI 53022
List $1

Plants of the Southwest
Route 6, Box 11A
Santa Fe, NM 87501
Catalog $2

Wildflower Seed Sources

Applewood Seed Company
5380 Vivian St.
Arvada, CO 80002

Bowman's Hill Wildflower Preserve Association, Inc.
Washington Crossing Historic Park
P.O. Box 103
Washington Crossing, PA 18977
Catalog $1

Green Horizons (provides seeds of Texas
wild flowers)
218 Quinlan 571
Kerrville, TX 78028
(Send SASE.)

Maver Seeds
Rt. 2, Box 265B
Asheville, NC 28805
(Also for perennials, grasses, woody
plants. Complete list, $5; wildflower
list only, $1.)

Midwest Wildflowers
Box 64
Rockton, IL 61072
Catalog $.50

Native Gardens
Rt. 1, Box 494
Greenback, TN 37742
Catalog $1

Natural Habitat Nursery
4818 Terminal Rd.
McFarland, WI 53558

Plants of the Southwest
Route 6, Box 11A
Santa Fe, NM 87501
Catalog $2

Clyde Robin Seed Company
25670 Nickel Place
Hayward, CA 94545-3222

Yerba Buena Nursery
19500 Skyline Blvd.
Woodside, CA 94062

Broom Corn Seed Source

Seeds Blum
Idaho City Stage
Boise, ID 83706

Sources for Garden Supplies

Catalogs are free unless otherwise noted.

Bean-Threshing Bags

Good Seed Company
P.O. Box 702
Tonasket, WA 98855
Catalog $1

Biological Insect Controls

Earlee, Inc.
726 Spring St.
Jeffersonville, IN 47130

Natural Gardening Research Center
Hwy. 48, P.O. Box 149
Sunman, IN 47041
Catalog $1

Peaceful Valley Farm Supply
11173 Peaceful Valley Rd.
Nevada City, CA 95959

Rincon-Vitova Insectaries Inc.
P.O. Box 95
Oakview, CA 93022

Botanical Insect Controls

Earlee, Inc.
726 Spring St.
Jeffersonville, IN 47130

Natural Gardening Research Center
Hwy. 48, P.O. Box 149
Sunman, IN 47041
Catalog $1

Necessary Trading Co.
663 Main St.
New Castle, VA 24127
Catalog $2

Peaceful Valley Farm Supply
11173 Peaceful Valley Rd.
Nevada City, CA 95959

Seed-Saving Supplies (including small zip-lock bags and color-indicating silica gel)
Crystal Springs Packaging Company
P.O. Box 2924
Petaluma, CA 94952
(Send SASE for price list.)

Cover Crops

Bountiful Gardens
5798 Ridgewood Rd.
Willits, CA 95490
(Also grains)

Johnny's Selected Seeds
299 Foss Hill Rd.
Albion, ME 04910
(Also grains)

Mellinger's Inc.
2310 W. South Range Rd.
North Lima, OH 44452

Natural Gardening Research Center
Hwy. 48, P.O. Box 149
Sunman, IN 47041
Catalog $1

Necessary Trading Company
663 Main St.
New Castle, VA 24127
Catalog $2

Cytokinin Standardized Seaweed (SM-3 brand)

Atlantic and Pacific Research, Inc.
P.O. Box 14366
North Palm Beach, FL 33408

Garden Irrigation Equipment

Gardeners Supply Company
128 Intervale Rd.
Burlington, VT 05401

A. M. Leonard Inc.
6665 Spiker Rd.
Piqua, OH 45356

Mellinger's Inc.
2310 W. South Range Rd.
North Lima, OH 44452

Smith and Hawken
25 Corte Madera
Mill Valley, CA 94941

Greenhouse Supplies

Charley's Greenhouse Supplies
1569 Memorial Hwy.
Mt. Vernon, WA 98273
Catalog $2

Great Lakes IPM
10220 Church Rd., NE
Vestaburg, MI 48891

Gro-Tek Greenhouse Supplies
RFD 1, Box 518A
S. Berwick, ME 03908

Peanut Seed Inoculant

Park Seed Company
Cokesbury Rd.
Greenwood, SC 29647

Seaweed Extract

Necessary Trading Company
663 Main St.
New Castle, VA 24127

North American Kelp
Cross St.
Waldoboro, ME 04572

Also available from many mail-order seed and garden supply catalogs and from local garden supply stores.

Seed-Starting Supplies

Gardeners Supply Company
128 Intervale Rd.
Burlington, VT 05401

Johnny's Selected Seeds
299 Foss Hill Rd.
Albion, ME 04910

Mellinger's Inc.
2310 W. South Range Rd.
North Lima, OH 44452

Necessary Trading Company
663 Main St.
New Castle, VA 24127
Catalog $2

Soybean Seed Inoculant

(Pea/bean seed inoculant is widely available.)

Mellinger's Inc.
2310 W. South Range Rd.
North Lima, OH 44452

Research Seeds Inc.
P.O. Box 1393
St. Joseph, MO 64501

Glossary

Afterripening—Changes that take place in the seed after harvest, making it possible for the seed to sprout when conditions are right.

Allelopathy—The inhibition of one kind of plant by substances produced by a different plant growing nearby.

Annual—A plant that lives for only one year or growing season.

Anther—The pollen-containing tip of the stamen in a flower.

Auxin—A growth hormone produced by plants.

Biennial—A plant that blooms, bears seed, and usually dies the year after it is planted.

Blocking out—The practice of cutting around plants in a flat a week before transplanting them into the garden row.

Bolt—To send up a seed stalk, when vegetative growth is preferred.

Carpel—The individual female part of the flower, corresponding to the male stamens.

Cell—The smallest structural unit of an organism.

Chelate—To chemically grab and hold molecules of metal from the soil.

Cloche—A protective cover of glass or plastic, often a bottomless glass jug, used to protect growing plants from cold weather.

Cold frame—A plastic- or glass-covered frame used to protect plants from cold weather.

Companion planting—The practice of making purposeful adjacent plantings of plants that seem to enhance the growth of the other plants or confer some disease or insect protection.

Complete flower—One that contains both stamens and pistil.

Control—In experiments, the control is the untreated, standard plant (seed, soil sample, etc.) used to check performance of treated samples.

Cotyledon—The seed leaves or first leaves that emerge from a germinated seed, different in form from the later true leaves.

Crop rotation—The practice of planting a succession of different plants on a certain piece of ground to promote soil nutrient balance and prevent disease and insect buildup.

Cross-pollination—The transfer of pollen from the anthers of one kind of flowering plant to the stigma of a different variety of that plant.

Cutworm—A soil-dwelling beetle larva that encircles and nips off seedlings at the soil surface.

Cytokinin—A growth-promoting plant hormone found in kelp.

Dessicant—A drying agent; a substance that absorbs moisture.

Dormant—Alive but inactive and, in some cases, incapable of growth until certain conditions (light, temperature, time, for example) have been fulfilled.

Electrode—One of the two terminals of an electric source.

Embryo—The rudimentary plant contained in a seed.

Endosperm—The stored plant nourishment surrounding the embryo in a seed.

Enzyme—An organic substance produced by a plant that causes chemical changes in other substances.

Exoskeleton—The external supporting structure of an insect.

Exudate—A substance that is produced and given off by a plant, as in root exudate.

Fertilization—The union of the male cell in the pollen with the ovule, or female cell.

Flea beetle—A tiny, very active black beetle, about $1/16$-inch long, that eats small holes in the leaves of plants.

Fluorescent lamp—A glass tube coated on the inside with phosphorescent powder, which glows when exposed to a stream of electrons from the electrode.

Fruit—Botanically, a ripened, seed-containing ovary.

Gibberellin—A growth hormone produced by plants.

Hardening off—The process of exposing young plants gradually to the stresses of outdoor life.

Heterosis—Hybrid vigor, exceptional vitality sometimes seen in a first-generation cross.

Hormone—A substance made by plant tissue that has the effect of stimulating certain plant functions.

Hotbed—A glass- or plastic-enclosed frame that is heated by buried manure or an electric soil cable and is used for raising early plants.

Hybrid—A plant grown from seed obtained by cross-fertilizing two different plant varieties.

Imbibition—The absorption of water by a seed.

Incandescent lamp—A bulb that produces light (and some heat) when the filament it contains receives an electric charge.

Kelp—A sea plant used in fertilizer.

Loam—A well-balanced soil consisting of approximately 40 percent sand, 40 percent silt, and 20 percent clay.

Metabolism—The chemical and physical processes necessary to maintain a living organism.

Mutation—A change in the gene pattern (and therefore characteristics) of a plant, which can be inherited by succeeding generations.

Open-pollinated—Referring to nonhybrid plants or seeds.

Ovary—The hollow chamber at the base of the pistil, containing one or more ovules.

Ovule—The female cell, or egg.

Pathogen—A disease-producing microorganism.

Perennial—A plant that bears flowers and fruit every year, surviving the winter. Some perennials live for 30 years or more; others die after 5 to 15 years.

Perlite—Volcanic rock that has been "popped" (heat expanded).

Photoperiodism—The influence of the length of the daily period of darkness on the blooming habit of plants. Some plants need short nights in order to bloom; others need long nights; others (including many vegetables) are neutral—not sensitive one way or the other.

Photosynthesis—The formation, by the living plant, of carbohydrates from water and carbon dioxide through the action of sunlight on the chlorophyll in the leaves.

Phototropism—The tendency of plants to grow toward a light source.

Phytochrome—The coloring matter in plants.

Pistil—The female part of a flower, consisting of an ovary containing at least one ovule, topped by a style and stigma. A carpel is a simple pistil. Compound pistils contain multiple carpels.

Pollen—Minute grains formed by the flower, which fertilize the ovule to produce the seeds of a new plant generation. Pollen is the male element in plants.

Pollination—The transfer of pollen from anther to stigma. Precedes fertilization.

Respiration—The energy-releasing process carried on by all living cells, in which oxygen is taken in and combined with carbohydrates to form carbon dioxide and water. The chemical reaction is the opposite of what happens in photosynthesis.

Scarifying—The practice of scratching or notching the seed coat to hasten germination.

Seed—A fertilized, ripened plant ovule. A living embryonic plant.

Self-fertile—Referring to a plant's ability to produce fruit after accepting its own pollen.

Self-incompatible—Referring to the uneven maturation of pollen and ovule, sometimes necessitating cross-pollination if the plant is to bear fruit.

Self-unfertile—Referring to a plant's inability to set fruit from its own pollen.

Shattering—In seed saving, the prompt dispersal of seeds as soon as they are ripe.

Stamen—The male part of the flower, bearing on its tip the pollen-containing anther.

Sterile—Referring to a plant's failure to bear fruit or viable seed.

Sterilize—To kill all living microorganisms (bacteria, fungi, and so forth), as by heat.

Stigma—The pollen-receptive tip of a flower pistil.

Stratification—Chilling seeds to promote germination. It is desirable for the seed to have absorbed some water before chilling.

Style—The slender part of the pistil, rising from the ovary and terminating in the stigma.

Succession planting—The practice of sowing a second crop to closely follow the harvest of the first crop.

Synthesize—To combine separate elements into a new form.

Trace elements—Elements that are necessary for growth but in very small amounts. Boron, manganese, copper, and zinc are trace elements.

Transpiration—The evaporation of internal water from plant leaves.

Ultraviolet—Light rays with short wavelength found just beyond the violet band in the visible spectrum.

Vermiculite—A form of heat-expanded mica used in soil mixes.

Vernalization—The practice of chilling young plants to induce early flowering.

Viable—Capable of germinating.

Watt—A unit of electric power, measuring a current of one ampere under one volt of pressure.

Zygote—The single cell formed by the union of the male and female plant cells.

Bibliography

ARTICLES

Allen, Judy. "Undercover Report." *National Gardening,* September 1986, pp. 18–19.

Armstrong, Colleen. "Giving Your Greenhouse Fresh Air." *Farmstead,* n.d., pp. 26–28.

———. "How to Grow European Cucumbers." *Farmstead,* n.d., pp. 39–40.

Ascher, Amalie A. "It's Safe to Mulch Your Garden in Color." *Green Scene,* May 1984, pp. 27–29.

Atwater, Betty Ransom. "Germination, Dormancy and Morphology of the Seeds of Herbaceous Ornamental Plants." *Seed Science and Technology,* August 1980, pp. 523–73.

Bartok, John, Jr. "High Intensity-Discharge Lighting." *Horticulture,* November 1986, pp. 60–62.

Bassett, James. "Saving Your Own Vegetable Seed: A Pollination Primer." *Horticulture,* August 1978, pp. 18–25.

Berenbaum, May. "That Devilish Herb." *Horticulture,* July 1980, pp. 23–24.

Beste, C. E. "Co-Cropping Sweet Corn and Soybeans." *HortScience,* June 1976, pp. 236–38.

Biran, I., and A. M. Kofranek. "Evaluation of Fluorescent Lamps as an Energy Source for Plant Growth." *Journal of American Society for Horticultural Science,* November 1976, pp. 625–28.

Boland, Maureen, and Bridget Boland. "Old Wives' Planting Lore." *Country Journal,* April 1977, p. 68.

Booth, E. Frost. "Resistance and Insect Pests: Seaweed Has a Two-Way Benefit." Reprinted from *The Grower,* November 27 and December 4, 1965.

Brody, Jane E. "Upstate Scientist Is Trying to Breed a Tomato that Can Stand the Cold." *New York Times,* 14 May 1977.

Bubel, N. "Making Your Own Soil Mix." *Country Journal,* March 1983, pp. 12–14.

———. "When Should You Water Your Garden?" *Country Journal,* July 1985.

———. "Garden Watering Devices." *Country Journal,* August 1985, pp. 14–15.

———. "The Elegant Eggplant." *Horticulture,* February 1986, pp. 26–28.

———. "Tennessee Red and Georgia Jet." *Horticulture,* May 1986, pp. 54–60.

"Cauliflower." *National Gardening,* September 1986, pp. 13–16.

Cox, Jeff. "Azotobacter: The Soil Bacteria that Can Increase Your Yields." *Organic Gardening and Farming,* April 1976, pp. 144–50.

"Crop Covering Materials of Polyvinyl Alcohol." *HortIdeas,* n.d.

Dawkins, T. C. K.; P. D. Hebblethwaite; and M. McGowan. "Soil Compaction and the Growth of Vining Peas." *Annals of Applied Biology,* October 1984, pp. 329–43.

DeBaggio, Thomas. *Gerard's Garden,* January/February 1984.
———. *Gerard's Garden,* March/April 1984.
———. *Gerard's Garden,* September/October 1984.
———. *Gerard's Garden,* March/April 1985.
———. *Gerard's Garden,* September/October 1985.
———. "Growing Artichokes as Annual Vegetables." *Gerard's Garden,* November/December 1985.

———, ed. *Gerard's Garden,* November/December 1983.

Dorschner, Cheryl. "Okra." *National Gardening,* February 1986, pp. 38–41.

Estabrook, Barry, and Billie Milholand. "Lunar Yeomancy." *Hdrrowsmith,* n.d., pp. 9–12.

Gauss, James. "Green Peppers." *Horticulture,* April 1981, pp. 23–26.

Goc, Michael. "Spunbonded Row Covers." *Rodale's Organic Gardening,* December 1985, pp. 82–92.

Hayward, Gordon. "Asparagus: A Vegetable Worth Waiting For." *Horticulture,* May 1981, pp. 27–31.

Horticultural Abstracts 55. Abstract 3374. United Kingdom: Farnham Royal, Slough, SL2, May 1985.

Kane, Mark. "Make the Most of the Pea Season." *Organic Gardening,* April 1984, pp. 32–39.

———. "Leeks: The Neglected Onion." *Organic Gardening,* August 1984, pp. 49–52.

———. "Beets: The Sooner the Better." *Organic Gardening,* November 1985, pp. 55–60.

Lafavore, Michael. "Digging for Gold." *Organic Gardening,* September 1984, pp. 24–29.

Lima, Patrick. "Stalwart Herbs." *Harrowsmith,* n.d., pp. 42–51.

Meeker, John. "Fennel: A Good Late-Season Vegetable." *Organic Gardening,* July 1979, pp. 63–65.

"Microwave Sterilization of Soil." *HortIdeas,* March 1984, p. 21.

Niklas, Karl J. "Aerodynamics of Wind Pollination." *Scientific American,* July 1987, pp. 90–95.

Rhodes, Landon H. "Mycorrhiza Is Fungus Plus Root." *Plants Alive,* September 1977, pp. 28–29.

Robinson, Frank. "Under the Black Walnut Tree." *Horticulture,* October 1986, pp. 30–33.

Roughgarden, Rocky. "The Modular Cold Frame." *Farmstead,* Spring 1977, p. 32.

Shimizu, Holly. "Herbs." *Gardens for All News,* August 1985, pp. 1–2.

Shultz, Warren, Jr. "Better Winter Lettuces." *Organic Gardening,* November 1983, pp. 34–36.

Swain, Roger. "Cloches." *Horticulture,* April 1986, pp. 20–24.

Tenga, A. G., and D. P. Ormrod. "Responses of Okra Cultivars to Photo-period and Temperature." *Scientia Horticulturae 27,* December 1985, pp. 177–87.

Tresemer, David. "Watch Out for Wetting Agents." *Organic Gardening,* July 1984, p. 55.

Upchurch, Woody. "Warm Soil Needed for Okra Planting." *North Carolina State University Bulletin*. Raleigh, N.C.: North Carolina State University, n.d.

White, John W. "Crops for a Passive Solar Greenhouse." *Horticulture*, February 1980, pp. 32–36.

Wilkes, H. G. "The World's Crop Plant Germplasm—An Endangered Resource." *Bulletin of the Atomic Scientists*, February 1977, pp. 8–16.

Wolf, Ray. "The Best Staking Methods for Tomatoes." *Organic Gardening*, April 1979, pp. 90–100.

BOOKS

Allard, R. W. *Principles of Plant Breeding*. New York: John Wiley and Sons, 1960.

Amaranth: Modern Prospects for a Modern Crop. A report by a panel of the Advisory Committee on Technological Innovation; the Board on Science and Technology for International Development; the Office of International Affairs; and the National Research Council. Washington, D.C.: National Academy Press, 1984.

Barton, Lela. *Seed Preservation and Longevity*. New York: Interscience Publishers, 1961.

Bebe, Miles. *Bluebells and Bittersweet*. New York: Van Nostrand Reinhold, 1970.

———. *Wildflower Perennials for Your Garden*. New York: Hawthorne Books, 1976.

Bubel, Nancy. *The Country Journal Book of Vegetable Gardening*. Brattleboro, Vt.: Country Journal Publishing Co., 1983.

Carr, Anna. *Good Neighbors: Companion Planting for Gardens*. Emmaus, Pa.: Rodale Press, 1985.

Cherry, Elaine C. *Fluorescent Light Gardening*. Princeton, N.J.: D. Van Nostrand, 1965.

Crocker, William, and Lela Barton. *The Physiology of Seeds*. Waltham, Mass.: Chronica Botanica, 1957.

Cruso, Thalassa. *Making Things Grow*. New York: Alfred Knopf, 1971.

Editors of *Organic Gardening*. *The Encyclopedia of Organic Gardening*. Emmaus, Pa.: Rodale Press, 1978.

Editors of Rodale Press. *How to Grow Vegetables and Fruits by the Organic Method*. Emmaus, Pa.: Rodale Press, 1961.

Edmond, J. B.; T. L. Senn; and F. S. Andrews. *Fundamentals of Horticulture*. New York: McGraw-Hill, 1964.

Fell, Derek. *Annuals: How to Select, Grow and Enjoy*. Tucson, Ariz.: HP Books, 1983.

Gardening with Wild Flowers. Brooklyn Botanical Gardens Handbook no. 1, 1976.

Gessert, Kate. *The Beautiful Food Garden*. New York: Van Nostrand Reinhold, 1983.

Harper, Pamela, and Frederick McGourty. *Perennials: How to Select, Grow and Enjoy*. Tucson, Ariz.: HP Books, 1985.

Hartmann, Hudson T., and Dale E. Kester. *Plant Propagation*. Englewood Cliffs, N.J.: Prentice-Hall, 1975.

Hemphill, John, and Rosemary Hemphill. *Herbs: Their Cultivation and Usage*. New York: Sterling Publishing Co., 1983.

Hills, Lawrence D. *Comfrey Report*. Essex, England: Henry Doubleday Research Assoc., 1974.

————. *Save Your Own Seed.* Essex, England: Henry Doubleday Research Assoc., n.d.

Johnston, Vernon, and Winifred Carriere. *An Easy Guide to Artificial Light Gardening for Pleasure and Profit.* New York: Gramercy Publishing Co., 1964.

Jones, Henry A., and Louis Mann. *Onions and Their Allies.* New York: Interscience Publishers, 1963.

Kingman, A. R. *Plant Growth Responses to Extracts of Ascophyllum Nodosum.* Clemson, S.C.: The South Carolina Experiment Station, Clemson University, 1975.

Knott, James Edward. *Handbook for Vegetable Growers.* New York: John Wiley and Sons, 1962.

Kozlowski, T. T., ed. *Seed Biology,* vols. 1–3. New York: Academic Press, 1972.

Kraft, Ken, and Pat Kraft. *Garden to Order.* Garden City, N.Y.: Doubleday, 1962.

Krantz, Frederick H., and L. Jacqueline. *Gardening Indoors under Lights.* New York: Viking Press, 1971.

Lathrop, Norma Jean. *Herbs: How to Select, Grow and Enjoy.* Tucson, Ariz.: HP Books, 1981.

Lorenz, Oscar A., and Donald N. Maynard. *Knott's Handbook for Vegetable Growers.* New York: John Wiley and Sons, 1980.

McDonald, Elvin. *The Complete Book of Gardening under Lights.* Garden City, N.Y.: Doubleday, 1965.

Mayer, A. M., and A. Poljakoff-Mayber. *The Germination of Seeds.* New York: Pergamon Press, 1963.

Myers, Amy. *A Manual of Seed Testing.* Sydney, Australia: New South Wales Department of Agriculture, 1952.

National Council of State Garden Clubs. *Directory to Resources on Wildflower Preparation.* Missouri: John S. Swift Co., 1981.

Nikolaeva, M. G. *Physiology of Deep Dormancy in Seeds.* Translated by Israel Program for Scientific Translations. Jerusalem: 1969. Washington, D.C.: National Science Foundation.

Patent, Dorothy Hinshaw, and Diane Bilderback. *Garden Secrets.* Emmaus, Pa.: Rodale Press, 1982.

Phillips, Harry R. *Growing and Propagating Wildflowers.* Chapel Hill, N.C.: University of North Carolina Press, 1985.

Raymond, Dick. *Down-to-Earth Vegetable Gardening.* Pownal, Vt.: Storey Communications, 1982.

Reilly, Ann. *Park's Success with Seeds.* Greenwood, S.C.: Geo. Park Seed Co., 1978.

Rice, Elroy L. *Pest Control with Natural Chemicals.* Norman, Okla.: University of Oklahoma Press, 1983.

Rickett, Harold William. *Botany for Gardeners.* New York: Macmillan Co., 1971.

Roberts, E. H. *Viability of Seeds.* Syracuse, N.Y.: Syracuse University Press, 1972.

Schopmeyer, C. S., ed. *Seeds of Woody Plants in the United States.* United States Department of Agriculture. Washington, D.C.: Agricultural Handbook no. 459 of the U.S. Forest Service, 1974.

Schultz, Peggy. *Growing Plants under Artificial Light.* New York: M. Barrows and Co., 1955.

Seed Savers Exchange Publications: 1980 Spring/Summer Ed.; 1981 Fall Harvest Ed.; 1981 Spring/Summer Ed.; 1982 Fall Harvest Ed.; 1982 Winter Yearbook; 1983 Winter Yearbook; 1984 Winter Yearbook; 1985 Fall Harvest Ed.

Senn, T. L., and A. R. Kingman. *A Report of Seaweed Research.* Clemson, S.C.: The South Carolina Experiment Station, Clemson University, 1975.

Simonds, Calvin. *The Weather-Wise Gardener.* Emmaus, Pa.: Rodale Press, 1983.

Smith, Shane. *The Bountiful Solar Greenhouse.* Sante Fe, N.Mex.: John Muir Publications, 1982.

Sperka, Marie. *Growing Wildflowers.* New York: Harper & Row, 1973.

Stefferud, Alfred. *The Wonders of Seeds.* New York: Harcourt, Brace and Co., 1965.

Stephenson, W. A. *Seaweed in Agriculture and Horticulture.* London: Faber and Faber, 1968.

Tannahill, Reay. *Food in History.* New York: Stein and Day, 1974.

Tiedjens, Victor. *The Vegetable Encyclopedia.* New York: Avenel Books, 1943.

United States Department of Agriculture. Agricultural Publication no. 450 of the U.S. Forest Service. Washington, D.C., 1974.

United States Department of Agriculture. *Seeds, the Yearbook of Agriculture.* Washington, D.C.: U.S. Government Printing Office, 1961.

Weatherwax, Paul. *Indian Corn in Old America.* New York: Macmillan Co., 1954.

Whitson, John; Robert John; and Henry Williams, M.D., LID. *Luther Burbank, His Methods and Discoveries and Their Practical Application.* New York: Luther Burbank Press, 1914.

PAMPHLETS

Coleman, Eliot. *The Use of Ground Rock Powders in Agriculture.* Harborside, Maine: The Small Farm Research Association, n.d.

Fletcher, Robert F., and J. O. Dutt. *Vegetable Varieties for Pennsylvania.* Including correspondence with Professor Fletcher. University Park, Pa.: Pennsylvania State University Press, 1977

Growing Vegetable Transplants. University Park, Pa.: Penn State University Extension Service, n.d.

Growing Vine Crops. University Park, Pa.: Penn State University Extension Service, n.d.

"Guide to Indoor Garden Lighting with Sylvania Grow-Lux Lamps." Danuois, Mass.: Sylvania, 1979.

Harrington, James F., and P. A. Minges. *Vegetable Seed Germination.* University of California Extension Service, n.d.

Johnston, Rob. *Growing Garden Seeds.* Albion, Maine: Johnny's Selected Seeds, 1976.

Lawrence, Eleanor. *The Conservation of Crop Genetic Resources.* New York: International Board for Plant Genetic Resources, 1975.

New Alchemy Technical Bulletin, no. 3. East Falmouth, Mass.: New Alchemy Institute, 1982.

New Alchemy Technical Bulletin, no. 22. East Falmouth, Mass.: New Alchemy Institute, 1985.

Plant Growth Lighting. Cleveland, Ohio: General Electric, n.d.

Sheldrake, Raymond, and James Boodley. *Commercial Production of Vegetable and Flower Plants.* Ithaca, N.Y.: Cornell University Press, 1974.

UNPUBLISHED PAPERS

Brain, K. R., et al. 1973. "Cytokinin Activity of Commercial Aqueous Seaweed Extract."

Brain, K. R., and D. C. Williams. n.d. "Plant Growth Regulatory Substances in Commercial Seaweed Extracts."

Race, Susan. 1963. "Seaweed in Horticulture."

Rosenour, Herbert. 1958. "Seaweeds—Soil and Plant Food." Chicago: Sea-Born Corp.

Recommended Reading

For more information about gardening specialties mentioned in this handbook, you might want to consult one of the following books.

Abraham, George "Doc," and Katy Abraham. *Organic Gardening Under Glass.* Emmaus, Pa.: Rodale Press, 1975.

Allard, R. W. *Principles of Plant Breeding.* New York: John Wiley and Sons, 1960.

Ball Blue Book. Muncie, Indiana: The Ball Corporation, n.d.

Campbell, Stu. *Let It Rot! The Gardener's Guide to Composting.* Charlotte, Vt.: Garden Way, 1975.

DeKorne, James. *The Survival Greenhouse.* El Rita, N.Mex.: Walden Foundation, 1975.

Editors of *Organic Gardening* magazine. *The Encyclopedia of Organic Gardening.* Emmaus, Pa.: Rodale Press, 1978.

Farb, Peter. *The Living Earth.* New York: Harper & Row, 1969.

Fisher, Rick, and Bill Yanda. *The Food and Heat Producing Solar Greenhouse.* Santa Fe, N.Mex.: John Muir Publications, 1977.

Jabs, Carolyn. *The Heirloom Gardener.* San Francisco: Sierra Club Books, 1984.

Jeavons, John. *How to Grow More Vegetables than You Ever Thought Possible on Less Land than You Can Imagine.* Palo Alto, Calif.: Ecology Action of the Midpeninsula, 1974.

Lieth, Helmut. *Phenology and Seasonality Modeling.* New York: Springer-Verlag, 1974.

Logsdon, Gene. *Small-Scale Grain Raising.* Emmaus, Pa.: Rodale Press, 1977.

McCullagh, James C. *The Solar Greenhouse Book.* Emmaus, Pa.: Rodale Press, 1977.

MacLatchie, Sharon. *Gardening with Kids.* Emmaus, Pa.: Rodale Press, 1977.

Raymond, Dick. *Down-to-Earth Vegetable Gardening.* Charlotte, Vt.: Garden Way, 1975.

Riotte, Louise. *Companion Planting.* Charlotte, Vt.: Garden Way, 1975.

———. *Planetary Planting.* New York: Simon and Schuster, 1975.

370

Skelsey, Alice, and Gloria Huckaby. *Growing Up Green.* New York: Workman Publishing Co., 1973.

Swanson, Faith H., and Virginia B. Rady. *Herb Garden Design.* Hanover, N.H.: University Press of New England, 1984.

Welsh, James R. *Fundamentals of Plant Genetics and Breeding.* New York: John Wiley and Sons, 1981.

Whealy, Kent, ed. *The Garden Seed Inventory.* Decorah, Iowa: The Seed Savers Exchange, 1985.

Wilson, Charles Morrow. *Roots: Miracles Below.* Garden City, N.Y.: Doubleday, 1968.

Yepsen, Roger, ed. *Organic Plant Protection.* Emmaus, Pa.: Rodale Press, 1976.

Young, James A., and Cherly G. Young. *Collecting, Processing and Germinating Seed of Wildland Plants.* Portland, Oreg.: Timber Press, 1986.

Index

Page numbers in boldface indicate information in tables.